THE BOGOMILS

THE BOGOMILS

A STUDY IN
BALKAN NEO-MANICHAEISM

BY

DMITRI OBOLENSKY
M.A., Ph.D.

CAMBRIDGE
AT THE UNIVERSITY PRESS
1948

PUBLISHED BY THE PRESS SYNDICATE OF THE UNIVERSITY OF CAMBRIDGE
The Pitt Building, Trumpington Street, Cambridge, United Kingdom

CAMBRIDGE UNIVERSITY PRESS
The Edinburgh Building, Cambridge CB2 2RU, UK
40 West 20th Street, New York NY 10011 - 4211, USA
477 Williamstown Road, Port Melbourne, VIC 3207, Australia
Ruiz de Alarcón 13, 28014 Madrid, Spain
Dock House, The Waterfront, Cape Town 8001, South Africa

http://www.cambridge.org

© Cambridge University Press 1948

This book is in copyright. Subject to statutory exception
and to the provisions of relevant collective licensing agreements,
no reproduction of any part may take place without
the written permission of Cambridge University Press.

First published 1948
First paperback edition 2004

A catalogue record for this book is available from the British Library

ISBN 0 521 58262 8 hardback
ISBN 0 521 60763 9 paperback

To
MY MOTHER

PREFACE

The Bogomil movement has come to be recognized as one of the major problems of south Slavonic and Byzantine history. The influence it has exercised on the history of the Balkan peoples—on their church and state, on their society and literature, on their religion and folk-lore—make the study of Bogomilism essential for Byzantinist and Slavist alike. To scholars and students in other fields Bogomilism still offers many unexplored, or half-explored, possibilities. The theologian and the philosopher can find in Bogomilism one of the most interesting examples of the growth on European soil in the Middle Ages of a pattern of thought and a way of life which may be termed 'dualistic'. A detailed study of Bogomilism should help Western medievalists to shed new light on the still somewhat obscure problem of the historical connections between Asiatic Manichaeism and the dualistic movements of western Europe, particularly of the Italian Patarenes and of the Cathars or Albigenses of southern France. This connection, if successfully established, would in its turn enable Church historians to regard the Bogomil sect as the first European link in the thousand-year-long chain leading from Mani's teaching in Mesopotamia in the third century to the Albigensian Crusade in southern France in the thirteenth. Moreover, the study of the Bogomil movement has its own, and by no means negligible, part to play in the investigation of the cultural and religious links between eastern and western Europe, the importance of which is increasingly perceived at the present time.

The study of Bogomilism has a fairly long, but not uniformly successful, history. In the eighteenth century Bogomilism began to attract the attention of German scholars. Some, like J. C. Wolf, regarded the Bogomils as heretics, while others, like J. L. Oeder, tried to prove that they were the bearers of a 'pure' Christianity and were unjustly persecuted by a corrupt Byzantine Church. Their investigations were necessarily limited by their ignorance of the non-Byzantine sources, which led them to take a view of Bogomilism at its best incomplete and in some cases false. In

England, about the same time, Gibbon was able to dismiss the Bogomils in a peremptory footnote of his *Decline and Fall* as 'a sect of Gnostics, who soon vanished'.

Bogomil studies received a fresh impetus and a new orientation in the second half of the nineteenth century, owing to the discovery of Slavonic documents which conclusively pointed to Bulgaria as the original home of the Bogomil sect. The study and publication of these manuscripts was carried out with great success by the Russian scholar M. G. Popruzhenko. In 1869 and 1870 the Croatian scholar F. Rački published his *Bogomili i Patareni*, a work which still remains an indispensable guide, although a number of its conclusions now stand in need of correction or revision. More recently, Slavonic scholars have shown a keen interest in the subject and have approached Bogomilism from several new angles, but, on the whole, they have tended to limit themselves to specific details and have not attempted to re-examine the whole problem from the historical point of view. However, the important place occupied by Bogomilism in the history of Bulgarian literature is stressed by Prof. I. Ivanov, who has analysed and edited the literary monuments of the Bulgarian Bogomils.

The study of Bogomilism has, in my opinion, suffered unduly from the preconceived or erroneous notions of many investigators. For example, several modern Balkan historians have over-emphasized the political significance of Bogomilism, often to the detriment of its importance as a religious movement, by regarding it primarily as a nationalistic attempt of the Slavs to resist the encroachments of Byzantine imperialism. The present study, it is hoped, may serve to show that this view, though justifiable within certain limits, has sometimes been grossly exaggerated. Moreover, the historians of Bogomilism have, for the most part, considered Bogomilism as a static phenomenon, and have unhesitatingly attributed to the sect at the very beginning of its history features which in fact only developed at later periods. At the same time they have usually failed to draw sufficiently clear distinctions between the Bogomils and other medieval Balkan sects, a failure which often leads them to erroneous conclusions regarding the former. Hoping to remedy these mistakes, I have decided to abandon the traditional plan, in which the history, doctrines and customs of the Bogomils were divided into separate chapters, and have adopted the method of studying the different aspects of Bogomilism simultaneously, period by period. In this manner a clearer picture may perhaps be obtained of the gradual evolution of the doctrines, ethics, ritual, customs and organization of the sect under the influence of historical circumstances. Moreover, in order to dispel the confusion often made between the Bogomils and other contemporary Balkan sects, I have been obliged to deal at some length with the latter, particularly with the Paulicians and the Massalians.

Like most other medieval sects, the Bogomils are known to us very largely from the evidence of their enemies. This would seem to render the task of impartial criticism a delicate one, particularly since the number of sources directly concerned with the Bogomils is not large. And yet the information of Orthodox Churchmen on the subject is, on the whole, fairly reliable: a comparison between the evidence supplied by writers widely separated in space and time reveals almost unanimous agreement on the essential features of Bogomilism. In these circumstances, an objective reconstruction of the doctrines and practices of the Bogomils is by no means impossible.

The present book was, in substance, completed in 1942 and its publication has been delayed mainly by the circumstances of war. The same circumstances prevented me from having any knowledge of the works by Mr S. Runciman on the Manichaean movement and by MM. H.-C. Puech and A. Vaillant on Cosmas's treatise against the Bogomils, until both books were in proof form. I am indebted to Mr Runciman and M. Puech for permission to consult the proofs of their books.

My thanks are due first of all to Trinity College for enabling me to pursue the research which led to the writing of this book. I owe a special debt of gratitude to Dr Elizabeth Hill, without whose encouragement and help this book would not have been written, to the Rev. Prof. F. Dvorník, who has unstintingly allowed me to benefit from his knowledge of Byzantine and Slavonic history, and to Prof. Sir Ellis Minns, who read the work in manuscript and made many valuable suggestions. I am much indebted to my wife for her help in compiling the index and reading the proofs. I wish also to thank the staff of the British Museum for innumerable kindnesses and the Syndics of the Cambridge University Press for their assistance in the publication of this book.

In order to simplify typographical problems in the quotations in Old Church Slavonic I have substituted for the letters Ѧ, Ѫ, Ѭ, the modern я, у, and ю respectively, I have omitted all accents, and transcribed abbreviated words in their complete form.

DMITRI OBOLENSKY

TRINITY COLLEGE
CAMBRIDGE

29 *October* 1946

CONTENTS

Abbreviations	*page* xiii
Chapter I. The Manichaean Legacy	1
II. Neo-Manichaeism in the Near East	28
III. The Rise of Balkan Dualism	59
IV. Bogomilism in the First Bulgarian Empire	111
V. Byzantine Bogomilism	168
VI. Bogomilism in the Second Bulgarian Empire	230

Appendices

I. The Chronology of Cosmas	268
II. The *pop* Jeremiah	271
III. The Date of the Bogomil trial in Constantinople	275
IV. Bogomilism in Russia, Serbia, Bosnia and Hum	277
V. Bogomils, Cathars and Patarenes	286
Bibliography	290
Index	305

ABBREVIATIONS

Abh. bayer. Akad. Wiss.   *Abhandlungen der bayerischen Akademie der Wissenschaften.* München.
Abh. böhm. Ges. Wiss.   *Abhandlungen der böhmischen Gesellschaft der Wissenschaften.* Prague.
Abh. preuss. Akad. Wiss.   *Abhandlungen der preussischen Akademie der Wissenschaften.* Berlin.
Bull. Acad. Belg.   *Bulletin de l'Académie royale de Belgique.* Bruxelles.
B.   *Byzantion.* Paris, 1924– .
B.I.B.   *Bŭlgarska Istoricheska Biblioteka.* Sofia, 1927– .
B.Z.   *Byzantinische Zeitschrift.* Leipzig, 1892– .
C.E.H.   *Cambridge Economic History (The).* Cambridge, 1941– .
C.M.H.   *Cambridge Medieval History (The).* Cambridge, 1911–36.
C.S.H.B.   *Corpus scriptorum historiae byzantinae.* Bonn, 1828–97.
Denkschr. Akad. Wiss. Wien.   *Denkschriften der Akademie der Wissenschaften in Wien.*
D.T.C.   *Dictionnaire de Théologie Catholique.* Paris, 1923– .
G. Soc. Asiat. Ital. (n.s.)   *Giornale della Società Asiatica Italiana.* (Nuova serie.) Firenze, 1887– .
G.S.U.   *Godishnik na Sofiyskiya Universitet.* Sofia, 1904– .
I.R.A.I.K.   *Izvestiya Russkogo Arkheologicheskogo Instituta v Konstantinopole.* Odessa, 1896–1912.
J.A.   *Journal Asiatique.* Paris, 1822– .
J.R.A.S.   *Journal of the Royal Asiatic Society.* London, 1823– .
Kh. Ch.   *Khristianskoe Chtenie.* St Petersburg, 1821–1918.
Mansi.   J. D. Mansi, *Sacrorum conciliorum nova et amplissima collectio.* Florence, 1759–98.
Mém. Acad. Belg.   *Mémoires de l'Académie royale de Belgique.* Bruxelles.
M.G.H.   *Monumenta Germaniae historica.* Hanover, 1826– .
P.G.   J. P. Migne, *Patrologiae cursus completus. Series graeco-latina.* Paris, 1857–66.
P.L.   *Patrologiae cursus completus. Series latina.* Paris, 1844–55.
P.O.   *Pravoslavnoe Obozrenie.* Moscow, 1860–91.
P.S.   *Periodichesko Spisanie na Bŭlgarskoto Knizhovno Druzhestvo.* Braila, 1870–6.
Rad   *Rad Jugoslavenske Akademije Znanosti i Umjetnosti.* Zagreb, 1867– .
R.E.   *Realencyklopädie für protestantische Theologie und Kirche.* Leipzig, 1896–1913.
Rec. Univ. Gand.   *Recueil de travaux de la Faculté de Philosophie et Lettres, Université de Gand.* Gand, 1888– .
R.E.S.   *Revue des Études Slaves.* Paris, 1921– .
R.H.R.   *Revue de l'Histoire des Religions.* Paris, 1880– .
R.Q.H.   *Revue des Questions Historiques.* Paris, 1866– .
S.B.A.N.   *Spisanie na Bŭlgarskata Akademiya na Naukite.* Sofia, 1911– .
S.B. bayer. Akad. Wiss.   *Sitzungsberichte der bayerischen Akademie der Wissenschaften.* München.

ABBREVIATIONS

S.B. preuss. Akad. Wiss.   *Sitzungsberichte der preussischen Akademie der Wissenschaften.* Berlin.

S.L.   *Sbornik statey po slavyanovedeniyu, sostavlenny i izdanny uchenikami V.I. Lamanskogo.* St Petersburg, 1883.

S.N.U.   *Sbornik za Narodni Umotvoreniya, Nauka i Knizhnina (i Narodopis).* Sofia, 1889–1936.

S.R.   *Slavonic and East European Review (The).* London. 1922– .

V.V.   *Vizantiysky Vremennik* (Βυζαντινὰ Χρονικά). St Petersburg, 1894–1928; 1947– .

Wiss. Mitt. Bosn. Herz.   *Wissenschaftliche Mittheilungen aus Bosnien und der Hercegovina.* Vienna, 1893– .

Zh.M.N.P.   *Zhurnal Ministerstva Narodnogo Prosveshcheniya.* St Petersburg, 1834–1917.

CHAPTER I

THE MANICHAEAN LEGACY

The problem of Evil: the Judaeo-Christian and the dualist views. Manichaeism and neo-Manichaeism. Was Zoroastrianism a dualistic religion? Manichaeism in Syria, Armenia and Asia Minor before the seventh century. Sectarian movements in Asia Minor: Gnostics, Massalians, Encratites, Montanists, Novatians. Dualistic and Christian asceticism; Eustathius of Sebaste and the Desert Fathers. Influence of Manichaeism on Christian sects and its adaptation to Christianity.

Among the ever-recurring problems which have confronted human reason throughout the ages one of the most complex is that of the nature and origin of Evil. Whenever man seeks to support his religious faith by rational thinking, sooner or later he is inevitably led to the problem of reconciling the absolute qualities he attributes to God with the obviously limited and contingent character of the world he lives in. The importance and urgency of this problem is easily perceived by both speculative and non-speculative minds. The metaphysician and the theologian must explain the possibility of any relation between the Infinite and the finite, between the perfection of the Creator and the imperfection of the creature, between God and the world; and those men who, without being philosophers, believe that God is the source of all perfection and goodness and that He has created the world, cannot but recognize that in this world moral and physical evil—suffering, cruelty, decay, death—is abundantly present. How then can God, the Supreme Good, be the cause of Evil? Is it possible to escape the following seemingly logical conclusion: either God *is* the creator of Evil, in which case He is not the source of all perfection and hence not truly God; or else He is *not* the creator of Evil, and the origin of Evil must be sought outside God in some agent distinct from and opposed to Him? In the many solutions to the problem of Evil attempted by the human reason two main attitudes of mind, completely opposed to each other, are clearly distinguishable.

The first is based on the belief in a fundamental relation between God and the world created by Him; it was above all the faith of

the Jewish people that the world, created by God, is good (Gen. i).
The Book of Genesis describes the creation as an act of God's
omnipotence, explains the appearance of Evil as a result of man's
disobedience to the will of God, but gives no philosophical theory
of the relation of the creature to the Creator. The direct contact
between the Infinite and the finite, the Absolute and the contingent, has all the reality of a fact willed by God, but remains
essentially a mystery, incomprehensible to the human reason.
Judaism, throughout its history, always emphasized the profound
nature of this relation between God and creation, recognized the
work of Divine Providence in the world by stressing the positive
importance of human history in preparing the Kingdom of God
on earth and thus proclaimed the ultimate value and significance
of this life. The Judaic view of life received a supreme confirmation
and an all-embracing significance by the Incarnation of the Word,
whereby God became flesh and entered human history. Christianity, by accepting and teaching the fundamental reality of God-man, recognized that the gulf between the Infinite and the finite
had been finally bridged and that the created world into which
the Creator Himself had entered was not only of positive value
but even capable of sanctification. Henceforth to those who on
account of the incommensurability of God and the material world
denied the possibility of contact between them Christianity was
able to reply that God created the world, became man and will
raise up the flesh. Taking their stand on the mystery of the
Incarnation, Christian theologians gradually built up a rational
solution of the problem of Evil. Starting from the proposition that
it is useless to seek for the origin of something without first defining
its nature, they showed that the origin of Evil can be logically
deduced from its nature. Everything that is, that has being, is
good; and since everything that is derives its substance from God,
it follows that Evil, as the opposite of Good, has neither substance
nor being, nor positive reality (otherwise it would be good). Evil
exists merely as a possibility of disorder: Evil is merely an accident
of the substance, the *privation* of Good. Evil as the opposite of Good
is not created by God, since nothing can generate its opposite; it is,
strictly speaking, non-being. But Evil as the privation of Good,
to exist at all, depends on the existence of substances in which this
privation can become operative, substances which have being and

hence are good. Thus, Evil exists in Good and depends for its existence on Good. Hence the cause of Evil is found to be in Good. This good is man's free will, which is a gift of God. Man's abuse of his free will, made possible by his finite condition, his state of inferiority as a creature in relation to his Creator, has resulted in his separation from God. This separation resulted in a state o privation, which has brought about disorder, suffering, corruption and the other manifestations of Evil.

In complete contradiction to the Christian view of Evil, which follows from the belief in the Hypostatic Union and the consequent value attributed to this life and to the body, we find another conception, already existing in many respects before the rise of Christianity. This conception, positing a fundamental opposition between Good and Evil, denied that God, who is essentially good, can be the author or the cause of Evil. The origin of Evil must be sought outside God. The seat of Evil is the visible, material world where disorder and suffering are dominant. The origin of Evil lies in Matter itself, whose opaqueness and multiplicity are radically opposed to the spirituality and unity of God. This view, which attributes to Evil the same positive and ultimate quality as is possessed by Good, thus leads to an inevitable *dualism* between God and the opposite principle of Matter. It seems that this dualistic cosmology was accepted, implicitly or explicitly, by most of the Greek philosophers before Plato. Plato himself, by tracing the origin of Evil to Matter, regarded as independent of God and outside His causality, could not escape at least a strong measure of dualism. But it is above all in Gnosticism, which arose in Asia Minor in the first century of our era, that we find the first systematic attempt to solve the problem of Evil in a strictly dualistic sense.[1] Behind the numerous discrepancies in the teachings of the different Gnostic sects there lies the basic idea that Matter, which is essentially evil, cannot be the creation of God. The Gnostics explained the origin of Matter either by regarding it as eternally evil in itself or by positing an intermediary between God and Matter, the

---

[1] It has hitherto been customary among historians and theologians to trace systematic dualism back to the Zoroastrian tradition of Persia. But, as it will be shown below, Zoroastrian 'dualism' differs from the Gnostic variety in some important respects and even contains several features incompatible with true dualism.

Demiurge, one of the emanations (aeons) of God, whose nature had been basically corrupted by a transgression which caused his expulsion from the divine pleroma; this Demiurge created the material world, which consequently shares in his essentially evil nature. Man himself, in Gnosticism and in every truly dualistic theory, mirrors this fundamental dualism: his soul is of divine origin, his body ineradicably evil. The ancient Greek myth of the soul, come down to earth from its heavenly abode and imprisoned in the darkness of a material body from which it is ever seeking to escape in order to return to its home, is present, at least implicitly, in every form of dualism. The body is 'the tomb of the soul', the instrument whereby the Demiurge seeks to imprison light in the darkness of Matter and to prevent the soul from ascending back to the heavenly spheres. Every truly dualistic conception must see the origin of all misfortune in life in this world: for the birth of a man is the imprisonment of a divine or angelic soul in an unredeemable body. The only final redemption is in death, the escape of the soul from its prison and the return of a particle of light to the One Uncreated Light. This redemption, this escape is not the repentance for the moral evil committed by man: man cannot be really responsible for the guilt of sin if Evil is not due to the abuse of his free will but is rooted in his material body and is thus the inevitable concomitant of life itself. But though he is not responsible for the existence of Evil and has thus ultimately no free will, man can and must collaborate in the work of God in striving by his knowledge and his actions to purify his soul from the contagion of its material envelope. Purification as understood and practised by the consistent dualist implies forbearance from all actions which further the soul's imprisonment in Matter (especially from marriage and the procreation of children, which strengthen the power of Matter in the world) and a rigid asceticism, based not on the desire to discipline the flesh but on a radical hatred of the body.

In the history of the Christian Church dualism plays a particularly important part. It was largely the necessity of refuting the doctrines of the dualists that led the Christian theologians to formulate in a comprehensive manner their own teaching on the problem of Evil. Moreover, dualism gave rise to a large number of sects which during the whole of Christian antiquity and until the very end of the Middle Ages were fierce and dangerous enemies of the Church, and against which both in eastern and western Europe the Church was compelled to wage an almost ceaseless war.

The most rigid and classical form of dualism in historical times is to be found in Manichaeism, invented in Babylonia in the middle of the third century A.D. by the Persian Mani. Mani's celebrated teachings spread, in the course of the thousand years after their first appearance, over large parts of Europe and Asia, extending from the Pacific to the Atlantic Oceans. Their main tenets, which were to exercise an astonishingly potent influence on human thought, may be briefly summarized as follows.[1]

From all eternity there exist two opposite and mutually independent principles, God and Matter, represented respectively on the physical plane by two 'natures', Light and Darkness. Our present world appeared as a result of an invasion of the realm of Light by Darkness, or Matter, and is a 'mixture' of both natures, an amalgam of divine particles of Light imprisoned in a material envelope. The future, or final, state of all things will come about as the result of the complete restoration of the original dualism by the absolute separation of both principles, which will render Darkness for ever incapable of further aggression. The present, in so far as it is a preparation for the future, consists in a gradual liberation of the particles of Light, consubstantial with God, which are the souls of men, from the prison of Matter, of the body. The separation of Light from Darkness is the work of God Himself, who desires that those elements which He lost when they became 'mixed' with Matter should return to their true abode, and is furthered by a series of 'evocations' (hypostatized divine attributes) which God sends into the world. One of these 'evocations', the Demiurge, created our visible world from materials belonging to the realm of Darkness: the purpose of this world is to be a prison for the powers of Darkness and a place of purification for the souls

---

[1] The best accounts of the Manichaean doctrines are to be found in the following works: P. Alfaric, *L'Évolution intellectuelle de Saint Augustin* (Paris, 1918), pp. 95–213; H. H. Schaeder, 'Urform und Fortbildungen des manichäischen Systems', *Vorträge der Bibliothek Warburg* (1924–5), pp. 65–157; H. J. Polotsky, article 'Manichäismus' in Pauly-Wissowa, *Real-Encyclopädie der classischen Altertumswissenschaft* (1935), Supplementband VI.

of men, a kind of machine for the distillation of Light. In order to counteract this gradual liberation of the Light and to strengthen the fetters which bind the souls to Matter, the powers of Darkness created man who, by the duality of the sexes and his instinct of self-propagation inherited from the demons who generated him, is intended to perpetuate the imprisonment of the particles of Light in his own body. Man is thus in a microcosmic form the image of the macrocosmic 'mixture'. A further counter-measure on the part of God then became necessary: this is another 'evocation': Jesus, a Divine Being, descended from the realm of Light into the world and appeared on earth to bring the true teaching to man.[1] He gave to man the knowledge of his dual nature by showing him that his soul is one with the Divine Light which suffers in the whole world from its 'mixture' with Darkness and taught him the path of salvation; this path consists in carefully avoiding all those actions which harm the particles of Light contained in man and further the imprisonment of the soul in Matter. The method by which man must effect within himself the gradual separation of Light from Darkness, the breaking up of the 'mixture', forms the object of Manichaean ethics, based on a radical hatred of the unredeemable flesh and extreme asceticism.[2]

The history of Manichaeism, which until the latter part of the last century remained almost exclusively the domain of Church historians, has now become a subject which no scholar investigating conditions in the later Roman Empire and the Middle Ages can afford to neglect. The influence exerted by Manichaeism over the entire Mediterranean world and its repercussions on the religious, political and social life of medieval Europe are questions which—though still obscure in many respects—are increasingly

[1] The Manichaean conception of Jesus is typically docetic: if His role is to enable man to effect within himself the liberation of the particles of Light from the tyranny of the unredeemable flesh, He clearly cannot Himself have assumed a material body and been born of woman.

[2] The followers of Mani were divided into two main groups: the elect, or 'righteous', bound to a rigid observance of the ethical precepts of Manichaeism, and the catechumens, or 'hearers', who could make some concessions to the weakness of the flesh. To the elect, who alone were regarded as true Manichaeans, sexual intercourse, the eating of any animal food and the drinking of wine were strictly forbidden; the 'hearers' were allowed to marry, to eat meat and to drink wine.

attracting the attention of historians. A particularly important aspect of Manichaeism is its direct connection with the problem of the Oriental influences exerted on medieval European civilization: Manichaeism was the last of these 'Oriental religions' which from the third century B.C. to the fifth century of our era penetrated into the Graeco-Roman world from the Near East—from Asia Minor, Egypt, Syria, Persia and Babylonia. Their common features were an alliance of faith and reason for the pursuit of ultimate knowledge, a strong syncretism ever ready to assimilate the most diverse religious and philosophical teachings and to adapt itself to the doctrines of other nations, and an earnest striving for moral purity by means of asceticism and mortification.[1]

In the course of the present century two important discoveries of original Manichaean sources have resulted in a considerable development of Manichaean studies.[2] Between 1899 and 1907 excavations and searchings carried out in the oasis of Turfan, in Chinese Turkestan, by Russian, German, British, French and Japanese missions led to the discovery of a large number of manuscripts, identified as Manichaean.[3] In 1930 a collection of Coptic papyrus codices was discovered in Egypt and identified as the remains of a Manichaean library, probably of the fifth century.[4]

---

[1] See F. Cumont, *Les religions orientales dans le paganisme romain* (4th ed.; Paris, 1929).

[2] An impetus was given to Manichaean studies in the second half of the nineteenth century by the publication of two oriental sources containing valuable information on Manichaeism, the *Fihrist* of the Arab writer An-Nadim and extracts from the writings of the Syrian Theodore bar Khonai. This led to the works of G. Flügel, *Mani, seine Lehre und seine Schriften* (Leipzig, 1862), and F. Cumont, *La Cosmogonie manichéenne d'après Théodore bar Khôni* (Bruxelles, 1908; *Recherches sur le Manichéisme*, vol. 1).

[3] See F. W. K. Müller, 'Handschriften-Reste in Estrangelo-Schrift aus Turfan, Chinesisch-Turkistan'. (I) *S.B. preuss. Akad. Wiss.* (1904), pp. 348–52; (II) *Abh. preuss. Akad. Wiss.* (1904); C. Salemann, 'Ein Bruchstük manichaeischen Schrifttums im asiatischen Museum', *Zapiski imperatorskoy akademii nauk* (ist.-fil. otd.) (1904), vol. VI; W. Radloff, *Chuastuanit, das Bussgebet der Manichäer* (St Petersburg, 1909); A. von Le Coq, 'A short account of the origin, journey and results of the first Royal Prussian expedition to Turfan in Chinese Turkestan', *J.R.A.S.* (1909), pp. 299–322; E. Chavannes and P. Pelliot, 'Un traité manichéen retrouvé en Chine, traduit et annoté', *J.A.* (1911), pp. 499–617; (1913), pp. 99–199, 261–394.

[4] See C. Schmidt and H. J. Polotsky, 'Ein Mani-Fund in Ägypten. Originalschriften des Mani und seiner Schüler', *S.B. preuss. Akad. Wiss.* (1933), pp. 4–90.

The investigation and publication of these documents, which contain much historical, doctrinal and liturgical material of the greatest value—including some works attributed to Mani himself—is still far from completed.[1] But there can be no doubt that the study of these newly discovered sources will shed much new light not only on the teachings of Mani—which have already been investigated in some detail—but also on the far less known question of the spread and development of the Manichaean sect in the territories of the Roman Empire.

Manichaean dualism penetrated into Europe in two waves, separated by an interval of some three centuries. The first wave, that of primitive Manichaeism, spread between the third and seventh centuries over the whole of the Mediterranean world, extending from Syria, Asia Minor, Judaea to Egypt, northern Africa, Spain, southern Gaul, Italy, and penetrated into the two centres of Roman Christian civilization, Rome and Byzantium.[2] The second wave was that of a revived and in many respects modified Manichaeism, sometimes known as 'neo-Manichaeism'.[3] It appeared in Europe with the dawn of the Middle Ages, and between the ninth and the fourteenth centuries swept over all southern and part of central Europe, from the Black Sea to the Atlantic and the Rhine. A comprehensive history of the neo-Manichaean movement as a whole has yet to be written, and before any such attempt can be made it will be necessary to study in greater detail than has yet been done its origin and development in each of the European countries where it found a home, particularly in Bulgaria, Serbia, Bosnia, northern Italy and southern France.

[1] The following documents have so far been published: *Manichäische Homilien* (*Manichäische Handschriften der Sammlung A. Chester Beatty*, Bd I), herausgegeben von H. J. Polotsky (Stuttgart, 1934); *Manichäische Handschriften der Staatlichen Museen Berlin*, herausgegeben in Auftrage der pr. Akad. der Wissensch., unter Leitung von C. Schmidt, Bd I, *Kephalaia* (Stuttgart, 1935–7); *A Manichaean Psalm-Book* (ed. by C. R. C. Allberry; Stuttgart, 1938).

[2] For this first spread of Manichaeism in the Near East, in Africa and Europe, see E. de Stoop, 'Essai sur la diffusion du manichéisme dans l'Empire romain', *Rec. Univ. Gand* (38ᵉ fasc., 1909); Alfaric, *Les Écritures manichéennes* (Paris, 1918), vol. I, pp. 55–71.

[3] The doctrinal and historical continuity between Manichaeism and 'neo-Manichaeism' has been denied by some scholars. An attempt is made in the following pages to prove this continuity and to justify the use of the term 'neo-Manichaeism' to describe this second wave of dualism.

The present book is concerned with the beginnings of neo-Manichaeism in Europe, with its penetration into the Balkans from the Near East in the ninth and tenth centuries and its development in Bulgaria between the tenth and fourteenth centuries, where its doctrine formed the basis of Bogomilism. A study of the Bogomil sect in Bulgaria may thus establish the first important link in the thousand-year-long chain leading from Mani's teaching in Mesopotamia in the third century to the Albigensian crusade in southern France in the thirteenth century.

Historians of neo-Manichaeism have generally taken the historical connection of this movement with the original teaching of Mani for granted. Evidence which points fairly conclusively to this connection is adduced in the following pages. On the other hand, in trying to establish the distant origins of neo-Manichaeism, some of these historians have not unnaturally been led to investigate the source and nature of those earlier dualistic theories which were accepted in the third century by Mani as the basis of his teaching. This question must now be briefly examined.

Unfortunately, the problem of the origins of Manichaeism proper, which has given rise to the most varied and even contradictory hypotheses,[1] though considerably clarified during the past twenty-five years, can still be solved only in a general manner. It is fairly certain that the dualistic doctrines which directly influenced Manichaeism arose in the Near East or, more precisely, in the borderland between the two great civilizations of the late classical period, the Hellenistic and the Persian. This borderland, stretching roughly from Egypt to Armenia, was already before our era the land *par excellence* of religious syncretism, and it seems an almost impossible task to trace with any degree of certainty the relations between the numerous dualistic sects in the highly intricate maze of the heretical movements in the Near East during the first centuries after Christ. It is, however, possible to identify the main currents of dualism which influenced the development

---

[1] Outlines of the history of Manichaean scholarship are given by U. Fracassini, 'I nuovi studi sul manicheismo', *G. Soc. Asiat. Ital.* (n.s., 1925), vol. I, pp. 106–21; H. S. Nyberg, 'Forschungen über den Manichäismus', *Z. Neutestamentliche Wiss....Kunde der älteren Kirche* (1935), vol. XXXIV, pp. 70–91; H. H. Schaeder, 'Der Manichäismus nach neuen Funden und Forschungen', *Morgenland. Darstellungen aus Geschichte und Kultur des Ostens* (1936), Heft XXVIII, pp. 80–109.

of Manichaeism and elements of which can be found in the later neo-Manichaean movement, and thereby to correct a number of errors and misconceptions regarding the origin of neo-Manichaean dualism which are still to be found in the works of some scholars.

According to the view prevalent among past historians and not infrequently upheld by present-day scholars, the origin of Manichaean dualism is to be sought in the ancient Zoroastrian tradition of Persia. The acceptance of this view has led many historians to regard neo-Manichaeism as a distant product of those doctrines which were taught in Iran at least six centuries before our era, and to which Manichaeism is supposed to have merely given a more definitely dualistic bent.

The close historical contact between Zoroastrianism and Manichaeism from the very time of appearance of the latter is undeniable. Mani himself was a Persian by birth, it was in Persia that he made his first public appearance as a religious teacher, gaining some success even in court circles, it was to Persia that after a long period of exile in central Asia he returned, to perish in A.D. 276 in the hands of the Zoroastrian priesthood.[1] Moreover, as it will be shown, there are some marked resemblances in doctrine between Zoroastrianism and Manichaeism. These factors, to which must be added the striking similarities revealed by the Turfan discoveries between the religious terminology of Zoroastrianism and that of Manichaeism in central Asia, have led many scholars to regard Manichaeism as an offshoot of the Iranian tradition, or at the most as a kind of reformation of Zoroastrianism in a more rigidly dualistic direction.

A detailed comparison of the Zoroastrian and Manichaean doctrines does not lie within the scope of this book. But evidence of a general character, based on the results of recent Iranian scholarship, may here be adduced to show that though several features of Zoroastrianism may appear to warrant the epithet 'dualistic' generally applied to this religion, Zoroastrian 'dualism' as a whole and in its basic philosophical and moral conclusions not only does not correspond to the general definition of dualism as given above, but is even opposed to it in more than one respect. If accepted, this view will lead to the conclusion that although a number of

[1] See A. V. W. Jackson, *Researches in Manichaeism* (New York, 1932), pp. 3–6; A. Christensen, *L'Iran sous les Sassanides* (Copenhagen, 1936), pp. 174–93.

Manichaean teachings (and particularly many elements of its cosmological myth) were in fact undoubtedly borrowed from the religion of Zoroaster, the basis of Mani's dualism is not of Iranian origin and that consequently the theory of the distant derivation of neo-Manichaeism from Zoroastrianism is not historically correct.

It cannot, however, be denied that several important features of Zoroastrianism appear to be 'dualistic'. Like Manichaeism, Zoroastrianism taught the duality of two co-existing 'principles', Light and Darkness, or Good and Evil, separated by a great gulf.[1] Like Manichaeism, it regarded the history of the universe as a cosmic drama with three successive acts: the primordial duality of the two 'principles', Ormazd, the creator of all that is good, and Ahriman, the personification of all that is evil—the state of 'mixture' which is that of our present world—and finally the separation of Good from Evil and the ultimate triumph of Light over Darkness.[2] This dualistic scheme appears to be so systematically developed that in a number of cases concepts relating to Ormazd and to Ahriman are expressed by a different vocabulary and by opposite sets of terms.[3] It may be added that Zoroastrian theologians of the Pahlavi period (third to ninth centuries of our era) criticize the Christian doctrine that Good and Evil have the same origin on the ground that to trace both Good and Evil to God is to deprive Him of His divinity.[4]

The presence in Zoroastrianism of doctrines which, in a certain sense of the word, can be called 'dualistic' is undeniable. But if we examine other teachings, fundamental to this religion, and compare them with the general definition of dualism attempted above, we must come, it seems, to the unavoidable conclusion that Zoroastrian 'dualism' is basically different from the Gnostico-Manichaean variety and is even completely opposed in more than one respect to every form of consistent dualism.

It is in the Zoroastrian teaching on the nature of man and in its moral and social consequences that we can find the most manifest

[1] See A. V. W. Jackson, *Zoroastrian Studies* (New York, 1928), pp. 28–30; H. S. Nyberg, *Die Religionen des alten Iran* (Leipzig, 1938), pp. 21 et seq.

[2] See H. S. Nyberg, 'Questions de cosmogonie et de cosmologie mazdéennes', *J.A.* (1931), vol. CCXIX, pp. 29–36; Jackson, *Zoroastrian Studies*, pp. 110–15.

[3] Jackson, op. cit. p. 29.

[4] See A. Christensen, *L'Iran sous les Sassanides*, p. 281, n.; M. N. Dhalla, *History of Zoroastrianism* (New York, 1938), pp. 384–91.

opposition to any dualistic view. For Zoroastrianism man was not a compound of a divinely created soul and an essentially evil body, but, together with the universe, wholly the product of Ormazd, the Supreme Ruler of the kingdom of Light, Good and Truth.[1] The dualism between Spirit and Matter, soul and body, which lies at the root of the Gnostic and Manichaean conceptions of life, appears to have been completely alien to the Zoroastrian view of the world:[2] the cosmic war between Ormazd and Ahriman is reflected in the struggle between good and evil in man; but owing to his free will, which is a gift of Ormazd, man has the power to choose between right and wrong and 'upon his choice his own salvation and his share in the ultimate victory of good will depend. Every good deed that man does increases the power of good; every evil he commits augments the kingdom of evil. His weight thrown in either scale turns the balance in that direction.... Responsibility accordingly rests upon man, and, because of his freedom of choice, he will be held to strict accountability hereafter'.[3] The emphasis on man's free will and personal responsibility which we find in Zoroastrianism[4] is in marked contrast with the more or less implicit determinism underlying all dualistic systems, which see the origin of moral evil not, as Zoroastrianism (and Christianity), in the abuse of man's free will, but in the very fact that he possesses a material body. The fundamental belief of Zoroastrianism that the body in itself is not evil explains its teaching on the renovation of the world and the resurrection of the body, which is strikingly similar to the Christian doctrine: although the physical constituents of the human being undergo dissolution at death, the dead will receive new bodies with the final restoration of all things, the establishment of 'a new heaven and a new earth' and the final reconciliation of the entire creation to its Creator.[5] The basic non-dualism of Zoroastrianism is particularly apparent in its moral and social teachings. True dualism, whether Gnostic or Manichaean, holding Matter to be the root of Evil, must, at least in theory, see in the

---

[1] Jackson, *Zoroastrian Studies*, pp. 110, 113, 133.
[2] H. S. Nyberg, *Die Religionen des alten Iran*, p. 22.
[3] Jackson, op. cit. p. 220.
[4] Jackson, op. cit. pp. 132-4, 219-44.
[5] Jackson, op. cit. pp. 143-51; Dhalla, op. cit. pp. 288-90, 423-33.

hatred of the flesh and in rigid asceticism the necessary conditions of salvation. On the contrary, nothing is further removed from the spirit of Zoroastrianism than any form of self-mortification: the body in itself is not evil, but can become an instrument of salvation, provided it is controlled and disciplined by the soul, which must rule over the body as a householder rules over his family or a rider rides his horse.[1] Zoroastrianism even condemned all forms of self-mortification and rejected not only the Manichaean but also the Christian conception of asceticism; celibacy it regarded as an evasion of man's religious and civic duties; 'even the priests were not to be celibates, for it is a cardinal point of the faith of every true Zoroastrian that he shall marry and rear a family'.[2] Monastic life was condemned and fasting held to be a sin.[3] Zoroastrianism strongly attacked the Manichaeans for their condemnation of material property, of agriculture and cattle-raising.[4] The insistence of Zoroastrianism on the importance of this world, the holiness of life and the value of the body is remarkably close to the Judaeo-Christian conception and completely opposed to every form of true dualism, which is always based, at least implicitly, on the hatred of the world and the denial of life.

The profound opposition between the Zoroastrian and the typically dualistic Manichaean views of man can be illustrated by the following quotation from the ninth-century *Denkart*, or 'Acts of the Zoroastrian Religion', which denounces, from the standpoint of orthodox Zoroastrianism, a series of Manichaean doctrines which were refuted in the fourth century by the Magian high priest Aturpat:[5]

'One contrary to that which the adorner of holiness, Aturpat, enjoined, [namely] to banish the fiend from the body. The fiend

[1] Dhalla, op. cit. pp. 342–4.
[2] Dhalla, op. cit. pp. 344–5; Christensen, op. cit. pp. 281–2, n.
[3] Dhalla, op. cit. pp. 345–6. It was only at the end of the sixth century, the century preceding the Arab conquest of Persia, when Zoroastrianism was on the decline, that it imbibed some features of asceticism under the influence of Gnosticism and Manichaeism. See Christensen, op. cit. p. 426.
[4] A. V. W. Jackson, *Researches in Manichaeism*, pp. 181, 207; Dhalla, op. cit. pp. 346–8; cf. E. G. Browne, *A Literary History of Persia* (Cambridge, 1928), vol. I, p. 161.
[5] See A. V. W. Jackson, 'The so-called Injunctions of Mani, translated from the Pahlavi of Denkart', 3, 200, *J.R.A.S.* (1924), pp. 213–27; and *Researches in Manichaeism*, pp. 203–17.

incarnate, Mani, falsely said mankind to be the body of the fiend.

One contrary to that which the adorner of holiness, Aturpat, enjoined, [namely] to make God a guest in the body. The fiend incarnate, Mani, falsely said God should not be a guest in the body, but He is a prisoner in the body.'

The problem of reconciling what Jackson called the 'dualistic traits' and the 'monotheistic tendencies' of Zoroastrianism has for long exercised Iranian scholars, but no really complete solution can be said to have been reached.[1] It would seem necessary to make a primary distinction between what would appear to be the essentially non-dualistic character of the Zoroastrian view of the world and the contingent nature of its 'dualism'. It is perhaps significant in this respect that modern Iranian scholars seem to attach great importance to a set of doctrines which were widespread in Persia from the very beginning of Zoroastrianism, if not still earlier, and are known as Zarvanism. Zarvanism taught that there is one Primordial Principle, the Supreme God, Zarvan. Zarvan begot twin sons, Ormazd, the spirit of Light and Good, and Ahriman, the spirit of Darkness and Evil. Ahriman was the first to issue from the bosom of his father and, as the elder of the two brothers, received temporary dominion over this world. Ormazd, the younger and beloved son of Zarvan, must struggle at first to assert his supremacy over his brother, but at the end will reign

---

[1] According to M. Haug Zoroastrianism was based on the 'idea' of the unity and indivisibility of the Supreme Being' and that the dualism commonly ascribed to Zoroaster's teaching is due to 'a confusion of his philosophy with his theology' (*Essays on the sacred language, writings and religion of the Parsis* (p. 303), edited by E. W. West; 3rd ed. London, 1884). For Jackson 'Zoroaster's dualism is a monotheistic and optimistic dualism' (*Zoroastrian Studies*, p. 31) and, according to Christensen, 'la religion de Zoroastre est un monothéisme imparfait...le dualisme n'est qu'apparent' (op. cit. p. 30). But these formulae posit the problem rather than solve it. The example of the Parsis of India who, with the small community of Gabars in Persia, are to-day the sole authentic descendants of the Zoroastrians, is sometimes invoked to support the view that Zoroastrianism was monotheistic. It is a fact that most present-day Parsis 'object to having dualism emphasized too strongly as a characteristic tenet of their faith' and 'in regard to theology they are strictly monotheistic' (Jackson, *Zoroastrian Studies*, pp. 34–5, 184–5). But this argument is inconclusive in itself, for, as Dhalla has pointed out, the modern Parsis have been considerably influenced by Christianity (op. cit. pp. 489–90).

alone.[1] It seems that Zarvanism was the most common form of orthodox Zoroastrianism in Persia during the Sassanian period (third to seventh century), so that it was with the Zarvanite form of Zoroastrianism that Mani himself must have been mostly familiar.[2] Moreover, recent research on Zarvanism has emphasized the close connections between this religion and Manichaeism.[3]

It can no longer be doubted that the true origin of Manichaean dualism does not lie in Zoroastrianism, which is basically anti-dualistic. No doubt certain Zoroastrian doctrines and concepts were borrowed by Manichaeism, and it is even possible that the Zarvanite cosmogony may have left some traces in certain forms of neo-Manichaeism;[4] but the fundamental source of Manichaean and neo-Manichaean dualism must be sought elsewhere. The preceding suggestions regarding the non-Iranian origin of Manichaeism are confirmed by the conclusions of modern scholars, who emphasize not the Zoroastrian but the Christian influences on Manichaeism. Christianity reached Mani by a Syrian channel and through the medium of religious thinkers influenced by Gnosticism,

[1] The best source of our knowledge of the Zarvanite cosmogony is the treatise 'Against the sects' of the fifth-century Armenian historian Eznik of Kolb (*Ausgewählte Schriften der armenischen Kirchenväter*, herausgegeben von S. Weber; München, 1927, Bd I, pp. 83-4); cf. F. Spiegel, *Erânische Alterthumskunde* (Leipzig, 1873), vol. II, pp. 176-87; J. Darmesteter, *Ormazd et Ahriman, leurs origines et leur histoire* (Paris, 1877), pp. 316-32; L. C. Casartelli, *La philosophie religieuse du Mazdéisme sous les Sassanides* (Paris, 1884), pp. 4-11; H. S. Nyberg, *Die Religionen des alten Iran*, pp. 22 et seq.; 'Questions de cosmogonie et de cosmologie mazdéennes', *J.A.* (1931) vol. CCXIX, pp. 71-82; A. Christensen, 'A-t-il existé une religion zurvanite?', *Le Monde Oriental* (Uppsala, 1931), vol. XXV, pp. 29-34; *L'Iran sous les Sassanides*, pp. 143 et seq., 430 et seq.; E. Benveniste, *The Persian Religion according to the chief Greek Texts* (Paris, 1929).

Dhalla (*History of Zoroastrianism*, pp. 331-3) regards Zarvanism as a sect which developed in opposition to orthodox Zoroastrianism and 'aimed at resolving the Zoroastrian dualism into monotheism'. This theory contradicts the opinion of the best authorities on Iranian religion like Nyberg and Christensen, according to whom Zarvanism is at least as old as Zoroastrianism proper and, until the Arab conquest of Persia (seventh century), was never incompatible with orthodox Mazdeism.

[2] A. Christensen, 'A-t-il existé une religion zurvanite?', *Le Monde Oriental*, pp. 33-4; *L'Iran sous les Sassanides*, pp. 144-5, 179 n. 1.

[3] Benveniste, op. cit. pp. 76-90.

[4] See infra, p. 95 and n. 4.

especially Marcion and Bardaisan.[1] It is generally accepted to-day that Manichaeism was not an oriental religion of Persian or Babylonian origin, but a form of Christian-hellenistic Gnosis, more simple and consistent than the previous Gnostic systems.[2] From Bardaisan Mani appears to have derived the basis of his cosmogony and from Marcion his opposition between the Old and the New Testaments, his ethical dualism and the principles of the organization of his sect.[3] Like Gnosticism, Manichaeism appears to have attempted to rationalize Christianity by subjecting the mysteries of faith to a preconceived philosophical interpretation of the universe.

We must now consider the grounds for the assertion, made above, that there is a definite historical connection between Manichaeism and those later dualistic movements which have been called neo-Manichaean. These grounds can be found on the one hand in the history of Manichaeism in the Near East and, on the other, in the presence in the same region of various heretical or distorted forms of Christianity.

From its birthplace in Babylonia Manichaeism spread in two main directions: eastwards to Persia, Turkestan, India and China and westwards to the Roman Empire.[4] The westward movement alone concerns us here.

[1] This theory of the derivation of Manichaeism from the teachings of Marcion and Bardaisan was already put forward by one of the earliest writers concerned with Mani, St Ephraim the Syrian. St Ephraim lived in Edessa a century after Mani and had a direct knowledge of that Syrian world in which Mani himself had moved. His historical appreciation of Manichaeism, clearly based on first-hand information, is now accepted in the main by present-day scholars. See *St Ephraim's Prose Refutations of Mani, Marcion and Bardaisan*, vol. I (edited by C. W. Mitchell, London, 1912); vol. II (completed by A. A. Bevan and F. C. Burkitt, 1921).

[2] The close dependence of Manichaeism on Christian and Gnostic ideas and its relation to the teachings of Marcion and Bardaisan are stressed by the best modern authorities on Manichaeism: P. Alfaric, *Les Écritures Manichéennes*, vol. I, pp. 13–16, 21–2, 56; F. C. Burkitt, *The Religion of the Manichees* (Cambridge, 1925), pp. 71–104; H. H. Schaeder, 'Urform und Fortbildungen des manichäischen Systems', loc. cit. pp. 65–157.

[3] See O. G. von Wesendonk, 'Bardesanes und Mani', *Acta Orientalia* (1932), vol. x, pp. 336–63; Schaeder, 'Bardesanes von Edessa in der Überlieferung der griechischen und der syrischen Kirche', *Z. Kirchengeschichte* (1932), pp. 21–73; F. C. Burkitt, Introductory Essay to *St Ephraim's Prose Refutations* (vol. II, pp. cxlii–cxliv), and infra, pp. 45–7.

[4] See de Stoop, 'Essai sur la diffusion du manichéisme dans l'Empire romain', *Rec. Univ. Gand*, loc. cit. pp. 51 et seq.; Alfaric, *Les Écritures Manichéennes*, vol. I, pp. 55–91.

The beginnings of Manichaeism in the Roman Empire are still wrapped in mystery. In the legendary account of the apocryphal *Acta Archelai* Mani is said to have sent his personal disciples to preach in Egypt and Syria.[1] Although the story is regarded by scholars with suspicion, it may have some historical foundation, as we know that Manichaeism was rife in Syria and Egypt in the fourth century. In Syria and western Mesopotamia it was at that time a formidable rival of the Christian Church and was strongly attacked by St Ephraim of Edessa (d. 373), by Syrian bishops and by St John Chrysostom during his priesthood at Antioch.[2] The fourth century also witnessed the spread of Manichaeism in Asia Minor, particularly in Paphlagonia and Cappadocia; in the latter region it found an adversary in St Basil the Great.[3] This widespread influence of Manichaean doctrines in Syria and Asia Minor in the fourth century—and, in Edessa, at least until the first half of the fifth—is significant in view of the fact that these countries were the respective strongholds of two dualistic sects, the Massalian (between the fourth and the ninth centuries) and the Paulician (between the seventh and the ninth) both of which were to exercise a direct influence on Bogomilism. If the historical connection between Manichaeism and Massalianism is uncertain,[4] it cannot be doubted that the teaching of Mani had a considerable influence on the growth of the Paulician sect in Armenia in the second half of the seventh century.[5] Armenia, the original home and—together with Asia Minor—the subsequent stronghold of Paulicianism, was also visited by Manichaean missionaries, whose teachings were later revised and reformed by the Paulicians. One of the early Manichaean epistles enumerated by the tenth-century Arab historian An-Nadim was addressed 'to the Armenians'.[6] In the fifth century, the Armenian Bishop Eznik of Kolb showed himself acquainted with Manichaeism, the religion of the 'two roots'.[7] The Armenian historian Samuel of Ani describes the arrival in Armenia in 588 of heretics from Syria, 'men with words like honey', equipped with a library of 'false books' which they

[1] De Stoop, op. cit. pp. 51–9; Alfaric, op. cit. pp. 55–6.
[2] De Stoop, op. cit. pp. 60–3.   [3] Ibid. pp. 63–9.
[4] For the Massalian sect, see infra, pp. 48–52.
[5] Cf. infra, pp. 43–5.
[6] See G. Flügel, *Mani, seine Lehre und seine Schriften* (Leipzig, 1862), p. 103.
[7] *Ausgewählte Schriften der armenischen Kirchenväter*, Bd 1, p. 85.

translated into Armenian for the benefit of the local inhabitants.[1] The list of these heterodox works includes the famous *Living Gospel of Mani*[2] and two apocryphal scriptures, the *Liber Paenitentiae Adam* and the *Liber de infantia Salvatoris*, both known to have been used by the Manichaeans.[3] The importance of this source lies in the proof it provides of the presence of Manichaeism at the close of the sixth century in the same country which, only some fifty years later, became the centre of the newly appeared neo-Manichaean sect of the Paulicians.

The paucity and vagueness of the sources do not permit us to determine precisely the manner in which the Manichaean doctrines were preserved and transmitted in the Near East between the third and the seventh centuries.[4] But the successful survival and propagation of Manichaeism in this region can be explained by two main causes: on the one hand, a number of heretical trends and sectarian movements which appeared during the first centuries within the Christian Church very probably prepared the ground for and facilitated the spread of Manichaeism; on the other hand, Manichaeism itself, by a process of conscious borrowing of Christian concepts and terms, was attempting to adapt its dualistic teaching to the dogmas of the Church and thus undoubtedly gained adherents among many ill-instructed Christians. There can be very little doubt that this partial—though essentially artificial—contact between Manichaeism and Christianity increased the vigour and prolonged the existence of the Manichaean sect in the Near East.

The early history of Christian heresies in the Near East offers a bewildering picture of numerous movements and sects whose relations to each other can seldom be proved directly, but certain features of which frequently suggest points of contact with Manichaeism. In default of a proven historical connection between

[1] *J.A.* (1853), vol. II, pp. 430–1. Samuel mistook these heretics for Nestorians.
[2] Cf. Alfaric, op. cit. vol. II, pp. 34–43. [3] Ibid. pp. 151, 172–3.
[4] It is not proposed in the following pages to analyse the general causes and methods of the development of Manichaeism in the Roman Empire, which has been admirably done by de Stoop (op. cit.) and by Cumont ('La propagation du manichéisme dans l'Empire, romain', *Rev. hist. litt. religieuses*, 1910, pp. 31–43), but only to consider some particular factors which explain the survival of the Manichaean doctrines in the Near East until they found a new expression in neo-Manichaeism in the seventh century.

these movements and the teaching of Mani, it is more satisfactory to regard them as successive and more or less independent manifestations of the same spiritual tendency. Their common feature is a tendency either towards extreme asceticism, surpassing and distorting the ethical teaching of Christianity, or at least towards a greater moral rigorism than was compatible with the practice of the Church. Of these two tendencies, the ascetic and the rigorist, only the first could lead to an explicit dualism, but it is probable that the second as well contributed, at least indirectly, to the success of Manichaeism.

Already in the first century there appeared within the Christian Church a false conception of asceticism, based on the belief that complete continence, which Christ and St Paul regarded as a desirable path for a minority of chosen souls, is obligatory for all the faithful and a necessary condition of salvation.[1] It is not surprising to find that this distorted view of asceticism, which arose from an over-emphasis laid on certain moral precepts of the Gospels, often proved itself incapable of resisting the infiltration of a background of Gnostico-Marcionic dualism. In particular, those unenlightened Christians who, by an exaggerated interpretation of Christian ethics, held that sexual intercourse, the eating of meat and the drinking of wine by rousing the physical passions were an obstacle to the salvation of the soul could not always avoid accepting at least implicitly a dualistic metaphysic of Matter, which placed the origin of Evil in the flesh, in the material body belonging to the realm of the Demiurge.

It is a striking fact that these outbursts of dualistic asceticism were nowhere so persistent and widespread in the first centuries A.D. as in Asia Minor. In the west of the peninsula, in Lydia and Phrygia, Gnosticism was already rampant in the first century A.D. The dualistic sect of the Massalians spread over a large part of Asia Minor in the fourth and fifth centuries, from Cappadocia and Lycaonia to Pamphylia and Lycia.[2] Asia Minor was likewise the centre of the Christian Encratite sects which afford a good example of the penetration of dualistic ideas into religious com-

[1] St Paul in his first Epistle to Timothy, written from Phrygia, condemns those who 'depart from the faith...forbidding to marry, and commanding to abstain from meats' (I Tim. iv. 1–5).
[2] Cf. infra, pp. 50–1.

munities unduly preoccupied with extreme asceticism: the Encratites, while remaining formally in agreement with the dogmas of Christianity, were led to condemn marriage and the use of meat and wine as subjecting man to the power of evil Matter. Encratism, which developed in the second century, was still widespread at the end of the fourth in Phrygia and in central and southern Asia Minor.[1]

Asia Minor was also a particularly fertile ground for the development of several heretical movements within Christianity which, without falling into any formally dualistic view of the world, nevertheless developed an extreme ascetic or rigoristic moral teaching opposed in several respects to the doctrine or discipline of the Church. In the second half of the second century Montanism arose in Phrygia. Although we lack any very detailed information on the doctrines of the Montanists of Asia Minor, they undoubtedly practised a more extreme form of asceticism than that required by Christian discipline and arrogantly accused the Orthodox Church of laxity and mediocrity.[2] Montanism was not essentially dualistic and in some respects was even anti-dualistic;[3] but at least two of its tenets bear a great resemblance to two doctrines which can be found, the one in Gnosticism (a century before Montanism), the other in Manichaeism. Like the Gnostics, the Montanists divided the believers into two separate categories, the 'pneumatics', who alone followed the true spiritual life (for the Montanists these were the members of their sect, as opposed to the ordinary members of the Church), and the 'psychics', who were capable only of an inferior degree of understanding.[4] Moreover, the title of 'Paraclete', given by the Montanists to their founder Montanus, was assumed a century later by Mani who, like Montanus, was regarded by his followers as the manifestation of the Holy Spirit.[5] No doubt it is impossible from this slender

[1] See G. Bareille, 'Encratites' in *D.T.C.* vol. v.
[2] See P. de Labriolle, *La crise montaniste* (Paris, 1913); G. Bardy, 'Montanisme' in *D.T.C.* vol. x; A. Hollard, *Deux hérétiques: Marcion et Montan* (Paris, 1935).
[3] See Labriolle, op. cit. pp. 106, 110, 149.
[4] Ibid. pp. 138–43. Labriolle thinks that the Montanists derived their distinction between πνευματικοί and ψυχικοί not from the Gnostics but from St Paul.
[5] Ibid. pp. 131–5, 225–8, 324. The same title of Paraclete was later claimed by the Paulician leader Sergius. See infra, p. 37.

evidence to deduce any direct doctrinal or historical relation between the Montanist and Manichaean sects; but the extraordinary vitality of Montanism (only surpassed by that of Manichaeism itself), which spread from the third to the fifth century over a large part of Asia Minor and survived at least as late as the eighth century in the same regions where Manichaeism was rife, testifies to the strength of the ascetic and anti-ecclesiastical tendencies which could not but facilitate proselytism by the followers of Mani. What has just been said of the Montanists applies also largely to the sect of the Novatians, which arose in the middle of the third century and at first merely demanded a greater strictness in ecclesiastical (particularly penitentiary) discipline and the exclusion from the Church of all penitents guilty of grave sin. The Novatians were especially numerous in Asia Minor, and more particularly in Phrygia, where they gradually fused with the Montanist sect.[1]

A direct contact between Manichaeism and all these sects, the Gnostic, the Massalian, the Encratite, the Montanist and the Novatian, which flourished in Asia Minor between the first and the eighth centuries, cannot be proved historically. Prima facie such a contact, at least with some of these sects, is not improbable, if it is remembered that Manichaeism was rife in northern and central Asia Minor in the fourth century.[2] But whatever their actual connection with Manichaeism, the history of these sects shows that the boundary between Christian asceticism and a dualistic conception of Matter, though it is absolute in theory, could often in practice become very narrow. Harnack has justly observed of the Gnostic and the Manichaean sects that 'it was not easy for them to gain any adherents except where some Christianity had gone before them'.[3] When this 'Christianity' with which Manichaeism could come into contact was itself heretical, possessing dualistic or ascetic features, Manichaean propaganda would often fall on very receptive ground. For these reasons it is probable that the Christian sects of the Encratites, the Montanists and the Novatians were a medium through which the

---

[1] See E. Amann, 'Novatien et Novatianisme', *D.T.C.* vol. XI.
[2] Cf. supra, p. 17.
[3] *Die Mission und Ausbreitung des Christentums* (4th ed.; Leipzig, 1924), vol. II, pp. 928–9.

doctrines of Mani were preserved in the Near East at least until the end of the seventh century, when they found a new and powerful expression in Paulicianism.

This danger of heresy which exists in every distorted view of Christian asceticism can also be found in the early history of monasticism in the Near East. More particularly, the question has sometimes been raised of the relations between Manichaeism and early Christian monasticism. A detailed examination of this problem cannot be attempted here, but a few indications seem necessary.

Around A.D. 330 the Council of Gangra in Paphlagonia (northern Asia Minor) condemned the teachings of Eustathius of Sebaste, sometime disciple of Arius in Alexandria, founder of coenobitic monasticism in Armenia and Asia Minor, the friend, and later opponent, of St Basil. The Council of Gangra, of which we possess the canons and a synodal letter addressed to the bishops of Armenia,[1] condemned Eustathius and his disciples for their self-righteous and exaggerated asceticism: the concrete charges against them included teaching that married people cannot be saved, forbidding their followers to eat meat, preferring their own private gatherings to the liturgy of the Church and encouraging their female adherents to cut their hair short and to dress like men. There is nothing specifically Manichaean in these teachings, which are more suggestive of the exaggerations of the Encratites. But two other tenets condemned by the Council of Gangra would seem to be, in the opinion of de Stoop, typical of Manichaeism and would point to an influence of this sect on the school of Eustathius: the practice of fasting on Sundays and the right enjoyed by the ascetics, as saints (ὡς ἁγίοις), of receiving the firstfruits which should normally have been given to the churches.[2] Both the Sunday fast and the obligation incumbent on the 'hearers' of supplying the 'righteous' with food undoubtedly existed among the Manichaeans, and their condemnation at the Council of Gangra may perhaps be taken as an indication of some Manichaean influence on the perverted asceticism of Eustathius

[1] See Mansi, vol. II, cols. 1095–1114; C. Hefele and H. Leclercq, *Histoire des Conciles* (Paris, 1907), vol. I, pt 2, pp. 1029–45.
[2] De Stoop, 'Essai sur la diffusion du manichéisme dans l'Empire romain', *Rec. Univ. Gand*, 38ᵉ fasc., 1909, p. 64.

of Sebaste.[1] However, the relations between the Eustathians and the Manichaeans could not have been anything but very indirect, for the name of Mani is not mentioned in the acts of the Council, and de Stoop himself admits that Eustathius, who had many enemies, was never accused of Manichaeism.[2] In any case, the condemnation of Eustathius by the Council of Gangra shows that the danger of an exaggerated asceticism, leading to a false view of matter and thus to an implicit dualism, was not unknown among the monks of the fourth century.

The particular case of Eustathius of Sebaste leads to the general question of the possible relation between Manichaeism and Christian monasticism. This question must now be briefly considered, because some historians have been tempted to seek for Manichaean influences on the growth of early Christian monasticism.

In theory the difference between the Christian and the Manichaean conceptions of asceticism is clear-cut and absolute. Those historians who would wish to break down or to minimize this difference by arguing that both the Manichaean and the Christian monks were striving, by the mortification of the body, towards moral purity and the liberation of the soul from the fetters of the sinful flesh,[3] are guilty of a grave misunderstanding of the Christian conception of asceticism. This point of view fails to grasp the fundamental difference between the Christian attitude to the flesh, which is 'contrary' to the spirit (Gal. v. 17) only in so far as it is not brought into subjection by a reasonable discipline, and

---

[1] The fact that the Eustathians were also accused of insulting the memory of the martyrs (Mansi, vol. II, col. 1103), which puzzled Hefele (loc. cit. p. 1042), may perhaps become significant when related to the derogatory attitude of the Manichaeans towards the Christian martyrs. See A. Dufourcq, *Étude sur les Gesta Martyrum romains* (Paris, 1900), p. 334. But the statement of Dom E. C. Butler in his chapter on monasticism in the *Cambridge Medieval History* (vol. I, p. 527) that the monasticism of Eustathius had 'strongly developed Manichaean tendencies' seems exaggerated.

[2] Some historians, including L. Duchesne (*Histoire ancienne de l'Église*, 2nd ed.; Paris, 1907, vol. II, p. 382), have thought that the teachings enumerated in the acts of the Council of Gangra are to be imputed not so much to Eustathius himself as to his disciples. Hefele, however, regards the accusations levelled against Eustathius as justified (loc. cit. pp. 1044–5).

[3] In recent years this point of view has been put forward by K. Heussi, *Der Ursprung des Mönchtums* (Tübingen, 1936), pp. 287–90.

the Manichaean dualistic view of the body as intrinsically evil. The attitude of Christian asceticism towards the body is based on its view of the sacramental character of Nature—which itself follows from the Incarnation—and on the belief that the whole of Nature can be used *sacramentally*, provided it is used *sacrificially*, that is with discipline and renunciation. The ideas of sacrament and sacrifice, on the other hand, are fundamentally alien and even opposed to the Manichaean dualism, which condemns the whole of Nature in so far as it is material.

In fact the question whether there is any historical relation between Manichaeism and Orthodox monasticism is more complex. It cannot be denied that in Egypt in the fourth century the Manichaean doctrines did penetrate to some extent into Christian monastic circles.[1] Manichaeism seems to have been widespread in the valley of the Nile already in the time of St Anthony, though later accounts of its success among the monks are no doubt much exaggerated.[2] It is not surprising to find that Christian monasticism, in the period of its formation, was not secure from distortions and heretical deviations, as we have seen in the case of Eustathius of Sebaste:[3] this danger must have been particularly acute in Egypt, where the very large number of anchorites and monks who went into the desert seized with a sudden enthusiasm for the ascetic life, and not all of whom were well equipped for this vocation, was scarcely conducive to the maintenance of a uniformly high level of monastic life. But in the eyes of the Orthodox these could only be pernicious aberrations, and those monks who lapsed into dualistic heresy invariably found themselves in opposition to the Church and were denounced with firmness (and often irony) by the Desert Fathers and with vigour and precision by the theologians and the councils.[4]

There can be no doubt that if Manichaeism was able to thrive so successfully and to survive for so many centuries in the Near

---

[1] See de Stoop, op. cit. pp. 73–9; Alfaric, op. cit. vol. I, pp. 58–60.

[2] De Stoop, ibid.

[3] It is interesting to note that Eustathius received his monastic training in Egypt.

[4] For this reason Cumont's statement (*La propagation du manichéisme dans l'Empire romain*, loc. cit. p. 35) that 'on n'a pas assez considéré l'intervention certaine du facteur manichéen dans le développement de *l'idéal* monastique' [the italics are mine] seems historically false.

East, it was partly because it could utilize as a receptive ground for the propagation of its doctrines certain heretical movements within Christianity, such as the false asceticism of several sects and the exaggeration of perverted monasticism. But the success of Manichaeism was also due to another factor: alongside the influence of dualism on heterodox forms of Christian asceticism, an opposite process can be observed in those countries where Mani's followers were proselytizing in Christian surroundings—a gradual infiltration of Christian concepts and terms into Manichaeism. It seems that this infiltration was the result of direct borrowing, for already Mani himself, whose intention was to found a universal religion, consistently adapted his teaching to the existing beliefs and religious terminology of the civilizations and peoples he wished to convert. Thus the ideas of Jesus and of the Paraclete were borrowed by Mani himself from Christianity.[1] These early attempts to bring Manichaean dualism into harmony with the teaching of the Church were continually made by Mani's followers in Christian countries, and became later eminently characteristic of the neo-Manichaean sects in the Near East and in the Balkans, especially of the Paulicians and the Bogomils.

In the Near East, an interesting precursor of these methods of neo-Manichaean exegesis was the Manichaean Agapius, author of the *Heptalogos*, an extensive treatise of Manichaean theology.[2] We possess a short summary of this work by the Patriarch Photius,[3]

[1] The name of Paraclete was applied by the Manichaeans to Mani himself (Polotsky, *Manichäismus*, Pauly-Wissowa, Supplbd vi, col. 266). The exact position of Jesus in the Manichaean system is not altogether clear. The strong Christian influence which can be found in Manichaeism already at the time of its formation would explain the apparently central position occupied by Jesus in its cosmology. (See Burkitt, op. cit. pp. 38–43; E. Waldschmidt and W. Lentz, 'Die Stellung Jesu im Manichäismus', *Abh. preuss. Akad. Wiss.* (phil.-hist. Kl.), 1926; Schaeder, op. cit. pp. 150 et seq.) It seems, moreover, that the following generations of Manichaean missionaries in Christian lands, in their attempt to prove that the gospel of Mani was but a more profound and universal interpretation of Christianity, made strenuous efforts to smooth out the differences between their view of Jesus and orthodox Christology (see Polotsky, op. cit. cols. 268–70). However, in spite of these syncretistic attempts, the Manichaeans could never entirely conceal their essentially docetic conception of Christ, which invariably provoked the indignant denunciations of the theologians of the Church.

[2] See de Stoop, op. cit. pp. 66–9; Alfaric, op. cit. vol. ii, pp. 106–12.
[3] *Bibliotheca*, Cod. 179, *P.G.* vol. ciii, cols. 521–5.

but practically no information on the person of the author.[1] From Photius's account Agapius appears as an original if somewhat syncretistic thinker who attempted to reconcile Manichaeism with the Neoplatonic philosophy current in his time, and especially with the dogmas of the Christian Church. As a true Manichaean Agapius held that there is an evil principle, self-subsistent and from eternity opposed to God, that the body is opposed to the soul, the latter being consubstantial with God, rejected the Old Testament and the Mosaic Law, condemned sexual intercourse, the eating of meat and the drinking of wine and taught other characteristically Manichaean doctrines. And yet he publicly professed a number of Christian dogmas basically opposed to Manichaeism, such as the belief in the Trinity, the Incarnation, the Baptism, Crucifixion and Resurrection of Christ, the Resurrection of the Dead and the Last Judgement.[2] This he was able to do, says Photius, by 'altering and translating almost all the terms of piety and of the Christian religion into other meanings, either strange and abominable, or monstrous and foolish' and by teaching 'perversely behind the names of our dogmas quite different things'.[3] There is no doubt that Agapius was a forerunner of those neo-Manichaeans—particularly the Paulicians and the Bogomils—who excelled in the art of professing adherence to the very Christian dogmas which most blatantly contradicted their dualistic tenets, while interpreting them in accordance with their own beliefs by a free use of the allegorical method. The same accusation of nominally 'agreeing with the words of the pious while barking at the things they designate'[4] was later constantly and angrily levelled by the Orthodox against the neo-Manichaeans. De Stoop is probably right in tracing back to Agapius those ingenious tricks by which the Paulicians, for reasons of personal safety, would subscribe to the letter of the dogmas of the Church without abjuring their faith.[5]

[1] According to Photius, Agapius wrote his large work of twenty-three chapters, dedicated to his female disciple Urania, as well as a number of hymns, and was an opponent of the Arian Eunomius, bishop of Cyzicus in western Asia Minor. De Stoop thinks that Agapius lived in the fourth century or at the beginning of the fifth, probably in Asia Minor; Alfaric, on the other hand, who identifies Agapius with Aristocritus, author of 'Theosophy' (cf. infra, p. 43, n. 3) places his life in Egypt in the second half of the fifth century.
[2] Photius, ibid.   [3] Ibid. col. 524.   [4] Ibid.   [5] Cf. infra, pp. 40–1.

The preceding remarks concerning the spread of Manichaeism in the Near East between the third and the seventh centuries have dealt with some of the reasons for the continued success of the sect in this region. Another factor which no doubt facilitated the propagation of Manichaeism in the eastern provinces of the Roman Empire was the penetration into Armenia and Mesopotamia of Manichaeans from Persia. The thirteenth-century Syriac writer Barhebraeus mentions the arrival in Armenia and Syria in the reign of Justinian II (685–95) of heretics whom he calls 'Barburiani', 'who in Syriac are termed "Maliunaie" and are an offshoot of the Manichaeans; these heretics, expelled from Persia, came to Armenia and thence to Syria, where they invaded and started to inhabit those monasteries which they found'.[1] It is very probable that the frequent persecutions of Manichaeans by the Sassanian rulers compelled many of these heretics to seek refuge in the adjacent territories of the Roman Empire.[2] From there they could extend their influence over Syria, Armenia and Asia Minor.

The doctrines of Mani, by their continued appeal to sectarian movements within Christianity, by their superficial adaptation to the teaching of the Church and by the influx of Manichaeans from Persia, survived in the Near East at least as late as the seventh century. It was then that the remains of Manichaeism were adopted and transformed by the Paulician sect, that first step in the neo-Manichaean movement that was to carry dualism over the greater part of southern Europe, from the Black Sea to the Atlantic.

[1] Gregorius Barhebraeus, *Chronicon ecclesiasticum* (ed. J. B. Abbeloos and T. J. Lamy; Lovanii, 1872), t. I, cols. 219–22.
[2] See de Stoop, op. cit. p. 81.

CHAPTER II

NEO-MANICHAEISM IN THE NEAR EAST

The Paulicians of Armenia and Asia Minor. Peter of Sicily, a Byzantine ambassador among the Paulicians (A.D. 869). Beginnings of Paulicianism: Constantine of Samosata and the 'Church of Macedonia'. The Paulicians in the seventh and eighth centuries: persecutions and schisms. Sergius and the seven Paulician 'Churches'. Decline and fall of Paulician power. Doctrines of the Paulicians. Was Paulicianism a revival of Manichaeism? Marcionites, Massalians and Borborites. Origin of Paulicianism: legends and facts.

In the second half of the seventh century the Paulician sect spread over large areas of Armenia and Asia Minor. Forming the border populations of the Asiatic Themes of the Byzantine Empire, the Paulicians inevitably came into contact with the religious and political life of Byzantium.

The Byzantine ecclesiastical and secular authorities made many attempts to convert these heretical subjects of the Basileus. In the seventh century the Emperors Constantine IV Pogonatus and Justinian II prescribed coercive measures against them: the latter even condemned a number of obdurate Paulicians to the stake.[1] Under the Iconoclastic emperors of the eighth century they seem to have suffered no persecution and to have spread throughout Asia Minor, from Phrygia and Lycaonia to Armenia.[2] In the ninth century, after the reign of Nicephorus I (A.D. 803–11), apparently the last emperor to have shown toleration towards the Paulicians,[3] violence was used against them on an unprecedented scale. Michael I, under the influence of the patriarch of Constantinople, officially introduced capital punishment against the heretics. The persecutions continued under Theophilus and reached their peak when Theodora, in her efforts to extirpate the heresy, ordered a wholesale massacre of the Paulicians, who

[1] Petrus Siculus, *Historia Manichaeorum qui et Pauliciani dicuntur*, P.G. vol. CIV, cols. 1280–1.
[2] See J. B. Bury, *A History of the Eastern Roman Empire* (London, 1912), p. 276.
[3] See A. A. Vasiliev, *Byzance et les Arabes* (French ed. by H. Grégoire and M. Canard; Bruxelles, 1935), vol. I, pp. 229–30.

perished in thousands.¹ The violence of these methods is explained by the fact that in the ninth century the Paulicians were both a military and a religious menace to the Byzantine Empire. At the beginning of the century a great leader, Sergius, had arisen among them; he brought them unity, inspired them with missionary zeal and formed them into well-organized communities.² They were warlike and unruly subjects of the Empire and, in alliance with the Arabs, made frequent raids into Byzantine territory. They built themselves fortified towns near Melitene, in the region of Sivas, on the western borders of Armenia; the most important of these was Tephrice, which became their capital and the residence of their great military leaders, Carbeas and Chrysochir.³ Such was the strength of the Paulicians that the armies of Chrysochir were able to raid Nicaea and Nicomedia and, in 867, to capture and plunder Ephesus. In reply to the emperor's proposals of peace, Chrysochir proudly demanded that the imperial provinces east of the Bosporus should be abandoned to the Paulicians. This led to the campaigns of 871–2, in which the Byzantine armies were victorious, Tephrice was razed to the ground, Chrysochir slain, and the military power of the Armenian Paulicians destroyed for ever.⁴

In 869 an imperial ambassador, Peter of Sicily, was sent by Basil I to Tephrice.⁵ His instructions were to arrange an exchange of prisoners with the Paulicians⁶ and also to negotiate peace between the emperor and Chrysochir. Peter remained in Tephrice for nine months and was successful in the first of his two missions.⁷

[1] Bury, op. cit. pp. 40, 277–8; Vasiliev, op. cit. p. 230.
[2] Cf. infra, pp. 35–7.
[3] See Vasiliev, op. cit. pp. 231–2, n. 2, who refers to the result of recent excavations showing the exact situation of the medieval Paulician fortresses. Cf. W. M. Ramsay, *The Historical Geography of Asia Minor* (London, 1890), p. 342.
[4] See Vasiliev, op. cit. vol. III (*Die Ostgrenze des byzantinischen Reiches*), p. 60; A. Vogt, *Basile Ier...et la civilisation Byzantine* (Paris, 1908), pp. 322–5.
[5] Petrus Siculus, *Historia Manichaeorum*, P.G. vol. CIV, col. 1241. The author says of his arrival in Tephrice: ἐκεῖσε παραγενόμενος ἐν ἀρχῇ τῆς αὐτοκρατορίας Βασιλείου τοῦ...μεγάλου βασιλέως ἡμῶν. Basil succeeded Michael III in 867. The time of Peter's stay in Tephrice (869–70) has been established by Vasiliev, Византия и Арабы, *Zapiski ist.-fil. fakulteta imperatorskogo S.-Peterburgskogo Universiteta*, vol. LXVI (1902), pp. 25–9, especially p. 26, n. 3.
[6] Petrus Siculus, loc. cit. cols. 1241, 1304. [7] Ibid. col. 1304.

But his peace proposals must have met with complete failure, for he had to return to Constantinople in 870[1] bearing to his sovereign the exorbitant demands of Chrysochir for the cession of Asia Minor to the Paulicians.

During his nine months' stay in the Paulician capital, Peter had many occasions to study the doctrines and customs of the Paulicians[2] and decided to write a systematic treatise to expose and refute their heresy.[3] Disquieting news which he learnt in Tephrice made this work all the more urgent: he heard from the Paulicians themselves that they were planning to send missionaries to Bulgaria to spread their teaching in that country.[4] Peter decided, therefore, to write his treatise not only for the authorities in Constantinople, but also for the special use of the Bulgarian Church, and addressed the prologue to the archbishop of Bulgaria.[5] The work was completed *c.* 872.[6]

The value of Peter's *Historia Manichaeorum* as a source for our knowledge of the Paulicians has been the subject of much discussion. Until recently the majority of scholars were inclined to depreciate its importance and reliability.[7] Thanks, however, to

---

[1] Peter says he accomplished his mission ἐν τῷ δευτέρῳ ἔτει τῆς βασιλείας Βασιλείου καὶ Κωνσταντίνου καὶ Λέοντος (*Hist. Man.*, *P.G.* vol. CIV, col. 1304).

[2] πολλάκις αὐτοῖς διαλεχθείς (ibid. col. 1241).

[3] Ibid. col. 1240.   [4] Ibid. col. 1241.   [5] Ibid. col. 1244.

[6] See H. Grégoire, 'Sur l'histoire des Pauliciens', *Bull. Acad. Belg.* (classe des lettres, 1936), p. 224.

[7] The main detractors of the *Historia Manichaeorum* were Karapet Ter-Mkrttschian, *Die Paulikianer im byzantinischen Kaiserreiche und verwandte ketzerische Erscheinungen in Armenien* (Leipzig, 1893), and J. Friedrich, 'Der ursprüngliche bei Georgios Monachos nur theilweise erhaltene Bericht über die Paulikianer', *S.B. bayer. Akad. Wiss.* (philos.-phil. hist. Kl.) (München, 1896), pp. 67-111). Mkrttschian, in particular, asserted against all historical evidence that the *Historia Manichaeorum* was written at the time of Alexius Comnenus (op. cit. pp. 122, et seq.). The views of Mkrttschian and Friedrich influenced a whole generation of scholars, even, to some extent, Bury, who in 1902 merely summed up the position of the problem, without offering any final solution (Gibbon, *Decline and Fall* (ed. J. B. Bury), vol. VI, App. VI, pp. 562-6). Bury was sceptical of Peter's account of the danger presented by the Paulicians to the Bulgarian Church.

But the historical importance of the mission of Peter of Sicily to Tephrice and of the relations between the Armenian Paulicians and the Bulgarians was stressed already in 1898 by F. C. Conybeare (*The Key of Truth. A Manual of the Paulician Church of Armenia*, Oxford, p. cxxxvii) and in 1902 by Vasiliev (Византия и Арабы, loc. cit. pp. 27–8).

the work of Prof. Henri Grégoire, it seems that Peter of Sicily is now finally rehabilitated. In a penetrating study of the sources concerned with the Paulicians,[1] Grégoire has shown that the *Historia Manichaeorum* is the only fully authentic and reliable firsthand account we possess of the history, doctrines and customs of the Paulicians of Tephrice. The other sources hitherto regarded as important by scholars—particularly the treatise *Contra Manichaeos* by the Patriarch Photius,[2] the tract of the 'abbot Peter' on the Paulicians,[3] the passage in the chronicle of George Monachus dealing with the Paulicians,[4] the 24th chapter of the *Panoplia Dogmatica* of Euthymius Zigabenus[5] and the formula of abjuration for the use of those Paulicians who were received into the Church[6]— are, in reality, either derivative or of secondary importance. The first book of Photius's treatise, which alone is concerned with the history and doctrines of the Paulicians, is proved by Grégoire to be a tenth-century forgery. Grégoire's researches have convincingly shown that the *Historia Manichaeorum* of Peter of Sicily must be regarded as the fundamental, almost exclusive, source for our knowledge of the Paulicians of Tephrice.[7] The last twenty-two chapters of the *Historia Manichaeorum* (XXI–XLIII) are devoted to a history of the Paulician sect, approximately from 668 to 868, which is of the greatest value.

Peter of Sicily and subsequent Byzantine historians and theologians regarded the Paulicians as direct descendants of the Manichaeans: Manichaeism and Paulicianism are for them one

---

[1] 'Les sources de l'histoire des Pauliciens: Pierre de Sicile est authentique et "Photius" un faux', *Bull. Acad. Belg.* (classe des lettres, 1936), vol. XXII, pp. 95–114.

[2] *P.G.* vol. CII, cols. 16–264; according to Grégoire, Photius's homilies against the Manichaeans (books II, III and IV) may all be authentic, though the last one alone was certainly written by the patriarch.

[3] Πέτρου ἐλαχίστου μοναχοῦ Ἡγουμένου περὶ Παυλικιανῶν τῶν καὶ Μανιχαίων, published by J. Gieseler (Göttingen, 1849).

[4] *Chronicon* (ed. de Boor), vol. II, pp. 718–25; the accounts of the abbot Peter and of George Monachus are derived from Peter of Sicily.

[5] *P.G.* vol. CXXX, cols. 1189–1244. The *Panoplia Dogmatica* was written at the time of Alexius Comnenus and the chapter dealing with the Paulicians is based entirely on earlier documents.

[6] *P.G.* vol. I, cols. 1461–72.

[7] Cf. H. Grégoire, 'Autour des Pauliciens', *Byzantion* (1936), vol. XI, pp. 610–14.

and the same heresy.[1] A comparison between the Manichaean and the Paulician doctrines will show that Peter's view of the filiation of the latter from the former, if somewhat over-simplified, is substantially correct.

This filiation likewise appears in Peter's account of the origins of the Paulician sect. According to him, Paulicianism first appeared in Samosata, a town on the Euphrates, on the borders of Syria, Mesopotamia and Cappadocia. There, at a time which Peter does not specify, a Manichaean woman called Callinice brought up her two sons, Paul and John, in the Manichaean faith and sent them to proselytize among the inhabitants of the neighbouring regions. From the names of these two Manichaean teachers, Paul and John, arose the name of the new sect of the Paulicians.[2]

From this extremely vague account it is impossible to date the activities of Callinice and her two sons, nor can we tell whether the teaching of Paul and John differed in any way from primitive Manichaeism. Peter of Sicily becomes more precise when speaking of him who appears to have been the real founder of the Paulician sect. This is a certain Constantine, an Armenian, born in the village of Mananali on the upper Euphrates, in the reign of Constans II (641–68).[3] Originally a follower of Mani and of Paul and John of Samosata, he wished, according to Peter, to support his doctrines by means of the New Testament. Moreover, in order to escape the stigma attached to Manichaeism and to 'revive the evil' he rejected the 'Manichaean books' used by his co-religionists[4] (while retaining the doctrines they contained) and decreed that no books should be read except the New Testament. Constantine's reforms, according to Peter, were in no way a departure from the basic doctrines of Manichaeism, but a reclothing of them in a form apparently more acceptable to Christians. Thus, together with the Manichaean books, he rejected 'the blasphemies of Valentinus

---

[1] οὐ γὰρ ἄλλοι οὗτοι, καὶ ἄλλοι ἐκεῖνοι, ἀλλ' οἱ αὐτοὶ Παυλικιάνοι καὶ Μανιχαῖοι ὑπάρχουσιν (*Hist. Man.* loc. cit. cols. 1240–1); cf. ibid. col. 1300.

[2] Ibid. col. 1273.

[3] 'Constantine, grandson of Heraclius', must be Constans II. See S. Runciman, *The Medieval Manichee* (Cambridge, 1946), pp. 35, 37. See ibid. p. 37, n. 1, for the location of Mananali and pp. 35–44 for a brief history of the Armenian Paulicians.

[4] For the Manichaean books in Armenia in the late sixth and in the seventh centuries, see supra, pp. 17–18.

concerning the thirty aeons', 'the legends of Cubricus (Mani) about the formation of the rain' and other theories (καὶ ἄλλα τινά)—in other words, it would seem, that part of the Manichaean teaching (in particular the cosmological myths) which most flagrantly contradicted the Christian doctrines. On the other hand, Constantine borrowed some of 'the filth of Basilides' and of others (καὶ τῶν λοιπῶν ἁπάντων) and thus appeared as a 'new leader' (νέος τις ὁδηγός). From the time of Constantine, according to Peter of Sicily, the Paulicians, ignorant of these tricks, anathematize readily (προθύμως) Mani and the other Manichaean teachers.[1] Peter's account of Constantine's role in reforming Paulicianism, in spite of its rather abstract character, is clear and consistent. The reform (or, probably more accurately, the foundation) of Paulicianism by Constantine was based on an attempt to bring the old Manichaean doctrines into an apparent agreement with Christianity. Constantine was doubtless one of the instigators of the distinction, characteristic of Paulicianism (and, later, of Bogomilism), between an exoteric teaching consisting mainly of the New Testament for the use of the ordinary members of the sect and an esoteric one, whereby the Christian Scriptures were interpreted in accordance with dualistic teaching by the secret and oral tradition of the initiates.[2] Constantine, as described by Peter of Sicily, was essentially a reformer of primitive Manichaeism; it seems that Paulicianism, regarded as an attempt to reconcile the dualism of Mani with the teaching of the Gospel, dates from him. One of the most remarkable Christian features of Paulicianism was its great veneration for St Paul. Here also Constantine was an initiator. According to Peter, he assumed the name of Silvanus, the companion of St Paul (Acts xv-xviii; II Cor. i. 19; I Thes. i. 1; II Thes. i. 1), and under this name took up his residence in the fortress of Cibossa, near Colonea, on the frontiers of Armenia Minor and the Pontus. There, in Peter's words, he claimed 'to be the Silvanus mentioned in the epistles of the Apostle, whom Paul sent as his faithful disciple to Macedonia.

[1] Ibid. cols. 1276–7.
[2] Peter says that the Paulicians do not impart their mysteries to all the members of their sect, but only 'to those few among them whom they know to be more perfect in impiety'; at the same time 'the heresy of the Manichaeans is observed and honoured by them in deep silence'. (Ibid. col. 1252.)

He would show his disciples the book of the Apostle... saying: "you are the Macedonians, and I am Silvanus, sent to you by Paul"'.[1] The Paulician community of Cibossa took the name of the 'Church of Macedonia'.[2] This name was doubtless chosen because Cibossa was situated near Colonea: the Paulicians, who had an intimate knowledge of the Acts of the Apostles and of the Epistles of St Paul, could not fail to remember the passage in the Acts where the name Macedonia is coupled with the term κολωνία.[3] There can be little doubt that Constantine-Silvanus was the actual founder of the Paulician sect. The many Christian elements in Paulicianism—in particular the cult of St Paul—can be traced back to his reforms. He was also the instigator of the tradition, prevalent among his successors, of giving to the Paulician 'Churches' names associated with St Paul and to the leaders of these communities the names of those disciples of St Paul connected by history or tradition with the names of these churches.[4] Constantine-Silvanus was at the head of the 'Church of Macedonia' for twenty-seven years and was finally arrested and stoned to death by order of an imperial officer, Symeon, sent to Armenia by Constantine IV to stamp out the Paulician heresy. The death of the founder of Paulicianism thus coincided with the first general persecution of the Paulicians, instigated by Constantine IV (668–85) and Justinian II (685–95). Many Paulicians refused to be converted and were martyred for their faith. We may suppose that their courage in persecution impressed even their enemies: for Symeon himself, the leader of the Byzantine punitive expedition, after his return to Constantinople, renounced everything he possessed, secretly left the capital, returned to Cibossa and was received into the Paulician 'Church of Macedonia'. He assumed the name of Titus—another companion of St Paul, like Silvanus associated with Macedonia (II Cor. ii. 13; vii. 5–6, 13–15)—gathered the remaining disciples of Constantine-Silvanus and became the leader of the Armenian Paulicians.[5] Symeon-Titus was soon to pay with his life for his apostasy from orthodoxy: three years after

[1] *Hist. Man.*, *P.G.* vol. civ, cols. 1277–80. [2] Ibid. col. 1297.
[3] Acts xvi. 11–12: 'Ἀναχθέντες οὖν ἀπὸ τῆς Τρωάδος εὐθυδρομήσαμεν εἰς Σαμοθράκην, τῇ τε ἐπιούσῃ εἰς Νεάπολιν, ἐκεῖθέν τε εἰς Φιλίππους, ἥτις ἐστὶ πρώτη τῆς μερίδος τῆς Μακεδονίας πόλις, κολωνία. Cf. H. Grégoire, 'Les sources de l'histoire des Pauliciens', loc. cit. pp. 102–3.
[4] *Hist. Man.* col. 1277 and infra, p. 36. [5] Ibid. cols. 1280–1.

he became the leader of the Paulicians, a quarrel arose between him and a certain Justus, the adopted son of Constantine-Silvanus. The dispute arose about the interpretation of the well-known passage in Col. i. 16, which, from the Paulician standpoint, was doubtless hard to reconcile with any dualistic cosmology; it is not improbable that the quarrel was also due to personal jealousy. Justus appealed to the bishop of Colonea, to whom he revealed all the secrets of the Paulicians. The bishop promptly notified the authorities in Constantinople, and the Emperor Justinian II, seeing that this Armenian sect, supposed to have been stamped out by his father, was still in existence, ordered the arrest of all Paulicians. Those who persisted in their faith, including Symeon-Titus, were burnt alive (c. 690).[1]

The eighth century was one of mixed blessings for the Paulician sect. On the one hand the lack of imperial persecution of the Paulicians favoured their spread in Armenia and Asia Minor; on the other hand, the sect was weakened within by a series of schisms and internecine struggles, exploited by the Byzantines and the Arabs, who were doubtless only too glad of an opportunity to fish in the muddy waters of Paulician politics.[2]

A new era dawned for the Paulician sect at the beginning of the ninth century with the advent of the greatest of the Paulician leaders, Sergius. His activities were manifold: Sergius was a teacher, a reformer, a missionary and an organizer. For thirty-four years (801–35) he ruled over the Paulicians under the name of Tychicus, the disciple of St Paul (Col. iv. 7). A strong and earnest figure, he reinvigorated the moral life of the Paulicians, who had fallen into lax ways under his predecessor. Himself an ardent missionary, he inspired his followers with the zeal for spreading their faith. Peter quotes these words of Sergius, which have a remarkably apostolic ring: 'from the East to the West, to the North and to the South I have journeyed, proclaiming the Gospel of Christ, walking with my own knees'.[3] In these missionary journeys on foot Sergius traversed the whole breadth of Asia Minor, from north to south: he founded three new Paulician 'Churches', those

[1] Ibid. col. 1281.
[2] For the Paulician schisms in the eighth century, see ibid. cols. 1281–8.
[3] Ibid. col. 1293. The text, as given in Migne, reads: τοῖς ἐμοῖς γόνασι βαρήσας. Βαρήσας must be a mistaken reading of βαδίσας. I am indebted for this information to Prof. R. M. Dawkins.

of 'Laodicea', in Cynochorion near Neocaesarea in the Pontus, of 'Ephesus' in Mopsuestia in Cilicia, and of 'Colossae' in Argaoun (Argovan) near Melitene.[1] The names of all three churches are associated with St Paul, and the 'Laodicea' in the Pontus was deliberately identified by the Paulicians with the Laodicea in southern Phrygia mentioned in the fourth chapter of the Epistle to the Colossians, which also refers to Tychicus (Col. iv. 7, 16). Thus at the height of Sergius' missionary activity the Paulician 'Churches' in Armenia and Asia Minor comprised the following:

(1) The 'Church of Macedonia' in Cibossa, near Colonea, founded by Constantine-Silvanus and reconstituted after his death by Symeon-Titus.

(2) The 'Church of Achaia' in Mananali on the eastern branch of the Euphrates, founded by Gegnesius-Timothy.[2]

(3) The 'Church of Philippi' (situation unknown), founded by Joseph-Epaphroditus.[3]

(4) The 'Church of Laodicea' in Cynochorion, near Neocaesarea.

(5) The 'Church of Ephesus' in Mopsuestia.

(6) The 'Church of Colossae' in Argaoun (Argovan) near Melitene.[4]

The last three were founded by Sergius-Tychicus.

(7) Moreover, as Grégoire has pointed out, the Paulicians regarded as their mother church the Church of Corinth, founded directly by St Paul. Thus the sacred number of seven was completed—Corinth remained the Church of St Paul himself, the other six those of his 'reincarnate' disciples.[5]

[1] *Hist. Man.*, *P.G.* vol. civ, cols. 1288–97.
[2] Ibid. col. 1297. For the location of Mananali see Runciman (op. cit. p. 37, n. 1), who, on this point, follows Conybeare in preference to Photius and Grégoire. For the Paulician leader Gegnesius-Timothy, see *Hist. Man.* cols. 1281–5. Timothy, the disciple of St Paul, was noted for his missionary work in Achaia, particularly in Athens and Corinth (I Thes. iii. 1–2; II Cor. i. 1, 19; Rom. xvi. 21).
[3] Ibid. col. 1297; for Joseph-Epaphroditus, see ibid. col. 1285. Epaphroditus was sent by St Paul to the Philippians (Phil. ii. 25).
[4] This location of the Paulician 'Churches', except in the case of the 'Church of Achaia', is that of E. Honigmann and H. Grégoire (Grégoire, 'Les sources de l'histoire des Pauliciens', loc. cit. pp. 101–5).
[5] Καὶ πάλιν φησίν (ὁ Τυχικός)· ἔτι δὲ λέγω, τὴν ἐν Κορίνθῳ Ἐκκλησίαν ᾠκοδόμησε Παῦλος· τὴν δὲ Μακεδονίαν, Σιλουανὸς καὶ Τίτος...(*Hist. Man.* col. 1297); cf. Grégoire, ibid. pp. 102–4.

The prestige of Sergius among the Paulicians was so great that he was regarded by his followers, as Mani had been, as the Paraclete Himself. In his missionary work he appears to have been extraordinarily successful: Peter of Sicily relates that, to follow him, married people broke their conjugal ties, monks and nuns their monastic vows, even children and priests became his disciples. It was undoubtedly the highest peak ever reached by the dynamic power of the Paulician sect.[1]

But meanwhile, in the face of this serious menace to the orthodox faith and to the security of the Empire, forces were gathering in Byzantium which were destined to destroy for ever the Paulician power in Armenia and Asia Minor. Persecutions against the Paulicians were resumed by Michael Rhangabe, and Theodora's massacres dealt them a crippling blow. Many Paulicians crossed the borders of the Byzantine Empire and found refuge with the Arabs, who regarded them as useful allies against Byzantium. Sergius, with a group of followers, was befriended by the emir of Melitene, in whose territory was situated the 'Church of Colossae'. There Sergius was murdered in 835.[2]

After his death important changes took place in the organization of the Paulician sect: hitherto the sect had been organized on a hierarchical principle and the Paulician leaders, called συνέκδημοι,[3] seem to have enjoyed some sacerdotal prerogatives. But after Sergius's death, the Paulicians replaced the hierarchical organization of their sect by a democratic one: in Peter's words, Sergius's disciples assembled their followers in Argaoun and, 'mutilating the teaching of their master Sergius and of his predecessors, all became equal in rank; no more did they nominate any one teacher (διδάσκαλον) as before, but were all equal (πάντες ἴσοι ὄντες)'. The συνέκδημοι were replaced by the νοτάριοι who, though doubtless distinct from the laity, were equal in rank.[4]

We do not know the effect of this reform on the inner strength of the Paulician sect; in any case its military power reached its peak some thirty years after Sergius's death, when the Paulician

---

[1] Ibid. col. 1293.   [2] Ibid. col. 1301.

[3] 'Companions in travel', or travelling preachers; the term was used to designate the companions of St Paul (Acts xix. 29; II Cor. viii. 19).

[4] Ibid. col. 1301. The νοτάριοι were probably, as Conybeare suggests (op. cit. p. cxxiv), 'copyists of the sacred books'.

armies, commanded by Chrysochir,[1] extended their domination to the Propontis and the Aegean Sea. The heyday of the Paulician power was as short as it was spectacular: after the attempt to secure peace with the Paulicians through Peter of Sicily had failed, the Byzantine armies succeeded in 872 in finally crushing their military power.

Peter's account of the history of the Paulician sect is of vital importance for all students of the Balkan dualistic movements, for from among the Paulicians of Tephrice were drawn those missionaries who spread Paulicianism in Bulgaria and who thus directly contributed to the rise of Bogomilism.

Of equal importance is Peter's analysis of the Paulician doctrines[2] which he had many occasions to study in Tephrice. According to him, the most characteristic feature of Paulicianism is its underlying dualism. The Paulicians believed in two Principles, the one good, the other evil; the second is the creator and ruler of the present, visible world, the first the creator and lord of the world to come.[3] Holding the material world to be a creation of the evil Principle, the Paulicians could naturally not accept the Christian dogma of the Incarnation. For how could Christ, who came from heaven, have become man and taken the flesh which belonged to the realm of evil? They were thus led to postulate a Docetic Christology, according to which the 'body' of Christ was of heavenly origin, His Incarnation only 'seeming', and the maternity and virginity of Our Lady were denied:[4] she was not the Mother of Christ, but the 'heavenly Jerusalem'.[5] Heretical in their non-acceptance of the fundamental dogmas of Christianity, the Pauli-

---

[1] The career of this famous Paulician general is described by Runciman (op. cit. pp. 41–3).

[2] Summaries of the Paulician doctrines may be found in the articles by Bonwetsch in the *Realencyklopädie für protestantische Theologie und Kirche*, vol. xv (Herzog-Hauck), and by Janin in the *Dictionnaire de Théologie Catholique*, vol. xii.

[3] πρῶτον μὲν γάρ ἐστι τὸ κατ' αὐτοὺς γνώρισμα, τὸ δύο ἀρχὰς ὁμολογεῖν, πονηρὸν Θεὸν καὶ ἀγαθόν· καὶ ἄλλον εἶναι τοῦδε τοῦ κόσμου ποιητήν τε καὶ ἐξουσιαστήν, ἕτερον δὲ τοῦ μέλλοντος (*Hist. Man.*, P.G. vol. civ, col. 1253).

[4] τὸν θεῖον αὐτῆς τόκον ἐν δοκήσει καὶ οὐκ ἐν ἀληθείᾳ γεγενῆσθαι δογματίζουσιν (col. 1248); τὸ τὴν πανύμνητον καὶ ἀειπάρθενον Θεοτόκον μηδὲ κἂν ἐν ψιλῇ τῶν ἀγαθῶν ἀνθρώπων τάττειν ἀπεχθῶς ἀπαριθμήσει· μηδὲ ἐξ αὐτῆς γεννηθῆναι τὸν Κύριον, ἀλλ' οὐρανόθεν τὸ σῶμα κατενεγκεῖν (col. 1256).

[5] ἔλεγε δὲ ταύτην εἶναι τὴν ἄνω Ἱερουσαλήμ, ἐν ᾗ πρόδρομος ὑπὲρ ἡμῶν εἰσῆλθε Χριστός (col. 1284).

cians denied much of the written and oral tradition of the Church. From the canon of the Scriptures they rejected the whole of the Old Testament (the Prophets they branded as 'deceivers and thieves') and the Epistles of St Peter, whom they hated as having denied Christ.[1] The Paulician canon consisted of the four Gospels, the fourteen Epistles of St Paul, the Epistle of St James, the three Epistles of St John, the Epistle of St Jude and the Acts of the Apostles. The text was identical with that used by the Christian Church 'without any change in the words'.[2] The extreme veneration of the Paulicians for St Paul, which caused them to name their leaders after the disciples of the Apostle mentioned in his Epistles and their own communities after the Churches ministered to by St Paul, has already been mentioned. Apart from the Christian Scriptures, the Paulicians also used some epistles of their leader Sergius-Tychicus.[3] One of them, quoted by Peter of Sicily, was addressed to 'Leo the Montanist', and may imply the existence of some direct relations between the Paulicians and the Montanists of Asia Minor at the beginning of the ninth century.[4] Their

[1] Ibid. col. 1256.
[2] A marginal annotation 'antiqua manu' in the MS. of the *Historia Manichaeorum* published by Migne gives the following valuable information: the author of the scholium wrote: 'I do not know whether [the Paulicians in the days of Peter of Sicily] used the Epistle of James, or another Epistle, and the Acts of the Apostles. But the present-day ones use only the four Gospels—and especially the Gospel according to St Luke—and fifteen epistles of St Paul: for they have another epistle [addressed] to the Laodiceans' (ibid. cols. 1255–6). This scholium, according to Grégoire ('Sur l'histoire des Pauliciens', loc. cit. p. 226), is of the eleventh century. 'On voit que les Pauliciens *dédoublaient* l'épître aux Éphésiens, et que leur canon portait: (1) une épître aux Colossiens; (2) une aux Éphésiens; (3) une aux Laodicéniens, très pareilles et contenant toutes trois au moins un passage relatif à Tychikos.' (Grégoire, 'Les sources de l'histoire des Pauliciens', loc. cit. p. 104.) This evidence that the Paulicians laid particular emphasis on the Gospel of St Luke and used the Epistle to the Laodiceans confirms the striking resemblance between their canon and that of Marcion (cf. infra, p. 47). In Migne there is an error in the text of the scholium: the Paulicians used not τοῖς δύο...εὐαγγελίοις but τοῖς δ' εὐαγγελίοις (see H. Grégoire, 'Sur l'histoire des Pauliciens', loc. cit. p. 226, n. 1).
[3] *Hist. Man.* col. 1256.
[4] Ibid. col. 1297. Sergius accuses Leo of rending 'the true faith', which, as Runciman points out (op. cit. p. 61, n. 2), may imply that Sergius and Leo were 'officially of the same faith'. But Leo's surname of 'the Montanist' need not, perhaps, be taken too seriously (Runciman, ibid.).

attitude towards their own canon was governed by their dualistic cosmology and strongly tainted by rationalism. The text of the Holy Writ was sacrosanct, and they were careful not to adulterate a single word, for for them the Word of God was not the Incarnate Logos, but solely the teaching of Christ, set out in the Gospels. On the other hand, since they denied the possibility of contact between God and Matter, the Paulicians were obliged, in explaining certain events recorded in the Gospels, to resort to their own interpretation, opposed to the teaching of the Church. Those events which, according to Orthodox Christianity, are based on the sanctification of Matter, such as the institution of Baptism and of the Eucharist, were perforce interpreted by them in a non-material, figurative sense. The Paulician view of the Eucharist is particularly typical of this: according to Peter, the Paulicians rejected the sacrament of the Eucharist and held that the bread and the wine, given by Christ to His disciples at the Last Supper were, 'symbolically', His words.[1] Peter relates the significant episode of the interrogation of the Paulician leader Gegnesius-Timothy; it affords a good illustration of the allegorical method, so frequently practised by the Paulicians (and later by the Bogomils) in interpreting Christian dogma. Gegnesius was summoned to Constantinople by Leo the Isaurian to render an account of his faith, which the Byzantine authorities had every reason to suspect of being heretical. Accused by the Patriarch of denying the Orthodox Faith, the Cross of Christ, the Mother of God, the communion of the Body and Blood of Christ, the Catholic and Apostolic Church and Baptism, Gegnesius professed a firm belief in all these doctrines. But, says Peter, Gegnesius meant in reality by the Orthodox Faith 'his own heresy', by the Cross—the Person of Christ, who formed that figure with His arms outstretched,[2] by the Mother of God— the 'heavenly Jerusalem', by the Body and Blood of Christ —simply His words, by the Catholic and Apostolic Church— 'the communities of the Manichaeans' (i.e. of the Paulicians), by Baptism—Christ again, the giver of the 'living water'

[1] τὸ τὴν θείαν καὶ φρικτὴν τῶν ἁγίων μυστηρίων τοῦ σώματος καὶ αἵματος τοῦ Κυρίου καὶ Θεοῦ ἡμῶν μετάληψιν ἀποτρέπεσθαι... λέγοντες ὅτι οὐκ ἦν ἄρτος καὶ οἶνος, ὃν ὁ Κύριος ἐδίδου τοῖς μαθηταῖς αὐτοῦ ἐπὶ τοῦ δείπνου, ἀλλὰ συμβολικῶς τὰ ῥήματα αὐτοῦ αὐτοῖς ἐδίδου, ὡς ἄρτον καὶ οἶνον (Hist. Man., P.G. vol. CIV, col. 1256).
[2] The Paulicians spurned the material figure of the Cross (ibid. col. 1256).

(John iv. 10). Incredible as it may seem, this mystification appears to have been successful and Gegnesius returned home to Armenia provided with a safe-conduct from the emperor.[1] It can scarcely be doubted that Peter simplified the details of this story, which, as it stands, taxes somewhat one's credulity with regard to the gullibility of the supreme Byzantine authorities; yet the story itself is quite credible and shows how difficult it must have been for the Church to combat heretics who, when they were questioned, professed complete conformity with the orthodox teaching. From the above examples it can be seen that the Paulicians, in their dualistic rejection of matter as a vehicle for Grace, were led to oppose the whole of the sacramental teaching of the Orthodox Church. It is, however, not clear whether they repudiated the use of images: the rejection of images would seem consonant with their view of matter; but, since evidence is lacking on this point, this cannot be affirmed with any certainty.[2]

The Paulicians not only rejected the principal dogmas but also the entire organization of the Christian Church, in particular the Order of Priesthood. The word πρεσβύτεροι was, it seems, especially hateful to them, as it also designated the Jewish elders who formed the council against Our Lord.[3] Their own elders—the συνέκδημοι (replaced, after Sergius's death, by the νοτάριοι)—claimed no Apostolic Succession except the spiritual descent from St Paul. As guardians of the true faith, the Paulicians claimed for

[1] Ibid. col. 1284.
[2] *The Key of Truth* mentions the rejection of images (Conybeare, op. cit. pp. 86, 115), but, as we shall see, it is essentially an Adoptionist, not a Paulician document and cannot be regarded as an authentic 'manual of the Paulician Church of Armenia', which Conybeare considers it to have been. It has been asserted, with even less justification, that the Paulicians were Iconoclasts (Janin, 'Pauliciens', *D.T.C.* vol. xii). Conybeare goes as far as to call them 'the extreme left wing of the Iconoclasts' (op. cit. p. cvi), a statement echoed by the Vardapet T. Nersoyan in his article on the Paulicians (*Eastern Churches Quarterly*, vol. v, no. 12, 1944, p. 405). A refutation of this view, on historical grounds, can be found in E. J. Martin's *History of the Iconoclastic Controversy* (London, 1930), pp. 275-8. Grégoire, relying on Peter of Sicily, strongly denies it: 'Nos recherches, en établissant la valeur éminente et presqu'exclusive de Pierre de Sicile, nous permettent d'écarter de la doctrine de la secte son prétendu iconoclasme, auquel Pierre ne fait pas la moindre allusion. Cette accusation et plusieurs autres ne viennent que beaucoup plus tard' ('Autour des Pauliciens', *Byzantion*, vol. xi, p. 613).
[3] *Hist. Man.* col. 1257.

themselves exclusively the name of 'Christians', while the Orthodox were 'Romans'. According to Peter, they were fond of opening a conversation with the question: 'tell me, what is it that separates us from the Romans?'[1] The conception of the Church Catholic they did not reject, but applied it to their own communities.[2] The Paulicians appear to have had a particular aversion to monks: according to George Monachus, they held that the monastic garb was revealed by the Devil to St Peter, who then gave it to men.[3]

Peter emphasizes the extreme difficulty of distinguishing the Paulicians from the Orthodox Christians. Like their future descendants, the Bogomils, the Paulicians not only called themselves Christians, but in ethics and even in doctrine they simulated complete conformity with the teaching of the Church.[4] The accusation of hypocrisy, frequently levelled at them for this reason, should be conditioned by the remark that although their outward profession of Orthodoxy was no doubt a commonly used weapon of self-protection against persecution, it in no way contradicted the principles of their faith, according to which the Christian dogmatic formulas were accepted, but interpreted by them in a figurative sense. On the other hand the Paulicians were quite capable of accepting martyrdom when necessary,[5] and Peter himself testifies to their remarkable courage and self-abnegation.[6]

A study of the Paulician doctrines naturally leads to the following questions: Are the Paulicians to be regarded as authentic representatives of Armenian Manichaeism? If the Paulicians and the Manichaeans formed two distinct sects, were the former, nevertheless, derived doctrinally and historically from the latter? Can any other origin, outside Manichaeism, be found for the Paulician doctrines?

The first two of these questions can be answered largely by reference to the *Historia Manichaeorum*. Grégoire's brilliant vindication of Peter of Sicily has to-day dispelled the scepticism which had long reigned regarding the reliability of his treatise; the neglect

[1] *Hist. Man.*, P.G. vol. CIV, col. 1253.   [2] Cf. supra, p. 40.
[3] See Friedrich, 'Der ursprüngliche bei Georgios Monachos...', loc. cit. p. 73.
[4] χρηστὸν σχηματίζονται ἔχειν τὸ ἦθος, καὶ πάντα τὰ παρὰ τοῖς ὀρθοδόξοις Χριστιανοῖς δόγματα ἐπικυροῦσι δολίως καὶ ἀναφωνοῦσιν (*Hist. Man.* col. 1245).
[5] See ibid. col. 1280.
[6] εἰώθασι...πολλοὺς κόπους καὶ κινδύνους προθύμως ἀναδέχεσθαι πρὸς τὸ μεταδιδόναι τῆς οἰκείας λοίμης τοῖς παρατυγχάνουσι (ibid. col. 1241).

of this most important source has led many scholars into false conclusions about the history and the doctrines of the Paulician sect.[1]

The picture of Paulicianism which we derive from Peter of Sicily is thát of a reformed and simplified Manichaeism. Peter, who, of all the contemporary Byzantine historians of Paulicianism, was the only one who had a first-hand knowledge of the sect, categorically affirms that the Paulicians are descended from the Manichaeans, and there is no conclusive evidence for disbelieving him. Even if we disregard, on account of its chronological vagueness, Peter's description of the beginnings of Paulicianism under Paul and John, sons of the Manichaean Callinice, it remains quite clear from the rest of his narrative that the real founder of the Paulician sect in the second half of the seventh century, Constantine-Silvanus, based his teaching on Manichaeism, which he merely divested of those cosmological and mythological accretions which were particularly offensive to Christian ears. We know from another source that 'Manichaean books', whose contents were used by Constantine to elaborate his teaching, existed in Armenia in the seventh century.[2] Those who have tried to disprove the filiation of the Paulicians from the Manichaeans have often stressed the fact that the Paulicians, from the very time of Constantine-Silvanus, anathematized Mani and other Manichaean heresiarchs. But the argument is inconclusive: this behaviour was not unknown to the Manichaeans themselves: in a fifth-century Manichaean writing, Mani is described as 'a wicked man'.[3] It is perfectly possible to explain Peter's observation that the Paulicians anathematized Mani by one or several of the following reasons: an outward repudiation of any connection with the ill-famed founder of Manichaeism would have been fully compatible with the Paulician habit of simulating Orthodoxy when necessary, to avoid interference or persecution; secondly, by appearing to dissociate themselves from Mani, the Paulicians could hope to pursue more convincingly their attempt to bring the Manichaean doctrines closer to Christianity by glossing over their more obvious differ-

[1] The most prominent scholars who fell into grave error regarding Paulicianism were Mkrttschian, Friedrich and Conybeare. Grégoire even describes somewhat sweepingly Mkrttschian's work on the Paulicians as 'un livre faux d'un bout à l'autre'! ('Autour des Pauliciens', loc. cit. p. 610.)

[2] Cf. supra, pp. 17–18.

[3] Words quoted from Aristocritus's 'Theosophy' (P.G. vol. I, col. 1468).

ences;[1] finally, it may also have been that those Paulicians whom Peter of Sicily heard anathematize Mani belonged to the not fully initiated members of the sect who may have been ignorant of its true origin.

If we compare the doctrines of the Paulicians with those of the Manichaeans, there appears immediately a striking resemblance between them. With one exception, all the main Paulician tenets can already be found in Manichaeism: belief in two principles, denial of the Incarnation, rejection of the Old Testament, anti-sacramentalism, predilection for the Pauline Epistles, ostentatious parading of the name of Christians, the title of 'Paraclete' assumed by the leader of the sect. In one respect, however, the Paulicians differed from the Manichaeans: while the 'elect' of the latter sect were bound to abstain from sexual intercourse, meat and wine, we find no trace of any such asceticism among the Paulicians described by Peter of Sicily. Probably this difference in their ethics was due to the fact that the two sects followed different modes of life: the Manichaean ideal was primarily contemplative and monastic, while the Paulicians led a life of action and even war. But in spite of this one important difference, the connections, both doctrinal and historical, between Paulicianism and Manichaeism are beyond any doubt: Paulicianism was not identical with Manichaeism; and yet Manichaeism must be regarded in many respects as the direct ancestor of Paulicianism.[2]

[1] According to the Paulician formula of abjuration, the same Manichaean writer who denounced Mani, Aristocritus—identified by Alfaric (*Les Écritures manichéennes*, vol. II, pp. 107–12) with Agapius (cf. supra, pp. 25–6)—tried in his book 'Theosophy' to prove that 'Judaism, paganism, Christianity and Manichaeism are one and the same doctrine' (*P.G.* vol. I, col. 1468). This is clear evidence of that Manichaean syncretism which gradually imbibed more and more elements of Christianity. In this respect Paulicianism followed and surpassed Manichaeism.

[2] It was no doubt the failure to recognize the paramount importance of the evidence of Peter of Sicily that led so eminent an authority on Manichaeism as H. H. Schaeder to the unjustifiable conclusion that the Paulicians were in no way related to the Manichaeans, that they were falsely accused of Manichaeism by the entourage of the Patriarch Photius, and that they do not deserve the name of 'neo-Manichaeans' given to them in recent years ('Der Manichäismus nach neuen Funden und Forschungen', *Morgenland*, Heft XXVIII, p. 83). The same erroneous statement is made by A. Harnack, *Marcion: das Evangelium vom Fremden Gott* (2nd ed.; Leipzig, 1924), p. 383, n. 2: 'Mit dem Manichäismus haben die Paulicianer nichts zu tun.'

The fact that a number of Manichaean doctrines—particularly the cosmological myth—were not found in Paulicianism is sufficiently explained by Peter's account of the reform and simplification effected in the Manichaean teachings by the founder of the Paulician sect, Constantine-Silvanus, and doubtless continued by his successors.

Nevertheless, it would be false to think that Manichaeism was the only influence which affected the growth of Paulicianism. It must be admitted that the Byzantine theory that Paulicianism was simply a slightly modified continuation of Manichaeism is insufficient. Peter of Sicily admits himself that the Paulician heresiarchs 'added certain idle terms to the earlier heresies',[1] and that Paulicianism appeared as something new.[2]

Church historians have recently tended to emphasize the connections between Paulicianism and Marcionism.[3] Harnack, the greatest authority on Marcionism, remained very cautious on this point: in his authoritative book on Marcion he confessed that after long and careful study of the relations between Marcionism and Paulicianism he was unable to reach any certain conclusion on the matter. He merely supposed that the Paulicians were influenced by their contact with the Marcionites of eastern Asia Minor and that from the eighth century Marcionism in the Near East became in a large measure merged in Paulicianism.[4] No attempt has yet been made to go further than Harnack towards a solution of this problem. It is true that we have as yet no direct evidence of a historical filiation of the Paulicians from the Marcionites. But our present knowledge of Paulicianism now permits of an improvement on Harnack's cautious statement: circumstantial historical evidence and, above all, striking similarities between the doctrines of the Paulicians and the Marcionites clearly point to a very probable contact between the two sects.

There is every reason to suppose that the Marcionites and the Paulicians lived in close geographical proximity. Armenia, the

---

[1] *Hist. Man., P.G.* vol. civ, col. 1276.   [2] Ibid. col. 1277.
[3] J. K. L. Gieseler, *Lehrbuch der Kirchengeschichte* (Bonn, 1846), vol. II, pt 1, p. 14; A. Neander, *Allgemeine Geschichte der christlichen Religion und Kirche* (Gotha, 1856), vol. II, pt 1, p. 134; I. von Döllinger, *Beiträge zur Sektengeschichte des Mittelalters* (München, 1890), vol. I, pp. 2–3; Mkrttschian, op. cit. pp. 104–12.
[4] *Marcion*, pp. 382*–3*.

home of the Paulicians from the seventh century, was still in the middle of the fifth century infested with Marcionites[1] who were considered by the Orthodox to be closely related to the Manichaeans.[2] In the fourth century there began among the eastern Marcionites a general exodus out of the towns into the country, to escape persecution.[3] This movement, which led to the formation of 'Marcionic villages', enabled their communities to survive for several centuries in distant places. The mountains and high valleys of Armenia were a good protection against any very rigorous control by the Church, and there can scarcely be any doubt that by the second half of the seventh century the Marcionites and the Paulicians were in close contact and, fighting the same enemy, found themselves on common ground.

If we turn to the doctrines of the Marcionites and the Paulicians, their similarity—in some cases even identity—is remarkable. The following Paulician doctrines existed already in the teaching of Marcion: the dualism between the good God and the evil creator of the world,[4] Docetism and the rejection of the Incarnation,[5] and the special cult of St Paul. In this respect Paulicianism seems

---

[1] The Armenian bishop Eznik of Kolb devoted a whole chapter of his treatise 'Against the Sects', written between 441 and 449, to an exposition and a refutation of Marcionism. See *Ausgewählte Schriften der armenischen Kirchenväter*, Bd I, pp. 152–80. Moreover, Gieseler has pointed out that the eight Marcionic localities which Theodoret in the fifth century claimed to have converted were situated in those very districts of Armenia which two centuries later became the home of the Paulicians ('Untersuchungen über die Geschichte der Paulicianer', *Theologische Studien und Kritiken*, 1829, pt I, pp. 104–5).

[2] Harnack, op. cit. p. 158. [3] Ibid. pp. 158 et seq.

[4] The Paulician dualism seems to have differed somewhat from that of the Manichaeans, for the Paulicians opposed not God to Matter, nor Light to Darkness, but merely the good God, Lord of the next world, to the evil God, creator of this world. Mkrttschian rightly remarked that the Paulicians do not appear to have had any explicit doctrine of matter (op. cit. p. 107). On the other hand, the Paulician dualism resembles more that of Marcion: for Marcion formulated his dualism essentially in terms of two Gods—the good God and the just God—and, according to Harnack, matter, regarded as a principle, played no part in his Biblical teaching (*Marcion*, p. 161).

Moreover, it is interesting to compare Marcion's teaching on 'the foreign God' ('der fremde Gott', cf. Harnack, op. cit. pp. 4–5, 118–20) with these words of the Paulicians, reported by Peter of Sicily: 'they [the Paulicians] say to us: "you believe in the Creator of the world, but we believe in him of whom the Lord speaks in the Gospels, saying: Ye have neither heard his voice, nor seen his shape".' (*Hist. Man.* col. 1253.) [5] *Marcion*, pp. 124 et seq.

to be even closer to Marcionism than to Manichaeism; for although the Manichaeans, in their attempts to oppose the New Testament to the Mosaic Law, had recourse to St Paul more than to any other Christian writer, it was above all Marcion who regarded St Paul as the corner-stone of the true faith, second only to Christ Himself.[1] Perhaps the most striking similarity between the Paulicians and the Marcionites lies in the canon of Scriptures they used. Not only did both sects reject the Old Testament and lay particular stress on the Epistles of St Paul (the Manichaeans did likewise); but the Marcionites, like the Paulicians, especially honoured the Gospel of St Luke. For Marcion, St Luke's Gospel was, in its original form, of divine inspiration, and, in its present form, comparatively free from the 'falsifications' of the Jewish apostles.[2] Moreover, the apocryphal Epistle to the Laodiceans, falsely attributed to St Paul, and used by the Paulicians, is, as Harnack has shown, of Marcionic origin.[3]

These close similarities between the doctrines of the two sects have led some scholars to regard the Paulicians simply as descendants of the Marcionites and Paulicianism as a restoration of original, pure Marcionism.[4] But this view runs counter to the evidence of Peter of Sicily and cannot be substantiated. Harnack justly remarked that for the numerous similarities between Marcionism and Paulicianism there are important features in which they differ.[5] Thus the anti-sacramentalism of the Paulicians was no part of the teaching of Marcion, whose followers celebrated Baptism, the Eucharist and other rites of the Christian Church.[6] Moreover, we find no trace among the Paulicians of the dualistic asceticism of the Marcionites, who were bound by the rules of their sect to avoid sexual intercourse and the eating of meat.[7]

Thus, while substantially accepting the opinion of Peter of Sicily that Paulicianism was derived mainly from Manichaeism, we must recognize that Marcionism also played an important part in the rise and development of the Paulician doctrines. On the one

[1] Ibid. pp. 30 et seq., 198 et seq.
[2] Ibid. pp. 40 et seq., 249* et seq. Cf. supra, p. 39, n. 2.
[3] Ibid. pp. 134*–49*.   [4] In particular Mkrttschian (op. cit. p. 110).
[5] *Marcion*, p. 383*.
[6] Ibid. p. 144. In the Marcionic Eucharist, however, water was used instead of wine.
[7] Ibid. pp. 148–51.

hand it is impossible to deny that Manichaeism exerted a direct influence on Paulicianism, at least at the beginning of the history of the latter sect; on the other hand, the Marcionic character of several Paulician teachings is equally undeniable. Both the Byzantine theory of the filiation of the Paulicians from the Manichaeans and the view that Paulicianism was simply a revival of Marcionism are both insufficiently accurate when viewed in isolation and must be supplemented by each other. It seems probable that the doctrinal reforms carried out in the seventh century by the Paulician leader Constantine-Silvanus, and which Peter of Sicily describes as a rejection of some Manichaean tenets and a borrowing from other sources, were largely an assimilation of some Marcionic teachings; we may assume that the influence of Marcionism enabled the Paulicians to attenuate somewhat the original Manichaean cosmological dualism and to bring, at least outwardly, their teaching nearer to Orthodox Christianity.[1]

But apart from Manichaeism and Marcionism, we must take into account another, at least partly dualistic, movement which seems to have exerted some influence on the Paulician sect in Armenia and Asia Minor, and whose doctrines were later to have a direct and lasting effect on the development of Balkan neo-Manichaeism.

The doctrines of the Massalian sect, with one probable exception,[2]

[1] E. Amann regards Paulicianism as a simplified form of Manichaeism and thinks it possible that this simplification was due to the influence of Marcionism (A. Fliche and V. Martin, *Histoire de l'Église*, 1940, vol. VII, p. 436). This hypothesis seems not only possible, but highly probable.

It is interesting to note that in the twelfth century, in a formula of abjuration used by the Byzantine Church for converted Bogomils (who were directly descended from the Paulicians), the Bogomils are identified on the one hand with the Massalians or the Euchitae, and on the other with the Marcionites. (Ἔλεγχος καὶ θρίαμβος τῆς βλασφήμου καὶ πολυειδοῦς αἱρέσεως τῶν ἀθέων Μασσαλιανῶν, τῶν καὶ Φουνδαϊτῶν, καὶ Βογομίλων καλουμένων, καὶ Εὐχιτῶν, καὶ Ἐνθουσιαστῶν, καὶ Ἐγκρατητῶν, καὶ Μαρκιωνιστῶν: in J. Tollius, *Insignia itinerarii italici* (Trajecti ad Rhenum, 1696), p. 106.)

[2] The so-called *Spiritual Homilies of Macarius* have in recent years been ascribed not to the great Egyptian ascetic of the fourth century, but to a contemporary Massalian who, it is thought, disguised the doctrines of his sect under an orthodox name and terminology. See Dom L. Villecourt, 'La date et l'origine des "Homélies spirituelles" attribuées à Macaire', *C.R. Acad. Inscriptions Belles-Lettres* (1920), pp. 250–8. Cf. *Fifty spiritual homilies of St Macarius the Egyptian*, ed. by A. J. Mason, London, 1921.

NEO-MANICHAEISM IN THE NEAR EAST 49

are known to us solely from the evidence of their Orthodox opponents.[1] The name Massalians—or Messalians—is derived from a Syriac word meaning 'those who pray', of which the exact Greek equivalent is εὐχίται; in Greek sources both terms are used synonymously to describe them. Some of their doctrines were identical with those of the Paulicians. Thus they condemned the Christian Church and its hierarchy, interpreted the New Testament in an individualistic way, disbelieved in the Real Presence in the Eucharist, but partook of the sacrament in order to conform outwardly to the discipline of the Church. In other respects, however, they differed from the Paulicians: the basic doctrine of the Massalians was that in every man from his birth there dwells a demon who cannot be expelled by Baptism, but only through prayer.[2] As their name shows, the Massalians held that prayer was the most essential occupation of man and the necessary and sufficient condition of salvation. They claimed to follow the precept of St Paul: 'Pray without ceasing' (I Thes. v. 17), and maintained that they alone understood the true meaning of the Lord's Prayer. Sacraments were powerless and unnecessary. They believed that the effect of continual prayer was to bestow the gift of the Spirit; this gift created in the soul, purified of all passions, a visionary and prophetic state, in which they claimed to contemplate the Trinity with their bodily eyes. The Massalians taught that in this state the soul becomes possessed with a sacred delirium, which manifests itself by jumping, dancing and symbolically trampling under foot the vanquished demon. For this reason the Massalians were also called ἐνθουσιασταί[3] and χορευταί.[4] Of

[1] See in particular St Epiphanius, *Adversus Haereses*, lib. III, t. 2, *P.G.* vol. XLII, cols. 756–73; Theodoret, *Haereticarum Fabularum Compendium*, lib. IV, *P.G.* vol. LXXXIII, cols. 429–32; Timotheus, *De receptione haereticorum*, *P.G.* vol. LXXXVI (1), cols. 45–52; St John Damascene, *De haeresibus*, c. 80, *P.G.* vol. XCIV, cols. 728–37; Photius, *Bibliotheca*, Cod. LII, *P.G.* vol. CIII, cols. 88–92. All the extant Orthodox sources—Syriac, Greek and Latin—concerning the Massalians have been edited and translated by M. Kmosko, *Patrologia Syriaca*, pars I, t. 3, (Paris, 1926), cols. clxx–ccxciii.

[2] Like so many anti-ecclesiastical sects, the Massalians claimed to base their doctrines on the teaching of Christ. In this case, they took the words of Our Lord: 'This kind goeth not out but by prayer and fasting' (Matt. xvii. 21; Mark ix. 29), but interpreted them in an anti-sacramental sense. The same method was frequently resorted to by the Paulicians.

[3] Theodoret, loc. cit. col. 432. [4] Timotheus, loc. cit. col. 48.

those who had not reached this state of perfection, rigorous asceticism was required: they lived in complete poverty on public charity,[1] renounced all manual labour as an obstacle to contemplation and frequently assumed the monastic garb. Their favourite centres of proselytism were the Orthodox monasteries. For those of them, however, who had succeeded in finally driving out the demon, sin was no longer possible and any discipline or restriction became superfluous; this belief frequently drove them into the worst sexual excesses, which are so commonly associated with the Massalians by their Orthodox opponents. Extreme asceticism and extreme immorality thus appear as equally characteristic of the behaviour of these heretics.[2] Women sometimes held the position of teachers in the sect,[3] as they also did among the Manichaeans, the Montanists and the Marcionites.

The origin of the Massalians is generally placed in Mesopotamia and Osrhoene (particularly round Edessa). In the second half of the fourth century they spread in great numbers to Syria and Asia Minor, where their presence is attested in Pamphylia and Lycaonia.[4] They were expelled from Syria by order of Flavian, patriarch of Antioch,[5] but succeeded in corrupting several monasteries in Armenia Minor; the bishop of Melitene, having gained information from Flavian about this heresy, had the Massalian monasteries burnt and their heretical occupants expelled.[6]

But these measures had little effect, and in the fifth century the Massalians were more widespread than ever. They were particularly numerous in Syria after the death of Flavian (404). In Asia Minor, apart from the above-mentioned provinces, they invaded Lycia and Cappadocia.[7] In Armenia, their tenets were

---

[1] The Massalians have been called for this reason 'the first mendicant friars'. See A. Neander, *Allgemeine Geschichte der christlichen Religion und Kirche*, vol. I, pt 2, p. 544.
[2] Ibid. pp. 521–2.   [3] Timotheus, op. cit. *P.G.* vol. LXXXVI, col. 52.
[4] See Neander, op. cit. p. 514; J. G. V. Engelhardt, *Kirchengeschichtliche Abhandlungen* (Erlangen, 1832), pp. 197–8; J. Gieseler, *Lehrbuch der Kirchengeschichte*, vol. II, pt I, p. 402; G. Bareille, 'Euchites', *D.T.C.* vol. v, cols. 1456 et seq.
[5] Theodoret, op. cit. *P.G.* vol. LXXXIII, col. 432; Photius, *Bibliotheca*, *P.G.* vol. CIII, col. 88.
[6] Theodoret, ibid.; Photius, loc. cit. col. 89.
[7] See Gieseler, ibid.; *D.T.C.* loc. cit.

condemned by the Synod of Shahapivan (447).[1] The importance of the heresy can be judged by the fact that it was condemned at the Third Oecumenical Council at Ephesus in 431, which emphasized the false conception of asceticism held by the Massalians and their predilection for monasteries.[2]

Between the sixth and the tenth centuries our knowledge of the Massalians is derived principally from Armenian sources. They appear to have been numerous in Armenia, judging from their condemnation by the Catholicos John of Otzun in the eighth century and by Gregory of Narek in the tenth, both of whom identify them with the Paulicians.[3] In the eighth century the Massalians are mentioned by St John Damascene, again with the observation that they are to be found particularly in monasteries.[4] In the ninth century Photius speaks of them as still existing.[5]

The question of the existence of direct doctrinal or historical connections between the Massalian and Paulician sects remains obscure for the lack of sufficiently definite evidence. Here, as in the case of Encratism, Montanism and Novatianism, their common dualistic and anti-sacramental tendency may be largely due to the inheritance of certain basic ideas and of a spiritual frame of mind, outlined in the previous chapter, and which are older than Manichaeism itself. Nevertheless, the identification of the Massalians with the Paulicians by contemporary Armenian Churchmen, and the fact that several centres of Massalianism in the fourth and fifth centuries, such as Lycaonia, Cappadocia and western Armenia, contained in the eighth and ninth centuries large numbers of Paulician colonies, strongly suggest that at least in Armenia the two movements co-existed and even blended in some measure.

[1] See Mkrttschian, op. cit. pp. 42 et seq.; F. Tournebize, *Histoire politique et religieuse de l'Arménie* (Paris, 1900), pp. 320–1.

[2] '[Massaliani] convicti...non permittantur habere monasteria, ut ne zizania diffundantur et crescant' (*Definitio sanctae et oecumenicae synodi Ephesinae contra impios Messalianitas*, Mansi, op. cit. vol. IV, p. 1477.

[3] See Mkrttschian, *Die Paulikianer in byzantinischen Kaiserreiche*, pp. 39–47; F. C. Conybeare, *The Key of Truth*, pp. lvii, cvii–cviii. According to both scholars, the Armenian term equivalent to Massalianism, *mtslnéuthiun*, became in the eighth century a general term of abuse.

[4] Μασσαλιανῶν, τῶν ἐν μοναστηρίοις μάλιστα εὑρισκομένων. (*De haeresibus*, c. 80, *P.G.* vol. XCIV, col. 736.)

[5] καθὼς καὶ ἡμεῖς...πολλὴν σηπεδόνα παθῶν καὶ κακίας τὰς ἐκείνων ψυχὰς ἐπιβοσκομένην ἑωράκαμεν. (*Bibl.*, Cod. LII, *P.G.* vol. CIII, col. 92.)

But even apart from its probable influence on Paulicianism, the Massalian sect occupies a prominent position in the history of the European dualistic movements by its far-reaching repercussions on the development of Bogomilism in the Balkans, which will be examined in the following chapters.

Some influence may also have been exerted on Paulicianism by the rather mysterious Borborites, whose activities in Armenia in the first half of the fifth century were causing anxiety in the highest quarters in Byzantium. The Armenian Catholicos Sahak received a letter from the Patriarch Atticus of Constantinople, requesting him, in the name of the Emperor Theodosius II, to convert the Borborites or else to expel them from his diocese. Sahak found himself impelled to prescribe the death-penalty against them.[1] Although these Armenian Borborites may have been Massalians under a different name,[2] it seems more probable that they were a sect of Gnostics, for they are considered as such by contemporary heresiologists.[3] Their name, derived from the Greek βόρβορος (mud), was probably given to them on account of the reputed immorality of their lives and ceremonies. If the Armenian Borborites were Gnostics, they may well have transmitted some of the dualist tradition to the Paulicians.[4]

A few words must be said here about the Armenian Thonraki, who were almost certainly related to the Paulicians.[5] Their founder was a certain Sembat, who lived in the first half of the ninth century in the district of Thonrak, north of Lake Van, not far from Mount Ararat. The Armenian writers who describe them[6] seem, with one exception,[7] to have regarded them as distinct

---

[1] Moses Chorensis, *Histoire d'Arménie* (French tr. by P. E. Le Vaillant de Florival; Venice, 1841), t. II, pp. 154–7; V. Langlois, *Collection des historiens anciens et modernes de l'Arménie* (Paris, 1869), t. II, pp. 165–6.

[2] See Mkrttschian, op. cit. pp. 39–42.

[3] Epiphanius, *Adversus Haereses*, lib. I, t. II, p. xxvi, *P.G.* vol. XLI, cols. 336 et seq.; Theodoret, *Haeretic. Fabul. Compend.* lib. I, p. 13, *P.G.* vol. LXXXIII, cols. 361–4; cf. G. Bareille, 'Borboriens', *D.T.C.* vol. II.

[4] Runciman thinks that the Borborites 'may well have followed a simplified form of Gnosticism...which was developed into Paulicianism' (op. cit. p. 61).

[5] I have made much use of Runciman's description of the Thonraki (op. cit. pp. 51–9).

[6] Gregory of Narek, Gregory Magister, Paul of Taron. See Conybeare, op. cit. pp. 125–30, 141–51, 174–7.

[7] See Runciman, op. cit. p. 53.

from the Paulicians. But the Paulician and the Thonraki doctrines show some remarkable similarities. Thus, the Thonraki, while anathematizing Mani, believed in two Principles, claimed that the earth had been created by the Devil, rejected the cult of Our Lady, the Sacraments, the Cross and the Order of Priesthood, spurned churches, icons and relics, and asserted that Moses was inspired by the Devil. Moreover, they were said 'to love Paul and execrate Peter'. Like the Paulician Sergius—and Mani— who were regarded by their followers as the Paraclete, the leader of the Thonraki claimed to be Christ. The Thonraki flourished in Armenia between the ninth and the twelfth centuries and still existed in the nineteenth, when a doctrinal manual, *The Key of Truth*, was in use among them.[1] *The Key of Truth* is generally considered to date from the ninth century at the latest and was presumably used by the original Thonraki. However, in its extant form it displays marked Adoptionist features and, while confirming in many respects the picture of the Thonraki painted by medieval Armenian writers, contains such essentially non-dualist teachings as the recognition of the sacraments of Baptism and the Eucharist and the belief that God (and not the Devil) created the world. As an Adoptionist work, probably later taken over by the Thonraki, it cannot be regarded as a reliable source of information on the medieval Thonraki, still less on the Paulicians. The relation of the Thonraki to the Paulicians cannot, with our present knowledge of the former, be exactly determined, but contact between the two sects seems fairly certain, and the hypothesis of a common ancestry (perhaps Marcionite) is probable.[2]

It is necessary to examine another theory regarding the origin of the Paulicians, which is still sometimes put forward to-day. The importance of this theory lies not in its conclusions, which are

---

[1] The MS. of *The Key of Truth* was discovered by Conybeare in 1891 and edited by him, together with an English translation, in 1898. Conybeare considered this document to be 'a manual of the Paulician Church of Armenia' and tried to prove, from its undeniably Adoptionist features, that Paulicianism was a form of Adoptionist Christianity. His conclusions are aptly refuted by Runciman, who points out that '*The Key of Truth* is probably an ancient work of the Armenian Adoptionists, and was probably at some much later date taken over by the Thonraki, who found most of its teaching closely akin to their own; and its influence may have inclined them out of Dualism into Adoptionism' (op. cit. p. 57). [2] See Runciman, op. cit. pp. 59–60.

completely untenable, but rather in its premises, which raise the still unsolved problem of the origin of the name of Paulicians.

According to the opinion of Peter of Sicily, followed by the majority of Byzantine historians and theologians, the Paulicians derived their name from either one or both names of Paul and John, sons of Callinice of Samosata, the supposed founders of the Paulician sect.[1] Those, however, who accepted this view were faced with the difficulty of explaining the exact derivation of the Greek form Παυλικιάνοι. Thus, already in the tenth century, the Pseudo-Photius showed some hesitation to pronounce on the origin of this name and quoted the opinion, current in his time, that Παυλικιάνοι is a debased form of Παυλοϊωάνναι, which is more clearly derived from the names of Paul and John.[2] Modern scholars have shown themselves sceptical of this theory,[3] and the few sentences devoted to Paul, John and Callinice in the *Historia Manichaeorum* are certainly too vague to permit the etymology of the name Paulicians to be entirely based on this hypothetical account of their historical origin.

On the other hand, some historians have preferred to derive the name of the Paulicians from that of St Paul.[4] The special veneration of the Paulicians for this apostle and their custom of calling their leaders after the disciples of St Paul would seem to justify this derivation. However, this theory still does not explain the form Παυλικιάνοι, which is not a simple derivation from Παῦλος.

[1] Cedrenus, *Historiarum Compendium, C.S.H.B.* vol. I, p. 756; Anna Comnena, *Alexiad*, lib. XIV, cap. 8, *C.S.H.B.* vol. II, p. 297; Euthymius Zigabenus, *Panoplia Dogmatica*, tit. 24, *P.G.* vol. CXXX, col. 1189.

[2] *Contra Manichaeos*, lib. I, *P.G.* vol. CII, col. 17: Ἐκ θατέρου τοίνυν τῶν εἰρημένων, ὅτῳ Παῦλος ἦν ὄνομα, ἀντὶ τοῦ γινώσκεσθαι διὰ τῆς τοῦ Χριστοῦ παρωνυμίας τὴν τῶν Παυλικιάνων κλῆσιν οἱ τῆς ἀποστασίας ἐρασταὶ μετηλλάξαντο, οἱ δὲ οὐκ ἐκ θατέρου φασὶν, ἀλλ' ἐξ ἑκατέρου συναφθέντων ἀλλήλοις τῶν ὀνομάτων εἰς ἐκβαρβαρωθεῖσαν ἐπίκλησιν σύνθετον, καὶ ἀντὶ τοῦ Παυλοϊωάνναι καλεῖσθαι αὐτοὺς ὅπερ νῦν ὀνομάζονται.

[3] According to Gieseler, the derivation of the name of the Paulicians from Paul and John, sons of Callinice, is 'a later, Catholic, fiction' (*Lehrbuch der Kirchengeschichte*, vol. II, pt 1, p. 15, n. 4).

[4] See Gibbon, *Decline and Fall* (Bury's ed.), vol. VI, p. 112; I. von Döllinger, *Beiträge zur Sektengeschichte des Mittelalters* (München, 1890), vol. I, p. 3; Gieseler thinks that the Paulicians originally received this name from the Christians, on account of their perpetual references to St Paul. Cf. Runciman, op. cit. p. 49.

An attempt to explain the etymology of Παυλικιάνοι was made by Mkrttschian: according to him, the name is of Armenian origin and could only have arisen on Armenian soil; it is formed of the root Pol (Paul) and the derisive suffix 'ik'; 'Paulikiani' is thus the Armenian derivative of 'Pauliani' and would mean literally 'the followers (or sons) of the wretched little Paul'. This personage whom the Armenian Christians derisively referred to as *Polik* cannot have been St Paul, but was a heretic by the name of Paul, rightly or wrongly regarded as the original teacher of the Paulicians.[1] Mkrttschian's theory has not been successfully refuted and no more satisfactory explanation of the origin of the name of Paulicians has been offered. It seems therefore that the derivation of *Paulicians* from the Armenian *Polik* can be accepted.

Who was this *Polik*? According to Mkrttschian, he may have been the celebrated heretic Paul of Samosata, the third-century bishop of Antioch. To the credit of Mkrttschian it must be said that he regarded this relation between the Paulicians and Paul of Samosata as purely fictitious and the result of a confusion, since the 'teaching [of Paul of Samosata] has nothing to do with Paulicianism'.[2] Other scholars, however, in particular Conybeare, have put forward the view that Paulicianism was directly derived from the teaching of Paul of Samosata, bishop of Antioch. Conybeare has tried to prove at great length that the teaching of the Paulicians was a recrudescence of 'primitive' Adoptionist Christianity, of which Paul of Samosata was one of the most celebrated protagonists. More recently the same opinion was expressed by L. Petit, who pointed out that, while western Armenia came within the orbit of the theological school of Caesarea, the south-eastern part of the country remained for a long period under the influence of the ante-Nicene doctrines of Antioch, particularly of Adoptionism.[3]

Petit's argument, however, is valueless when applied to those Paulicians we are considering, i.e. those west of the Euphrates,

[1] 'Die Wurzel ist hier der abgekürzte volkstümliche Name Pol—Paul mit dem verkleinernden Suffix -ik, welches, wie auch in anderen Sprachen, im Sinne des Spottes gebraucht werden kann.... Auf solche Weise würde Polikianer einfach einen Anhänger des Pol, des vielleicht von dem Volke verspotteten Polik, bedeuten' (op. cit. pp. 63–4); cf. Conybeare, op. cit. pp. cv–cvi.
[2] Op. cit. p. 64.   [3] L. Petit, 'Arménie', *D.T.C.* vol. I, col. 1900.

who alone came into direct contact with the Byzantine Empire and the Balkans, since their home was not in south-eastern, but in western Armenia. Moreover, it is high time that the entirely mythical theory of the connections between the Paulicians and Paul of Samosata, which cannot be justified either doctrinally or historically, be finally abandoned.[1]

The doctrines of the Paulicians and those of the bishop of Antioch are in many respects in direct contradiction to each other. Thus the Adoptionist Christology of Paul of Samosata is diametrically opposed to the Docetic Christology of the Paulicians; both, no doubt, denied the full reality of the Incarnation, but for opposite reasons: Paul of Samosata, in his attempt to emphasize the unity of God, regarded Christ not as God-man, but as a human being who, on account of his absolute obedience and abundant virtue, received by Grace the name of Son of God:[2] the Paulicians, starting from a fundamental dualism between the heavenly God and the evil creator of this world, taught that Christ was a heavenly being, incapable of assuming the flesh which belongs to the realm of the wicked Demiurge. Several other instances could be found of the opposition between the doctrines of the Paulicians and those of Paul of Samosata: for example, the Judaic tendencies present in the teaching of the latter[3] are in contrast with the fundamental anti-Judaism of the former.

All the historical evidence likewise militates against any possible filiation of the Paulicians from Paul of Samosata. Bardy, in his authoritative work on Paul of Samosata, clearly shows that Paul never succeeded in founding a lasting school and that the 'Samosatean sect' was virtually extinct by the fifth century.[4] However, the memory of the heretical bishop of Antioch was for long kept alive by the frequent denunciations of his teaching by the Fathers

[1] The view that the doctrines of the Paulicians derive from those of Paul of Samosata was already expressed by the tenth-century Arab writer Mas'ūdī: '(Les Pauliciens) suivent l'hérésie de Paul de Samosate...; il professa des doctrines qui tiennent le milieu entre celles des Chrétiens et celles des Mages et des dualistes(!), car elles comportent la vénération et le culte de toutes les lumières selon leur ordre' (?). (*Le Livre de l'Avertissement*. Traduction par B. Carra de Vaux, Paris, 1896, p. 208.)

[2] G. Bardy, *Paul de Samosate* (Louvain, 1923), pp. 364, 370–80.

[3] Ibid. pp. 382–4, 442–3.

[4] 'La secte samosatéenne ne se répandit jamais en dehors de son pays d'origine et...dès les dernières années du quatrième siècle, au plus tard, elle s'y éteignait au milieu de l'indifférence universelle' (ibid. p. 443).

of the Church, especially by St John Chrysostom. This persistence of the memory of his name long after his doctrines had been forgotten caused Paul of Samosata to become 'le héros d'aventures légendaires auxquelles l'histoire n'a...rien à voir',[1] legends of which his supposed influence on Paulicianism is undoubtedly one.

According to Peter of Sicily, the Paulicians traced their origin to Paul and John, sons of Callinice, a Manichaean woman from Samosata. To identify this Paul with Paul of Samosata is impossible on chronological grounds, for Paul of Samosata became bishop c. A.D. 260,[2] when Mani was only beginning to preach his doctrine, and hence the mother of the bishop of Antioch could not have been a Manichaean.[3] On the other hand, this identification seems to have been made by the contemporary enemies of the Paulicians: at least as early as the ninth century, the Paulicians were probably accused of being the disciples of Paul of Samosata, for Peter of Sicily tells us that they anathematized Paul of Samosata together with Mani.[4] It seems probable that the imaginary connection between the Paulicians and Paul of Samosata was made owing to a confusion between the bishop of Antioch and another 'Paul of Samosata', the son of Callinice, who was, according to Peter of Sicily, one of the original founders of the Paulician sect. Whether the Paulicians really derived their name from Paul, son of Callinice, or whether *Polik* was some other heretic bearing the name of Paul, must remain, for the present, an unsolved problem, since the historical origins of the Paulician sect are still insufficiently known.[5]

[1] Ibid. p. 441. [2] Ibid. p. 169.
[3] Conybeare (op. cit. p. cv) remarked on this chronological impossibility, but unjustifiably used it as an argument to prove that Paulicianism was never subjected to any Manichaean influences. What it proves in reality is simply that Paul of Samosata, bishop of Antioch, and Paul of Samosata, son of Callinice, were two different persons.
[4] *Hist. Man.* loc. cit. col. 1245.
[5] Runciman (op. cit. pp. 48–9) thinks that the Paulicians derived their name from St Paul, but that some later Byzantine historian, led astray by the fact that one of Callinice's sons bore the same name, 'decided too impetuously that he had solved the problem'. Contrary to Mkrttschian, Runciman applies the contemptuous suffix 'ik' not to the name of Paul, but to that of the sectarians themselves. 'Paulicians' would mean not 'the followers of the wretched little Paul' (cf. supra, p. 55), but 'the petty followers of Paul' (i.e. St Paul). This name, according to Runciman, was applied to the Paulicians by 'their opponents in Armenia, tired of having St Paul continually thrust at them by these heretics'. The hypothesis is attractive, but as the evidence seems somewhat inconclusive, I prefer to follow Mkrttschian's interpretation.

When divorced from its legendary and imaginary accretions[1] the problem of the origin of Paulicianism cannot yet be said to have been finally elucidated. Some of its historical aspects still remain somewhat obscure. Nevertheless, it seems clear that Paulicianism arose under the combined influence of Manichaeism and Marcionism, and, to some extent at least, of Massalianism. While it appears impossible to deny the Manichaean origin of the Paulician dualism, the Marcionic character of several Paulician doctrines is also undeniable. Compared with original Manichaeism, Paulicianism is, at least outwardly, nearer to Christianity in many respects. In its strong consciousness of the New Testament it is closer to the teaching of Marcion. The difference between Manichaeism and Paulicianism is rather that between a non-Christian dualistic religion, gradually trying to adapt itself to Christianity (Manichaeism), and an attempt to 'reform' Christianity itself on a dualistic basis (Paulicianism).[2]

[1] Another fantastic theory was put forward by C. Sathas and E. Legrand (*Les Exploits de Digénis Akritas*, Paris, 1875, p. lxxviii), according to which Paulicianism was a revolt of 'hellenism' against the 'Roman traditions' of the Byzantine Church. There is not the slightest justification for this view.

[2] This twofold aspect of Paulicianism—Christian and dualistic—was noted by Mas'ūdī: 'Nous avons parlé ailleurs de la doctrine et des dogmes de Beïlaki [i.e. Paulicians], secte qui tient à la fois du christianisme et du magisme.' (*Les Prairies d'Or*. Texte et traduction par C. Barbier de Meynard et Pavet de Courteille, Paris, 1874, vol. VIII, p. 75.) Cf. supra, p. 56, n. 1.

CHAPTER III

THE RISE OF BALKAN DUALISM

I. *The appearance of Paulicianism in Bulgaria:* Penetration of the Paulicians into Bulgaria. The Bulgars, the Slavs and Byzantium. The pagan religion of the Bulgarians. Paulician proselytism in the ninth century.

II. *The rise of Slavonic Christianity:* Baptism of Boris and the growth of the Bulgarian Orthodox Church. Revolt of the boyars against Christianity. Temporary subjection of Bulgaria to the Roman See. Paulicians, Monophysites and Jews in Bulgaria. The 'dual faith' and the pagan revival. The beginnings of a Slavonic culture; the work of St Clement and St Naum. Spread of Massalianism to Bulgaria; growth of heresy at the beginning of the tenth century.

III. *The influence of Byzantium on Bulgaria:* Opposition of the heretics to Byzantine institutions. Byzantinization of Church and State under Symeon and Peter. Social and economic development of Bulgaria; its effect on the growth of heresy. Monasticism and heresy in tenth-century Bulgaria. A favourable ground for heresy.

Peter of Sicily, for all his concern for Orthodoxy in the Balkans, was probably unaware of the fact that there were Paulicians in Bulgaria long before 870.

In the eighth and ninth centuries, Byzantine foreign policy was aimed above all at safeguarding the Empire from the incessant threats of enemy forces on the eastern and northern frontiers. In the east the Arabs, frequently allied with the heretical Paulicians, were a source of never-ending trouble. The northern borderland of Thrace, the perpetual battlefield in the struggle of the Empire with the invaders from the north and which had been laid open from the sixth century to the devastations of the Avars and the Slavs,[1] became, in the eighth century, the road inevitably taken by the armies of the Bulgars, in their frequent raids into the heart of the Empire.[2]

Owing to this double necessity of defence in the east and in the north, several Byzantine emperors pursued the policy of transplanting groups of Armenian heretics into Thrace. This seemed the most effective way to break up the heretical communities, and,

[1] See L. Niederle, *Manuel de l'Antiquité Slave* (Paris, 1923), vol. I, pp. 59–66; F. Dvorník, *Les Slaves, Byzance et Rome au IXe siècle* (Paris, 1926), pp. 3–9.

[2] See S. Runciman, *A History of the First Bulgarian Empire* (London, 1930), pp. 38, 48 et passim.

by settling them in a region largely inhabited by Christians, to render them accessible to Orthodox influences. Moreover, those of them who were noted for their sturdy and warlike qualities (such as the Paulicians) would be, it was hoped, a buffer against invasions from the north.

The Byzantine policy of transplanting groups of Asiatic subjects to various European Themes was an old one, and Thrace was the traditional colonizing ground. Both Diocletian[1] and Heraclius[2] had transported groups of Asiatics into Thrace.

The Emperor Constantine V Copronymus, whose foreign policy was governed by his hostility towards Bulgaria,[3] on two occasions transferred eastern populations to Thrace. In 745, according to Theophanes, a large colony of Syrian Monophysite heretics was settled in Thrace.[4] In 757[5] Copronymus transferred a number of Syrians and Armenians to the same province from Theodosiopolis and Melitene. Their function was to repopulate the plague-stricken districts of Thrace, but they were, in fact, according to Theophanes, responsible for spreading the Paulician heresy there. Nicephorus states that the Emperor liberally provided for the needs of the new settlers.[6]

[1] *Incerti panegyricus Constantio Caesari*, 21 (ed. G. Baehrens; Teubner, Leipzig, 1911), p. 247.

[2] Sebêos, *Histoire d'Héraclius* (Paris, 1904), p. 54. The number settled in Thrace was 30,000 families.

[3] In the course of his reign (741–75) he carried out no less than eight campaigns against Bulgaria.

[4] Theophanes, *Chronographia* (de Boor ed.; Leipzig, 1883), p. 422. Cf. *C.S.H.B.* vol. I, pp. 650–1.

[5] The date of Constantine's second transplantation was fixed at 755 by A. Lombard (*Constantin V, empereur des Romains*, Paris, 1902, pp. 92–3), V. N. Zlatarski (История на българската държава прѣзъ срѣднитѣ вѣкове, Sofia, 1927, vol. I, pt 2, p. 62) and Runciman (op. cit. p. 35), and at 756 by Dvorník, who follows the chronology of Nicephorus (op. cit. p. 68, n. 3). But the correct date seems to be 757, since Theophanes places the event (*Chronographia*, *C.S.H.B.* vol. I, p. 662) in the first year of the pontificate of Paul I, who became Pope in April 757.

[6] Theophanes, *C.S.H.B.* vol. I, p. 662; de Boor ed. vol. I, p. 429: ὁ δὲ βασιλεὺς Κωνσταντῖνος Σύρους τε καὶ 'Αρμενίους, οὓς ἤγαγεν ἀπὸ Θεοδοσιουπόλεως καὶ Μελιτινῆς, εἰς τὴν Θρᾴκην μετῴκισεν, ἐξ ὧν ἐπλατύνθη ἡ αἵρεσις τῶν Παυλικιανῶν.

Nicephorus Patriarcha, *Opuscula Historica* (de Boor; Leipzig, 1880), p. 66: ταῦτα ἐπιτελῆ ποιήσας Κωνσταντῖνος ἤρξε δομεῖσθαι τὰ ἐπὶ Θρᾴκης πολίσματα, ἐν οἷς οἰκίζει Σύρους καὶ 'Αρμενίους, οὓς ἔκ τε Μελιτηναίων πόλεως καὶ Θεοδοσιουπόλεως μετανάστας πεποίηκε, τὰ εἰς τὴν χρείαν αὐτοῖς ἀνήκοντα φιλοτίμως δωρησάμενος. Cf. Nicephorus, *C.S.H.B.* p. 74.

THE RISE OF BALKAN DUALISM 61

The military aims of Copronymus in his colonization of Thrace are clearly apparent and outweigh any religious motive which he might have had.[1] This is shown by the ensuing events: the Syrians and Armenians were settled in a number of fortresses which Constantine was building along the Bulgarian frontier. The Bulgarians, seeing the aggressive intentions of the emperor, tried to reach an agreement with him concerning these fortresses.[2] This having failed, they overran Thrace as far as the Long Wall protecting Constantinople. Constantine, however, attacked suddenly and drove them back with heavy losses.[3]

The next transplantation of Asiatics into Thrace was effected by Leo IV the Khazar in 778, the colonists again being heretics—Syrian Jacobites—who had been captured by the Byzantine forces during their campaign round Germanicea (778).[4]

Although these Asiatic heretics were settled by Constantine V and Leo IV in towns and fortresses originally within the boundaries of the Empire, they soon penetrated into Bulgaria. This colonized borderland between Byzantium and Bulgaria was continually changing hands in the eighth, ninth and tenth centuries. Any Bulgarian attack pre-supposed the invasion of Thrace; thus the Bulgarian Khan Telets captured some of the Thracian frontier towns in 763. Kardam advanced as far as Versinicia, near Adrianople, in 796. Krum carried out terrible devastations of Thrace between 807 and 814 and in 813 he pitched his camp beneath the very walls of Constantinople. After the capture and destruction of Adrianople, he had its entire population, numbering, it was said, 10,000, transported to the northern shores of the Danube. His successor, Omortag, likewise marched through Thrace beyond Arcadiopolis and returned with prisoners and booty.[5]

It is very probable that among the numerous prisoners taken by

[1] Martin supposes that 'Constantine was himself probably a Monophysite' and infers that he transplanted Monophysites and Paulicians 'because they were...no friends to orthodoxy.' (*A History of the Iconoclastic Controversy*, p. 53, n. 5.) This may be so, but it remains true that the colonization of Thrace was effected primarily for a military purpose. Constantine was in need of strong loyal garrisons in his wars against Bulgaria both for offensive and defensive purposes.
[2] ἐζήτησαν...πάκτα διὰ τὰ κτισθέντα κάστρα. Theophanes, *C.S.H.B.* ibid. p. 662.   [3] Nicephorus, ibid.; Theophanes, ibid.
[4] Theophanes, *C.S.H.B.* vol. I, pp. 698–9: αἰχμαλωτεύσας τοὺς αἱρετικοὺς Ἰακωβίτας Σύρους πάλιν ὑπέστρεψεν ἐν τῷ κάστρῳ...Ἐπέρασεν δὲ καὶ τοὺς αἱρετικοὺς Σύρους ἐν τῇ Θρᾴκῃ, καὶ κατῴκισεν αὐτοὺς ἐκεῖ.
[5] See Zlatarski, op. cit. vol. I, pt 1, pp. 213, 243–6, 266–81, 297–8.

the Bulgarian armies in Thrace or forcibly removed to various parts of Bulgaria there were at least some Paulician heretics whose missionary zeal could now be exerted in a new country. Furthermore, it must be assumed that a large contingent of them was incorporated into Bulgarian territory towards the middle of the ninth century, when the great Thracian cities of Sardica and Philippopolis were annexed by the Bulgarians.[1] Their number must have been increased even more by the annexation of Macedonia in 864.

This policy of colonization was a complete failure with regard to its intended purpose. As a military force the Armenian and Syrian garrisons of Thrace did not justify the hope placed in them by the Byzantine authorities, since they proved to be incapable of stemming the frequent incursions of Bulgarian armies into the heart of Thrace and their advances up to the walls of Constantinople. Moreover, far from abandoning their heretical doctrines as a result of contact with the Orthodox, these colonists, and the Paulicians in particular, indulged in open proselytism and spread their heresy in Thrace. But in one respect they did contribute to the intended weakening of Bulgaria, though this was in a manner which the Byzantine emperors could not have foreseen or even desired. The gradual penetration of Paulicians into Bulgaria and the consequent spread of their heretical doctrines in that country became a serious menace to the establishment of Orthodox Christianity, and paved the way for several anti-ecclesiastical movements, which were destined to become for many centuries the bitter opponents of the Byzantine Church. The most important and dangerous of them was the Bogomil heresy.

What was the fate of these Paulician missionaries in Bulgaria? To answer this question, it is necessary to consider the religious, ethnical and social situation in Bulgaria at the time when these heretics began to penetrate into the country. Only thus will it be possible to understand the reasons for the success of Paulician proselytism in Bulgaria.

Its beginnings can be placed with some probability in the second half of the eighth century after the transplantations of Syrian and Armenian heretics into Thrace by Constantine Copronymus. Bulgaria then had existed as an organized state

---

[1] See Runciman, op. cit. pp. 87–8.

THE RISE OF BALKAN DUALISM 63

for barely a century. In 679, the Bulgars, a Turco-Tatar tribe[1] closely related to the Huns, under the leadership of their Khan Asperukh, left the steppes of southern Russia and crossed the Danube into the Byzantine province of Moesia.[2] The local population which they encountered there belonged to the eastern branch of the southern Slavs who had spread in considerable numbers over the Balkan peninsula from the sixth century onwards and had settled in a vast area, comprising the valleys of the Morava and of the Timok, Macedonia, Thrace, Albania, Epirus, Greece and even the Peloponnesus. The alliance of these two distinct racial elements, the Bulgar and the Slav, formed the basis of the Bulgarian nation. The former, however, although they were a small minority, imposed their customs and organization on the Slavs, who thus became at first a subject race. The Bulgar State was military and aristocratic: central and local power was vested in the boyars and the supreme authority belonged to the Sublime Khan.[3] In the course of the eighth and ninth centuries, the aristocratic Bulgar minority was gradually absorbed into the Slavonic element, which was continually growing in power and influence. The increasing importance of the Slavs in Bulgaria was due not only to this process of racial absorption, but also to the frequent support they received from the Khans, who were forced to rely on them to counterbalance the excessive strength of the Bulgar boyars, and also from Byzantium, which regarded the Slavs as the most convenient medium for extending its domination over the Balkans. For the Khan and the boyars, Byzantium remained nearly always the traditional foe. Asperukh's successors in the eighth century, Tervel (701-18), Kormisosh (739-56), Telets (761-4), Telerig (? -777), devoted much time and energy to waging wars with the Empire.[4]

[1] The origin of the Bulgars is discussed by K. Jireček, *Geschichte der Bulgaren* (Prague, 1876), pp. 136-8; L. Niederle, *Manuel de l'Antiquité Slave*, vol. I, pp. 100, 177; V. N. Zlatarski, История, vol. I,. pt I, pp. 21-122; I. Ivanov, Българитѣ въ Македония (2nd ed.; Sofia, 1917), pp. 25-6.

[2] See Niederle, op. cit. pp. 98-111. Cf. Dvorník, op. cit. pp. 12-16; Ivanov, op. cit. pp. 1-11.

[3] See Runciman, op. cit. p. 29 and ibid. App. V, where the question of old Bulgar titles is discussed.

[4] For the history of Bulgaria in the eighth century, see Zlatarski, op. cit. vol. I, pt 1; Runciman, op. cit. pp. 30-50; A. Pogodin, История Болгарии (St Petersburg, 1910), pp. 2-11.

In the ninth century the armies of Krum, the mightiest of the early Bulgar rulers (807–14), on several occasions made the foundations of the Empire tremble. But in spite of repeated efforts, they did not succeed in taking Constantinople. Krum did much to lessen the ethnic duality between Bulgars and Slavs by promulgating his celebrated code of laws, obligatory for all his subjects.[1]

The aggressive policy towards the Empire of most of the Khans from Asperukh to Krum rendered Bulgaria refractory to the civilizing influence of Byzantium. It was just this influence that the boyars, jealous of their ancient privileges, feared the most, and if a Khan showed himself friendly to Byzantium he incurred their distrust and the accusation of wishing to subject his country to the traditional enemy.[2]

But the situation changed after Krum's death. In the reign of his successor Omortag, in 815–16, a thirty years' peace was concluded between the Empire and Bulgaria and the strength and authority of the Khan were sufficient to overcome any restlessness or dissatisfaction of the boyars. In these circumstances it was inevitable that Byzantine influence should have made itself felt in Bulgaria, particularly since all writing had to be done in Greek, there being as yet no Slavonic alphabet.[3] The expansion of Byzantium, the greatest spiritual and cultural centre of eastern Europe, was then intimately connected with the spread of Christianity. The evangelization of those peoples who came into contact with the Empire was one of the principal aims of the Byzantine Orthodox

[1] Krum's laws were issued, it was said, in order to prevent hatred, collusion between thieves and judges, drunkenness and commercial fraud, apparently widespread vices in Bulgaria at that time. It is characteristic of the sweeping nature of these laws that one of them ordered all the vines in the country to be uprooted. It has been thought that this measure bears some relation to the view, later held by the Bogomils, that the vine was first planted by the Devil. (See infra, p. 128, n. 3.)

The only account of Krum's laws is given by Suidas, *Lexicon* (ed. by Adler; Leipzig, 1928), vol. I, pp. 483–4. Cf. Zlatarski, op. cit. vol. I, pt 1, pp. 283–9; Dvorník, op. cit. p. 35, n. 2; Runciman, op. cit. pp. 68–9; G. Kazarow, 'Die Gesetzgebung des bulgarischen Fürsten Krum', *B. Z.* (1907), vol. XVI, pp. 254–7.

[2] Thus Sabin in 766 was dethroned by the boyar party for entering into negotiations with Constantine Copronymus (see Zlatarski, ibid. p. 218).

[3] Omortag's celebrated inscriptions are written in Greek, though in a rough, ungrammatical language, as used by the Greek captives in Bulgaria.

Church. The Balkan Slavs in particular, since the days of the Emperor Heraclius, had been an object of special solicitude for its fertile missionary activity.[1]

In Bulgaria, the spread of Christianity was facilitated by the presence in Moesia and Macedonia of Christian nuclei, remnants of a time preceding the Slavonic invasions,[2] and also by the numerous Greek prisoners settled by the Khans in various parts of the country. Thus Christianity penetrated into Bulgaria in the same manner and approximately at the same time as Paulicianism. At the time of Omortag (815–31), Christianity was gaining ground in the country.

But a strong section in Bulgaria was viewing this development with grave concern. For many Bulgarians Christianity was synonymous with Byzantine domination, a foreign and hostile force. Omortag himself, probably for political motives, started to persecute the Christians. Four bishops were martyred, including Manuel, archbishop of Adrianople. We are also told that Omortag ordered his Christian subjects to eat meat in Lent; those who, out of loyalty to their religion, refused to comply with this order, were arrested and put to death.[3]

The resistance to Christianity in ninth-century Bulgaria was not limited to the anti-Byzantine political party. There is no doubt that among the masses there existed stubborn opposition to the new religion in the name of the traditional pagan beliefs. The character of these beliefs is a matter of some importance for a study of Paulicianism in Bulgaria, since the Paulician missionaries, as bearers of a new religion, were naturally brought into

---

[1] See the chapter on the Byzantine evangelization of the Slavs in Dvorník, op. cit. pp. 60–105. Cf. M. Spinka, *A History of Christianity in the Balkans* (Chicago, 1933), pp. 1–29.

[2] The Bulgarian envoys at the Council of Constantinople in 869 declared that their ancestors, after conquering Moesia, discovered Greek priests there: 'Nos illam patriam a Graecorum potestate armis evicimus, in qua...Graecos sacerdotes reperimus' (Guillelmus Bibliothecarius, *Vita Hadriani II*, J. S. Assemanus, *Kalendaria Ecclesiae Universae*, vol. II, p. 190). Cf. P. J. Šafařík, *Slovanské Starožitnosti* (Prague, 1837), p. 587; Dvorník, op. cit. p. 100.

[3] According to an eleventh-century source, Christianity under Omortag penetrated into the very family of the Khan. Omortag's eldest son Enravotas was converted by his slave, the Greek Cinamon, and was consequently martyred by order of his brother, the Khan Malamir. See Zlatarski, op. cit. vol. I, pt 1, pp. 293–7, 332–4; Dvorník, op. cit. pp. 100–2; Runciman, op. cit. p. 89.

close contact with the existing pagan religion of the Bulgarian masses. Moreover, the success of Paulician proselytism in the country largely depended on the extent to which paganism was capable of satisfying the masses and holding them within the bounds of their ancient traditions.

Unfortunately, however, it is not possible to obtain a clear picture of the pre-Christian cult in Bulgaria, as the relevant sources are scanty and vague and illustrate no more than the general character of southern Slavonic demonology and ritual. The earliest and fundamental account is that of Procopius, who describes the religion of those Slavs who invaded the Balkans in the sixth century: they worshipped one supreme God, creator of lightning and Lord of all things, and also honoured the spirits of rivers and woods, to whom they offered sacrifices in exchange for oracles.[1]

Apart from this general information on the beliefs of the early

[1] Procopius, *De Bello Gothico*, vol. III, p. 14; vol. II (Teubner, Leipzig), pp. 357–8: 'They recognize that there is one God, the maker of lightning and sole lord of all things, and they sacrifice to him cattle and all other victims. They do not know destiny, nor do they admit in any way that it has any power over men. But whenever death stands before them, when they are stricken with sickness or preparing for war, they make a promise that, if they escape, they will straight away make a sacrifice to the God in return for their life; and if they escape, they sacrifice what they had promised, and consider that their safety has been bought with this same sacrifice. They venerate, however, rivers, nymphs and some other spirits (δαιμόνια); they offer sacrifices to all these also, and in sacrificing expect oracles.' This passage has served as a basis for all researches into the pagan religion of the Slavs.

The evidence of Procopius is particularly important, as it clearly shows that the belief of the early Balkan Slavs was monotheistic. The νύμφαι mentioned by him are in all probability the Slavonic *vily*, the belief in whom is an essential characteristic of the pagan tradition of the southern Slavs, and particularly of the Bulgarians. See L. Niederle, *Slovanské Starožitnosti: Život starých Slovanů* (Prague, 1916), pt II, vol. I, pp. 59–60, and *Manuel de l'Antiquité Slave*, vol. II, p. 133. V. Mansikka, however, considers the *vily* to be of Turco-Tatar origin (*Die Religion der Ostslaven*, FF Communications, no. 43, Helsinki, 1922, p. 153).

Southern Slavonic paganism is also discussed by Jireček, *Geschichte der Serben* (Gotha, 1911), vol. I, p. 160 et seq.; L. Léger, *La Mythologie Slave* (Paris, 1901); L. Niederle, *Slov. Starož.* loc. cit. pp. 44–5.

Evidence concerning folk beliefs of the southern Slavs can be found in Phyllis Kemp's *Healing Ritual: Studies in the Technique and Tradition of the Southern Slavs* (London, 1935), which contains an extensive bibliography of the subject.

southern Slavs, we possess some evidence concerning pagan ritual in Bulgaria. We know of the existence of pagan feasts, particularly that of the summer solstice, celebrated on a day which became, after the introduction of Christianity, the eve of the feast of St John the Baptist (24 June),[1] and of the summer festival of the *rusalii*.[2]

The scanty information concerning Bulgarian paganism can be supplemented by the evidence of a few historical sources. We are told that Krum, preparing for a final assault on Constantinople, offered sacrifices of men and animals.[3] Omortag, in pledging himself to friendship with the Byzantine emperor, swore on his sword and on the entrails of sacrificed dogs.[4]

The singular paucity of historical evidence concerning early Slavonic paganism is, according to Niederle, not fortuitous. The pre-Christian Slavonic cult, when compared with that of other Indo-European peoples, appears to have been rather indefinite and poor.[5] This is particularly true in the case of the southern Slavs, who, unlike the Russian and Baltic Slavs, do not seem to have had a distinct 'cycle' of gods or an organized priesthood.[6] In Bulgaria, for 200 years after the arrival of Asperukh, the consolidation and unification of the pagan cult was furthermore prevented by the existing racial and religious duality between the pagan Slavs and the Bulgars, who were Shamanists.[7] Neither force was strong enough to absorb the other and the religious and racial dichotomy was only overcome after the introduction of Christianity. In these circumstances, it can scarcely be doubted

---

[1] L. Niederle, *Manuel*, vol. II, p. 166. This pagan feast was common to most Slavs. It is known in Bulgaria as Иванъ-день and in Russia as купало, купалы. Its most important features included jumping through fire and the ritual killing and burial of a human figure. In Bulgaria, in later times, it was connected by the Orthodox with the practice of Bogomilism (see infra, p. 247).

[2] Ibid. p. 55. The *rusalii* are undoubtedly of Romano-Byzantine origin and correspond to the Latin *rosaria, rosalia* and to the Greek ῥοδώνια (ἡμέρα τῶν ῥόδων).

[3] μιαρὰς καὶ δαιμονιώδεις θυσίας. Theophanes (de Boor), vol. I, p. 503; *C.S.H.B.* vol. I, p. 785; Symeon Magister, *Annales*, ch. 8, *C.S.H.B.* p. 612.

[4] Theophanes Continuatus, *Chronographia*, p. 31.

[5] *Manuel*, vol. II, pp. 126–7.

[6] See N. P. Blagoev, История на старото българско държавно право (Sofia, 1906), pp. 190–1.

[7] I. Ivanov, Богомилски книги и легенди (Sofia, 1925), pp. 364–7.

that the pagan cult of the Bulgarians lacked the force necessary to ensure the religious and cultural development of their country.

The Paulicians, on the other hand, whose culture was on a considerably higher level, were no doubt to some extent able to fulfil the role of teachers to the Bulgarian people. Their superiority lay largely in the fact that their teaching, for all its dualism, had borrowed many elements from Christianity. In view of the exclusive importance they attached to the New Testament, we may legitimately assume that in some cases the Paulicians were the first to bring the Gospel to the pagan Bulgarians. In spite of their heretical interpretation of the Scriptures, the Paulician missionaries were vested with a moral superiority over paganism which goes far to explain their undoubted success in Bulgaria. It is not clear whether the early Paulicians were antagonistic to the pagan beliefs or whether, on the contrary, they adapted them to their own teaching.[1]

[1] According to a theory developed in the nineteenth century and sometimes brought forward to-day, the beliefs of the pagan Slavs were dualistic and hence connected with the teachings of the dualistic sects in Bulgaria and with those of the Paulicians in particular. This theory is based primarily on the following description of the twelfth-century Polabian pagan Slavs by Helmold: 'The Slavs...have a strange delusion. At their feasts and carousals they pass about a bowl over which they utter words, I should not say of consecration but of execration, in the name of the gods—of the good one, as well as of the bad one (*boni scilicet atque mali*)—professing that all propitious fortune is arranged by the good god, adverse, by the bad god. Hence, also, in their language they call the bad god Diabol, or Zcerneboch, that is, the black god (*malum deum sua lingua Diabol sive Zcerneboch, id est nigrum deum, appellant*).' Helmoldi Presbyteri *Chronica Slavorum*, lib. I, c. 52, Pertz, *M.G.H. Ss.* vol. XXI, p. 52 (English tr. by F. J. Tschan, 1935, p. 159). By antithesis with *Zcerneboch*, the black god, to the Slavs was also attributed the worship of 'the white god'. (See J. Gieseler, 'Über die Verbreitung christlich-dualistischer Lehrbegriffe unter den Slaven', *Theologische Studien und Kritiken*, Hamburg, 1837, pt 2, pp. 357-66; C. Schmidt, *Histoire et doctrine de la secte des Cathares ou Albigeois*, vol. I, p. 7; vol. II, pp. 271 et seq.; D. Tsukhlev, История на българската църква, Sofia, 1910, vol. I, pp. 662-4.) This view is convincingly refuted by Ivanov (Богомилски книги и легенди, pp. 361-4). His main arguments are as follows: (1) Helmold's evidence is of the twelfth century and does not refer to the southern Slavs. (2) The existence of a 'white god' as opposed to the 'black god' is not mentioned by any old reliable source. (3) The 'black god' is probably a later conception which may well have developed among the Baltic Slavs under the influence of the Christian teaching concerning the Devil. (This is admitted by Gieseler, op. cit. pp. 360-2.) (4) The little information we possess on the pagan religion of the Slavs strongly

THE RISE OF BALKAN DUALISM 69

This initial advantage enjoyed by the Paulicians in Bulgaria was increased by the fact that, as far as is known, their teaching was not opposed by the State. To proselytize openly on Byzantine territory would have been unthinkable, but the Khans seem to have been fairly tolerant in matters of religion. They opposed Christianity because it represented for them Byzantine imperialism, but the Paulicians, though they also were foreigners, presented no such danger. We possess no direct evidence of Paulician proselytism in Bulgaria during the first half of the ninth century,[1] but it cannot be doubted that their doctrines spread there.[2]

suggests that they were monotheists; thus Helmold himself, in another passage, refers to Svantovit, the supreme god of the Slavs: 'Zvantevit deus terre Rugianorum inter omnia numina Sclavorum primatum obtinuerit....Unde etiam...omnes Sclavorum provincie, illuc tributa annuatim transmittebant, *illum deum deorum esse profitentes*.' (Op. cit. lib. II, c. 12, p. 97.) Above all, the evidence of Procopius (see supra, p. 66, n. 1) clearly shows that monotheism existed among the southern Slavs. (5) As for the popular dualistic legends, so widespread in Bulgaria in later times, they are not autochthonous, but originate from Asia. Ivanov thinks that they may have been brought to Bulgaria by the Paulician colonists (op. cit. p. 378).

It may be added that the term 'dualism' is frequently used in far too loose a sense; the belief in good and evil forces outside man, either benevolent or harmful, and the consequent worship of the first and avoidance or propitiation of the second, common to all religions with a developed demonology (such as Slavonic paganism), cannot be accurately described as 'dualism'. Dualism proper, applied, for instance, to Manichaeism and Paulicianism, is a metaphysical doctrine, according to which the visible, material world is the creation of an evil force outside God.

[1] We know, however, that the Paulicians formed part of the armies of the celebrated rebel Thomas, who in his unsuccessful attempt to capture Constantinople (820–3) directed his land operations against the capital from Thrace. They are referred to by Genesius (*Regum* l. II, *C.S.H.B.* p. 33) as 'ὅσοι τῆς Μάνεντος βδελυρίας μετεῖχον', and were in all probability Armenians. Thomas himself, who has often been thought to have been a Slav, was in reality, it seems, of Armenian origin. See A. Vasiliev, *Byzance et les Arabes* (French ed.), vol. I, pp. 22–49. Cf. also J. Laurent, 'L'Arménie entre Byzance et l'Islam depuis la conquête arabe jusqu'en 886', *Bibliothèque des Écoles Françaises d'Athènes et de Rome*, fasc. 117, Paris, 1919, p. 252; J. B. Bury, *A History of the Eastern Roman Empire*, pp. 86, 109.

[2] A signal error has been committed by a number of historians, due to the false interpretation of a passage of Georgius Hamartolus, who writes, alluding to the Paulicians: ἔχουσι δὲ καὶ γ' ἐκκλησίας ἐν τῇ ὁμολογίᾳ αὐτῶν, (α') τὴν Μακεδονίαν, ἥτις ἐστιν Κάστρον κολωνίας... (*Chronicon*, ed. Muralt, St Petersburg, 1859), p. 607; cf. de Boor ed. vol. II, p. 720. Assuming that τὴν Μακεδονίαν must refer to the Balkan region of that name, E. Golubinsky, Краткий очерк

The growth of Christianity in Bulgaria could not be arrested indefinitely by State persecution. The Gospel was preached in the country by the Orthodox missionaries and the Paulician heretics. Moreover, the constant contacts, cultural and diplomatic, of the Bulgarian Khans with the Empires of the East and of the West, the Byzantine and the Frankish,[1] had, no doubt, brought to their minds the necessity for Bulgaria of taking her place among the civilized nations of Europe. This could only be achieved by renouncing the pagan isolation and accepting Christianity. But the Christian missionaries came from Byzantium, and the Khans and the boyars were naturally loth to open the doors of Bulgaria to priests and institutions coming from Constantinople. The Khan Boris (852–89), for this reason, was inclined to seek Christianity from the West and in 862 he concluded an alliance with Louis the German.[2] This Franko-Bulgarian pact seems to have been directed at once against Byzantium, the traditional enemy of both the Bulgarians and the Franks, and against the young Moravian State, whose rapid political growth under its able rulers Mojmir and Rastislav was arousing the displeasure of its German and Slavonic neighbours. The Moravians and the Byzantines, who were equally interested in preventing the consolidation of the Franko-Bulgarian pact, promptly concluded an alliance of their own. This alliance, initiated by Rastislav's celebrated embassy to the Byzantine court

истории православных церквей болгарской, сербской и румынской (Moscow, 1871), p. 155, identifies Κάστρον κολωνίας with the locality of Colonia (or Staria), to the south-west of Kastoria, in southern Macedonia, and concludes that in the first half of the ninth century the Paulicians possessed an organized 'Church' in Macedonia. This opinion is repeated by such authorities on Bulgarian history as Jireček, *Geschichte der Bulgaren*, p. 175, and Zlatarski, op. cit. vol. I, pt 2, pp. 62–3.

This surprising mistake can only be due to an insufficient acquaintance with the *Historia Manichaeorum* of Peter of Sicily. As we have seen, Peter expressly tells us that the Paulician communities of Asia Minor were named after the various Christian churches associated with St Paul. Among these was the 'Church of Macedonia' situated in Cibossa, near Colonea (see supra, pp. 33–4) in Armenia Minor, and which is, beyond any doubt, the Κάστρον κολωνίας mentioned by Hamartolus, where the 'Church of Macedonia' was situated. Hamartolus himself in his narrative is clearly referring to the Paulicians of Asia, which makes Golubinsky's error all the more astonishing.

[1] On several occasions Omortag entered into negotiations with the Emperor Louis the Pious. (See Runciman, op. cit. pp. 81–3.)
[2] Dvorník, op. cit. pp. 184–7.

(862), was to have profound repercussions on the history of the Slavs: its immediate effects were the Byzantine mission to Moravia (863), with the consequent rise of Slavonic Christianity in central Europe, and, on the other hand, the far-reaching events which took place in Bulgaria the following year. To counter the Franko-Bulgarian pact, with its danger of Carolingian influence spreading to the Balkans, the Emperor Michael decided to strike. A Byzantine army entered Bulgaria, and Boris, whose military position was precarious, was forced to capitulate: he accepted all the emperor's conditions, renounced the Frankish alliance and agreed to receive baptism and to admit Greek missionaries into his country. In 864 Boris together with a large number of Bulgarian boyars was baptized, the Emperor Michael being his godfather.[1]

By accepting baptism in the name of all his people Boris did much to achieve the unification and centralization of his realm, for which his pagan predecessors had vainly striven. It had been a constant aim of the Khans to overcome the racial duality between Bulgars and Slavs. Some of them, like Tervel and Krum, had been partly successful, owing to their strong personalities and to the help of the Slavs. But under their weaker successors, who were often incapable of holding together the different elements in the country, their work was largely undone. The failure of paganism to unify and centralize was thus largely due to the fact that the pursuit of these aims was the sole prerogative of the Khan, on whose personality and strength of character its success, in the last resort, depended.

[1] The circumstances of Boris's baptism are discussed by Zlatarski, op. cit. vol. I, pt 2, pp. 18–31; Dvorník, *Les Slaves, Byzance et Rome*, pp. 187–9, and 'Les Légendes de Constantin et Méthode vues de Byzance' (*Byzantinoslavica*, Supplementa I, Prague, 1933), pp. 229–31; Runciman, op. cit. pp. 103–5.

The event was recorded by a number of Byzantine and Frankish chroniclers (see Dvorník, *Les Slaves, Byzance et Rome*, p. 186, n. 1). Its exact date has been variously fixed. Zlatarski, by a calculation based on the old Bulgar chronological cycle, and on the basis of an Albanian inscription, concluded that Boris was baptized in September 865 (ibid. pp. 29–31). This date is accepted by Dvorník, Runciman and Spinka. However, A. Vaillant and M. Lascaris, 'Date de la conversion des Bulgares', *R.E.S.* (1933), vol. XIII, fasc. 1, 2, pp. 5–15, criticize Zlatarski's conclusion by a detailed examination of his sources. In their opinion, the correct date is 864, which is accepted by Grégoire, *B.* vol. VIII (1933), pp. 663–8.

Christianity was able to achieve both: by recognizing the fundamental equality of all races in the State, it did much to destroy the ethnic duality, and, moreover, by its principle of the divine origin of authority, it sanctioned and legalized the supreme position of the autocrat. With the help of Christianity, Boris ceased to be a pagan Khan and became a Christian Slavonic prince, whose aim it was to unite the southern Slavs under his sceptre.

Above all, the effect of Boris's conversion was to link Bulgaria with Byzantium through the Orthodox Church. The new faith came from Constantinople and, together with the doctrines, ethics and ritual of Christianity, Byzantine political and social institutions could penetrate freely into Bulgaria.

The Patriarch Photius, who, as the principal inspirer of all missionary work among the barbarians, took a particular interest in the conversion of Bulgaria, now assumed the position of the spiritual father of the newly baptized ruler and his subjects. In 865 he sent Boris a long and learned letter setting forth with his customary force and lucidity the mysteries of the Christian faith and the duties of a Christian ruler.[1] As well as giving an exposition of the Nicene Creed and of the doctrines of the seven Oecumenical Councils, and an explanation of the principal Christian virtues, the patriarch warned Boris against all deviations from Orthodoxy and innovations in matters of doctrine.[2]

Photius's letter can serve as an illustration of the task which confronted the Greek clergy in Bulgaria in its mission of consolidating Christianity. After preaching the Gospel, its most urgent duty was to destroy paganism and heresy, and to bring all national customs and institutions into harmony with the Christian law. The essential instrument in the Christianization of Bulgaria was thus Byzantine canon and civil law, both of which entered into the composition of the Byzantine nomocanon. There is no doubt that the Greek clergy sent to Bulgaria after 864 was supplied

---

[1] Photius, *Epistolae*, lib. I, ep. 8, *P.G.* vol. CII, cols. 628–96. See also I. N. Valetta, Φωτίου τοῦ σοφωτάτου καὶ ἁγιωτάτου πατριάρχου Κωνσταντινουπόλεως ἐπιστολαί ('Εν Λονδίνῳ, 1864), 'Επιστολή 6, pp. 200–48.

[2] μήτε δεξιᾷ μήτε ἀριστερᾷ, μηδὲ ἐπὶ βραχύ, ταύτης ἀποκλίνειν. *P.G.* vol. CII, col. 656.

THE RISE OF BALKAN DUALISM 73

with the Byzantine nomocanon.[1] Thus, together with the canonical, penitentiary and service books, the foundation of Byzantine civil law, i.e. the *Ecloga* of the Isaurian emperors, was introduced into Bulgaria, where it replaced the old customary law and became the basis of the Bulgarian civil code.[2]

The ecclesiastical organization and administration of Bulgaria were at first left entirely in the hands of the Greeks. There could be as yet no local Bulgarian hierarchy, as Christianity was still in its infancy there and much had to be improvised, the suddenness of the Bulgarian conversion having surprised even the Greeks.[3] Photius himself assumed immediate authority of jurisdiction over the Bulgarian Church. This no doubt explains the absence of an Orthodox bishop in Bulgaria until 870.[4]

But the very suddenness with which baptism was decided on and the speed with which it was carried out, characteristic of so many of Boris's important acts, were a source of danger to the newly established Bulgarian Orthodox Church. Baptism had been

[1] See S. S. Bobchev, Римско и византийско право в старовремска България, *G.S.U.* (1925), vol. xxi, p. 77. The nature of the books of Byzantine canon and civil law which were sent to Bulgaria at that time forms the subject of an ingenious article by Zlatarski, Какви канонически книги и граждански закони Борисъ е получилъ отъ Византия, *Letopis na Bŭlgarskata Akademiya na Naukite* (Sofia, 1914), vol. i, pp. 79–116.

[2] Bobchev, loc. cit.; Zlatarski, ibid. p. 115. Cf. C. A. Spulber, *L'Éclogue des Isauriens* (Cernautzi, 1929), pp. 103–11; E. H. Freshfield, *Roman Law in the later Roman Empire* (Cambridge, 1932), p. 32.

[3] Photius, in his encyclical to the Eastern patriarchs of 867 (*P.G.* vol. cii, col. 724, ed. Valetta, p. 168), says that the Bulgarian people εἰς τὴν τῶν Χριστιανῶν παραδόξως μετενεκεντρίσθησαν πίστιν. Theophylact, archbishop of Ochrida, writing almost 250 years later, also stresses the unexpectedness of the event: 'Ρωμαῖοι δέ, τὸ μηδέποτε παρὰ Βουλγάρων ἐλπισθὲν αὐτοῖς τὸ περὶ τῆς εἰρήνης μήνυμα ἀσμένως δεξάμενοι, πάντα διὰ τάχους ἐτέλεσαν (*Historia Martyrii XV Martyrum, P.G.* vol. cxxvi, col. 200).

[4] Most of the Byzantine chroniclers who record the baptism of Boris mention a Greek bishop sent from Byzantium to perform the sacrament. Theophanes Continuatus, l. iv, cap. 13–15, pp. 162–5; Cedrenus, vol. ii, pp. 151–2; Zonaras, l. xvi, c. 2, *C.S.H.B.* vol. iii, p. 388—Genesius mentions several bishops (*Regum* l. iv, *C.S.H.B.* p. 97).

But there is no evidence that any bishop remained in Bulgaria after the baptism or was sent to organize the Bulgarian Church, as Zlatarski appears to think (op. cit. vol. i, pt 2, p. 28). On the contrary, there is good reason for supposing, as Tsukhlev has pointed out (История на българската църква, p. 274, n. 1), that the first bishop of Bulgaria was only appointed by the Patriarch Ignatius in 870. See Theophanes Continuatus, p. 342.

enforced on many Bulgarians, who thus became Christians only in name. Moreover, in all the social classes Orthodox Christianity had many enemies who both actively and passively resisted the introduction and enforcement of the new law: the pagan masses, still the vast majority, resented in the main the attempt of the Church to destroy their old traditions and beliefs; the Paulician heretics were actively spreading their anti-Orthodox teaching; the boyars were observing with alarm that Christianity was threatening to destroy their ancient privileges and dominant position in the State and, moreover, they were faced with a peaceful invasion of men and institutions from Byzantium, their hereditary foe. Finally, the Roman See had not abandoned the hope that the contact established between Boris and Louis the German in 862, so rudely interrupted in 864, would eventually bear fruit and lead to the attachment of Bulgaria to the Western Church.

All these factors, present in Bulgaria from the time of the baptism, were destined to influence at different moments its inner life for several centuries and caused Bulgaria to be the fighting ground for a number of anti-Orthodox movements throughout most of its medieval history.

The first one broke out as early as 866. A number of boyars, supported by some Bulgarians of lesser rank, rose in revolt against Boris. Their intention was to kill him and appoint another Khan. Boris, whose position seems to have been extremely precarious, gathered a handful of faithful followers and, by a timely intervention which contemporary sources describe as miraculous, attacked and defeated the rebels. They were punished with great severity: fifty-two of the ringleaders, together with their children, were put to death.[1] According to Hincmar of Rheims the leaders of the revolt were 'intra decem comitatus', which no doubt means that they were governors of provinces, into which Bulgaria was

[1] The event is described in Byzantine and Latin sources: Theophanes Continuatus, p. 164; Cedrenus, vol. II, p. 153; Zonaras, l. XVI, c. 2, vol. III, pp. 388–9; Theophylact. Bulg. *Hist. Mart. XV Martyr.*, *P.G.* vol. CXXVI, col. 200; Nicolai Papae *Responsa ad Consulta Bulgarorum*, *P.L.* vol. CXIX, 17, col. 988. Hincmar of Rheims (*Annales Bertiniani*, pars III, *sub anno* 866, Pertz, *M.G.H. Ss.* vol. I, pp. 473–4) gives the most detailed account.

Cf. Zlatarski, op. cit. vol. I, pt 2, pp. 44–59; Dvorník, *Les Slaves, Byzance et Rome*, p. 189.

then divided.¹ Pope Nicholas I states, furthermore, that these nobles belonged to two distinct classes, of which the upper alone was put to the sword.² It is thus very probable, as Zlatarski suggests, that the strongest resistance to Christianity came from the provinces, where, far from the vigilant eye and centralizing efforts of Boris and his government, the semi-independent aristocracy could foment a revolt in defence of its jeopardized privileges.

The sources do not tell us whether the rebellious boyars were particularly attached to their pagan beliefs or customs. Zonaras merely says that they were dissatisfied because Boris had abandoned the traditions of his fathers;³ Nicholas, more significantly, mentions in his letter to Boris that the boyars revolted, 'dicentes, non bonam vos eis legem tradidisse'.⁴ It would appear that the basic motive of the rebellion of 866 was social and political. As it has already been shown, a considerable section of the boyars were traditionally opposed to the 'lex Christiana'.⁵

Hardly had this serious threat to Orthodoxy been overcome when another event occurred of great importance for the newly baptized country. The very same year (866) Boris sent envoys to Rome with a request to Pope Nicholas I to instruct him and his people in the pure Christian faith. At the same time he sent a similar mission to Louis the German, asking him for bishops and

¹ See S. S. Bobchev, Симеонова България отъ държавно-правно гледище, *G.S.U.* (1926–7), vol. XXII, pp. 58, 79–80. M. S. Drinov, Южные славяне и Византия в X веке (Moscow, 1876), pp. 84–5.

² 'Qualiter...omnes primates eorum, atque majores cum omni prole sua gladio fuerint interempti; mediocres vero, seu minores nihil mali pertulerint' (*Responsa*, loc. cit.). Zlatarski has shown (loc. cit. vol. I, pt 2, pp. 49–51) that the distinction between *majores* on the one hand, and *mediocres seu minores* on the other (*primates* being merely a generic term equivalent to 'boyars') corresponds to the well-known distinction in Bulgarian and Byzantine sources between the 'Great Boyars' who held high military and administrative posts at Court or in the provinces, and the less important or 'little' boyars, employed in various branches of the civil service, or, also, between the βοιλάδες and the βαγαῖνοι.

³ ὡς τῆς πατρίου δόξης ἀποστάντος, *C.H.S.B.* vol. III, p. 388.

⁴ *Responsa*, loc. cit., *P.L.* vol. CXIX, col. 988.

⁵ This should explain the bitter enmity between the Greek clergy and the opposing boyar party. Those of the rebels who had been pardoned by Boris and were prepared to do penance were nevertheless refused absolution by the Greek clergy in Bulgaria (*Responsa*, 78, loc. cit. col. 1008; cf. Zlatarski, op. cit. vol. I, pt 2, pp. 57–8).

priests, and thus returning to his policy of 862. Boris's motive in turning to Rome was, it seems, above all, the hope of obtaining the independence of his Church, which the Byzantine patriarch was not prepared to grant.[1]

This was a wonderful opportunity for the Roman See. For centuries the Papacy had claimed its ancient rights over eastern Illyricum, a large portion of which was now of its own will returning to the fold.[2] Nicholas's first act was to compose a letter with answers to specific questions raised by Boris. These celebrated *Responsa Nicolai ad Consulta Bulgarorum* are remarkable for their clarity, practical sense and shrewdness.[3] The letter was sent to Bulgaria together with a mission, headed by two bishops, Paul of Populonia and Formosus of Porto. It arrived at Pliska, Boris's capital, in November. 866.[4] The next year, there arrived a group of German missionaries, headed by Ermenrich, bishop of Passau, and sent by Louis the German in reply to Boris's appeal; but, as the Latin clergy was already installed in Bulgaria, they were forthwith sent home.[5] The Latin bishops and priests rapidly set about their work of bringing Bulgaria to the Roman obedience; the Greek clergy was expelled and much of its work of the past two years consciously undone.[6]

Many of the practices instituted by the Greek priests were now roundly condemned by their Latin successors. Already Nicholas I had denounced several of them to Boris as unnecessary or absurd.[7] The Latin priests, rejecting Greek Confirmation, insisted that all

---

[1] See Zlatarski, op. cit. vol. I, pt 2, pp. 88–107; Dvorník, op. cit. p. 191.

[2] The question of Illyricum is discussed at length in Dvorník's *Les Légendes de Constantin et Méthode*, pp. 248–83. Cf. Tsukhlev, История на българската църква, p. 244.

[3] *P.L.* vol. CXIX, cols. 978–1016.

[4] Zlatarski, op. cit. p. 108.

[5] *Annales Fuldenses*, pars III, sub anno 867, Pertz, *M.G.H. Ss.* vol. I, p. 380.

[6] Anastasius Bibliothecarius, *Historia de Vitis Rom. Pont.*: 'Nicolaus I', *P.L.* vol. CXXVIII, cols. 1374–5: 'Gloriosus autem Bulgarorum rex fidei tanta coepit flagrare monitis hujus pii Patris illectus constantia, ut omnes a suo regno pellens alienigenas, praelatorum apostolicorum solummodo praedicatione usus missorum.' Zlatarski thinks that 'alienigenas' may refer not only to the Greeks but to the heretical teachers proselytizing in Bulgaria (ibid. p. 111).

[7] Such as the prohibitions to bathe on Wednesdays and Fridays (*Resp.* 6, col. 982), to take communion without wearing a belt (*Resp.* 55, col. 1000), to eat the flesh of an animal killed by a eunuch (*Resp.* 57, col. 1001), or the teaching that it is a grave sin for a man to stand in church without his arms folded on his chest (*Resp.* 54, col. 1000).

Bulgarians should be re-confirmed.[1] It is to be presumed that they attempted to destroy the confidence of the Bulgarians in the Byzantine Church by means of the arguments used by Nicholas I: the non-canonicity of Photius's election to the patriarchal throne,[2] the Papal primacy,[3] the inferior position occupied by the Church of Constantinople,[4] and Byzantine imperialism.[5]

It is not hard to imagine the confusion regarding Christianity which must have reigned in the minds of the Bulgarian people who, in the space of three years, had been confronted with Byzantine, Latin and German priests, and who now were being told that the Greek priests, whom they had been taught to obey, were no more than ambitious impostors.[6]

This confusion was, no doubt, further increased when, following the decision of the council held in Constantinople in 869–70, Bulgaria was once again attached to the Eastern Church. The Bulgarian Church having been granted a considerable measure of independence from Byzantium, a newly appointed archbishop, consecrated by the Patriarch Ignatius, arrived in Bulgaria in 870 with a number of bishops and priests; it was now the turn of the Latin clergy to be expelled and of the Greeks to refute the Latin teachings and practices in order to justify their position.[7] The

---

[1] See Photius's Encyclical to the Eastern patriarchs (*P.G.* vol. cii, col. 725). Cf. Zlatarski, op. cit. p. 109. The non-validity of Greek Confirmation was argued by the Latins on the ground that it was performed by priests, according to the custom of the Eastern Church, and not by bishops alone, according to the rule of the Western Church.

[2] See Nicolai Papae *Epistola ad Photium*, *P.L.* vol. cxix, col. 780.

[3] Nicolai Papae *Epistola ad Michaelem Imperatorem*, ibid. col. 773; *Resp.* 73, ibid. col. 1007; 92, cols. 1011–12.

[4] *Resp.* 92, col. 1012: 'Constantinopolis nova Roma dicta est, favore principum potius quam ratione.'

[5] *Epistola ad Hincmarum*, *P.L.* vol. cxix, col. 1153.

[6] Constantine Porphyrogenitus thus describes the religious instability of the Bulgarians at that time: τὸ γὰρ τοιοῦτον ἔθνος...ἀπαγὲς ἦν ἔτι πρὸς τὸ καλὸν καὶ ἀνίδρυτον, ὡς ὑπὸ ἀνέμου φύλλα ῥᾳδίως σαλευόμενον καὶ μετακινούμενον. (*De Basilio Macedone*, cap. 96: Theophanes Continuatus, p. 342.)

[7] Anastasius Bibliothecarius, op. cit. *Vita Adriani*, *P.L.* vol. cxxviii, cols. 1395–6. The causes and circumstances of Boris's return to the Byzantine Church are described by Zlatarski (op. cit. vol. i, pt 2, pp. 111–45). The main causes appear to be Boris's failure to appreciate the Roman conception of centralization and his disappointment resulting from the inflexible refusal of the Popes Nicholas I and Hadrian II to grant autonomy to the Bulgarian Church and, on the other hand, the energetic Slavonic policy of the Emperor Basil I, which caused the Balkan Slavs to gravitate into the orbit of Byzantium. (Cf. Spinka, op. cit. pp. 41–3.)

Bulgarians were now persuaded that the Latin missionaries were heretics.[1] On matters of ecclesiastical law and discipline many of the arguments used by the Greeks after 870 were the exact counterpart of those used by the Latins between 866 and 870.[2] In doctrinal matters, the Latin dogma of the Double Procession of the Holy Spirit (the 'Filioque') was the principal object of attack; it should be noted that the arguments probably used to refute it supply indirect evidence of the proselytism of the Paulicians in Bulgaria at that time.[3]

It cannot be doubted that this struggle between the Byzantine and Roman Churches in Bulgaria indirectly contributed to the growth of heresy. Polemics between the rival hierarchies were almost solely concerned with matters of discipline and ritual. In the doctrinal field the question of the 'Filioque' appears to have

[1] Photius, in his *Encyclical to the Eastern Patriarchs* (*P.G.* vol. cii, col. 724), denounces the Latin missionaries in Bulgaria as ἄνδρες δυσσεβεῖς καὶ ἀποτρόπαιοι...ἐκ σκότους ἀναδύντες...ἀπὸ γὰρ τῶν ὀρθῶν καὶ καθαρῶν δογμάτων ...παραφθείρειν τούτους [τοὺς Βουλγάρους], καὶ ὑποσπᾶν, κατεπανουργήσαντο, while Nicholas I complains that the Bulgarians 'utpote adhuc in fide rudes ...nos quasi noxios, et diversarum haereseon squaloribus respersos, vitent, declinent, atque penitus deserant'. (*Epistola ad Hincmarum*, *P.L.* vol. cxix, col. 1153.)

[2] See Photius's *Encylical*, loc. cit. cols. 724–36; Nicholas I, *Epistola ad Hincmarum*, loc. cit. cols. 1155–6.

[3] Photius, loc. cit. cols. 725–32. Nicholas, loc. cit. col. 1155. Photius's theological refutation of the 'Filioque' was certainly beyond the powers of comprehension of the vast majority of Bulgarians. A vulgarized criticism was necessary. Theophylact of Ochrida, attacking the 'Filioque', alleges that this dogma is based on a fundamental dualism, which is in his opinion frankly Manichaean: Δύο δὲ ποιοῦντες ὑμεῖς ἀρχάς, τοῦ μὲν Υἱοῦ τὸν Πατέρα, τοῦ δὲ Πνεύματος τὸν Υἱόν, ἄλλην τινὰ μανίαν μανιχαϊκὴν μαίνεσθε. *Vita S. Clementis Bulgar. Archiep.* *P.G.* vol. cxxvi, col. 1209.

It seems legitimate to suppose that a similar argument was used after 870 by the Greeks *ad usum Bulgarorum*, as the Bulgarians were quite familiar with 'Manichaeism', which the Paulician heretics were spreading in their midst.

The probability of this hypothesis is increased by the knowledge that the accusation of Manichaeism was undoubtedly put forward by the Greek clergy in Bulgaria against the Latin rule concerning the celibacy of priests, which was said to imply a general hatred of marriage. See Photius, loc. cit. cols. 724–5: τοὺς ἐνθέσμῳ γάμῳ πρεσβυτέρους διαπρέποντας...οὗτοι, τοὺς ὡς ἀληθῶς Θεοῦ ἱερεῖς, μυσάττεσθαί τε καὶ ἀποστρέφεσθαι, παρεσκεύασαν· τῆς Μανοῦ γεωργίας, ἐν αὐτοῖς, τὰ σπέρματα κατασπείροντες. Cf. Nicholas, loc. cit. col. 1155: 'dicunt ...nos abominari nuptias, quia presbyteros sortiri conjuges prohibemus.'

been the only subject of disagreement. But for the Bulgarians, 'adhuc in fide rudes', the distinction between the fundamental principles of Christianity and their external expression in the ethical and social spheres was a difficult one. Faced with far-reaching transformations in almost every aspect of their private and public lives, they naturally tended to confuse the more important and the less important.[1] This confusion was in itself a favourable ground for heresy; the Paulician teachings, in particular, spread the very same confusion by neglecting the unity of the Christian tradition and by unduly emphasizing certain of its aspects to the detriment of others.

Moreover, the embittered polemics between the representatives of the two rival Churches in Bulgaria undermined the prestige of both in the eyes of the people, who, ignorant of the true significance of the changes, must have noticed above all the contradictions between the teachings and the quarrels of the hierarchs.[2] It is likely enough that those feuds were exploited by the Paulicians, who, in their bitter hostility to the Byzantine and the Roman Churches, were not likely to miss such an opportunity to discredit both.

Apart from these general considerations, there is positive evidence that at the time when Boris became a Christian and imposed his faith on his people the Paulicians were actively proselytizing in Bulgaria. The manner in which they penetrated from Armenia to Thrace in the eighth century, and from Thrace to Bulgaria in the eighth and ninth centuries, has already been described.[3] Though we do not hear of them directly until 866, it is highly

---

[1] Such confusions were numerous in Boris's questions to the Pope; judging from the *Responsa* of Nicholas they seem to have contained an astonishing mixture of the essential and the trivial. This was due to a correct understanding of some Christian principles and to a complete ignorance of others.

[2] A similar state of confusion, due to the preaching of Christianity in different forms to a still largely pagan Slavonic population, had arisen towards 862 in Moravia before the arrival of Constantine and Methodius; this can be seen from the following words of Rastislav's ambassadors to Byzantium, who said, referring to the Frankish, Latin and Greek missionaries in their country: Учяште ны различь ('they instruct us in different ways'). See F. Dvorník, *Les Légendes de Constantin et Méthode*, p. 385 and *Les Slaves, Byzance et Rome*, p. 158, n. 4.

[3] See supra, pp. 59–62.

probable that they were active in Bulgaria during the first half of the ninth century.

The *Responsa ad Consulta Bulgarorum* of Pope Nicholas I, that fundamental source of our knowledge of the inner conditions in Bulgaria in the third quarter of the ninth century, contains several allusions to the proselytism of heretical teachers on Bulgarian soil, some of whom were almost certainly Paulicians. We learn from the words of the Pope that Boris had asked his advice on the proper method of treating those whose teachings did not conform to the Apostolic commands.[1] He had also complained of the arrival in Bulgaria of numerous 'Christians' from various countries, who taught much and differently from one another, in particular Greeks and Armenians.[2] These Armenian 'Christians' were, in all probability, Paulicians. It is not impossible, however, that some of them were Monophysites who had come directly from Armenia; but it should be noted that those Monophysites transplanted into Thrace in 745 by Constantine Copronymus were Syrians, while the Armenians settled there by the same Emperor in 757 were, according to the evidence of Theophanes, Paulician heretics. It will be remembered that the Paulicians called themselves Christians, which would justify Boris describing them by that name to the Pope.

In support of this hypothesis, there is evidence showing that Bulgaria was at that time a centre of Paulicianism. A letter written by Stylianus, bishop of Neocaesarea, to Pope Stephen V, probably in 886 or 887,[3] mentions a certain 'Manichaean', Santabarenus, the father of Theodore Santabarenus, the well-known supporter of the Patriarch Photius. This Santabarenus, who, according to Stylianus, was a magician, seeing that his practices were discovered and that he was threatened with arrest, fled from Byzantine territory to Bulgaria, where he abjured Christianity.[4] These accusations of Manichaeism and magic must not, in all probability,

[1] 'Consulentibus...vobis, quid de eo faciendum sit, qui super praecepta apostolica se efferens praedicare tentaverit.' (*Resp.* 105, *P.L.* vol. CXIX, col. 1015.)

[2] 'Asserentes quod in patriam vestram multi ex diversis locis Christiani advenerint, qui...multa et varia loquuntur, id est Graeci, Armeni, et ex caeteris locis.' (*Resp.* 106, ibid. col. 1015.)

[3] See A. Vogt, *Basile Ier, empereur de Byzance*, p. 235, n. 4.

[4] J. Mansi, *Sacrorum conciliorum nova et amplissima collectio*, vol. XVI, col. 432. According to Friedrich ('Der ursprüngliche...Bericht über die Paulikianer', *S.B. bayer. Akad. Wiss.*, philos.-hist. Kl., the flight of Santabarenus to Bulgaria took place some time between 842 and 846.

be taken too seriously, as Stylianus, who, as a polemical writer, is not noted for objective accuracy, is trying in this letter to discredit Photius, and may simply be repeating unauthenticated rumours about the father of one of the patriarch's principal adherents. Yet the epithet 'Manichaean' was, at that time, generally applied to the Paulicians, and it can perhaps be inferred from this account that Santabarenus escaped to Bulgaria in order to seek protection among his co-religionists there, and that he was, indeed, received into their community.

At this very time (870) we possess the evidence of Peter of Sicily that the Paulicians of Tephrice were planning to send missionaries to Bulgaria.[1] Whether this was done or whether there had been any previous missions is not known for certain; but the covetous eyes which the Armenian Paulicians cast on Bulgaria strongly suggest that they were in contact with their co-religionists in that country.

An Old-Bulgarian manuscript which describes in legendary form the origin of the Bulgarian Paulicians[2] supports the evidence of the Greek sources. It shows that according to a medieval tradition current in Bulgaria the Paulician heresy was brought there by missionaries from the East. The names of the Paulician missionaries, Subotin and Shutil, are, according to Prof. Ivanov, of Eastern origin.[3] Moreover, the legend asserts that the Paulicians came to Bulgaria from Cappadocia,[4] and this is substantiated by our knowledge that the Paulician heresy was rife in Cappadocia in the ninth century.[5] This, together with the fact that the legend clearly refers to the early days of Paulicianism in Bulgaria,[6]

[1] See supra, p. 30.
[2] Произходъ на Павликянитѣ споредъ два български рѫкописа, S.B.A.N. (1922), vol. XXIV, pp. 20–31. The MS., published by I. Ivanov, is in the Bulgarian National Library in Sofia.
[3] Loc. cit. p. 30.
[4] И прѣидоста в бльгарьскои земли ѡт Кападоки, и прѣтвориmе си имена апостольска Павель, Іѡань. И оучаху людіе ѡт князѣ да коего человѣка увѣриши, ѡни ему закон пріимлаху. И тизи люди наричетсе Павликѣне (loc. cit. p. 22). Cf. Ivanov, Богомилски книги и легенди, p. 11.
[5] See Vasiliev, op. cit. p. 230. Bury, op. cit. p. 277. Cf. Conybeare, *The Key of Truth*, p. lxxiii.
[6] The MS. is undated. Ivanov places the composition of the story not earlier than the twelfth century. Its authorship, ascribed in the title to St John Chrysostom, is clearly apocryphal, as the saint died more than three centuries before we know of any Paulicians in the Balkans. But it shows that the story refers to a very early period.

confirms the evidence of Paulician infiltration from Asia Minor into Bulgaria in the ninth century.

It is thus possible to state with certainty that the Paulician heresy was a strong and dangerous force in Bulgaria in the third quarter of the ninth century at the time of its Christianization and the struggles between the Eastern and Western Churches.[1] Moreover, its teachers came from the Armenian colonies in Thrace and also directly from Asia Minor.

It is probable that many of the latter arrived together with Armenian merchants. These were numerous in Bulgaria in the ninth century. Some of them had remained in the commercial centres of Anchialus and Develtus after Krum had captured them from the Greeks. Beside the Greek and Bulgarian merchants, the Armenians acted as carriers in the lively trade between Byzantium and Bulgaria, especially during the Thirty Years' Peace (816–46). It should be remembered that Bulgaria was the main emporium in the trade between central and northern Europe on the one hand and Byzantium on the other.[2] Along the trade routes leading through Bulgaria,[3] Armenian merchants carried Transylvanian salt to Moravia and the industrial products of Constantinople and Asia Minor to central Europe. They brought their faith, Paulician or Monophysite, with them. Their mobility made them useful intermediaries between the Empire and Bulgaria. Armenians were particularly numerous in Thessalonica, whence they could easily penetrate into Macedonia.[4] Armenian communities are attested in Macedonia from the tenth to the fourteenth century.[5]

[1] The same conclusion was reached by Prof. Grégoire, merely through a critical study of the sources of Peter of Sicily: 'il n'y a...aucune raison de douter que, vers 872, les Pauliciens ne fussent nombreux et dangereux en Bulgarie' ('Autour des Pauliciens', *B.* (1936), vol. XI, p. 611). Cf. A. Lombard, *Pauliciens, Bulgares et Bons-Hommes en Orient et en Occident* (Geneva, 1879), pp. 11–21.

[2] See F. Dvorník, *Les Légendes de Constantin et Méthode*, pp. 222–6, who has brought to light many new facts showing the importance of Byzantine trade with central Europe from the sixth to the ninth century. Towards the middle of the ninth century, Bulgaria became the principal intermediary between Byzantium and the Moravian Empire.

[3] See K. Jireček, *Die Heerstrasse von Belgrad nach Constantinopel und die Balkanpässe* (Prague, 1877).

[4] See G. L. F. Tafel, *De Thessalonica ejusque agro* (Berlin, 1839), pp. xv–xix.

[5] For the tenth century see infra, p. 147; for the eleventh, Theophylact of Ochrida, *Epistolae, P.G.* vol. CXXVI, cols. 344–9; for the twelfth, infra, p. 223;

Apart from the danger presented by the Paulician and Monophysite Armenians, the young Orthodox Church of Bulgaria had to fight the proselytism of various non-Christian religions, particularly that of the Jews, colonies of whom were settled in several large Balkan towns and were frequently aggressive towards the Christians.[1] Their presence in Bulgaria accentuated the state of religious confusion and swelled the number of those who 'multi ex diversis locis...advenerint, qui...multa et varia loquuntur'.[2] The *Responsa Nicolai* tell us that a certain Jew had baptized many Bulgarians and that Boris, in ignorance whether he was a Christian or a pagan, and doubting the validity of his baptism, had asked the Pope for guidance on the matter.[3]

According to the same document, Bulgaria had been open to Moslem influences in the past: Boris had asked Nicholas what he should do with those books which his people had received from the Arabs.[4] There is, however, no clear indication that these books had any great success or that Islam was preached on any large scale in Bulgaria at that time.[5]

for the thirteenth and fourteenth, K. Jireček, *La Civilisation Serbe au Moyen Âge* (Paris, 1920), p. 63; and 'Staat und Gesellschaft im mittelalterlichen Serbien', *Denkschr. Akad. Wiss. Wien* (1912), vol. LVI, pt 2, p. 52.

[1] See G. L. F. Tafel, *De Thessalonica*, p. xiv. D. Tsukhlev, История на българската църква, p. 243. A large Jewish colony ('mercatorum genti') had existed already in Thessalonica in the first century A.D. Like the Armenians, they probably spread from there to Macedonia and to other parts of Bulgaria long before 864. Zlatarski, however, thinks that Jewish missionaries came to Bulgaria from southern Russia; Jewish colonies were widespread round the Sea of Azov even before our era, and from the eighth century they showed considerable proselytizing activity among the populations of the northern shores of the Black Sea, even succeeding in the ninth century in converting the Khazar Khan and nobles (История, vol. I, pt 2, pp. 65–6). Cf. F. Dvorník, *Les Slaves, Byzance et Rome*, pp. 138–41, and *Les Légendes de Constantin et Méthode*, pp. 148–211.

[2] See supra, p. 80, n. 2.

[3] 'A quodam Judaeo, nescitis utrum Christiano, an pagano, multos in patria vestra baptizatos asseritis, et quid de his sit agendum consulitis.' *Resp.* 104, *P.L.* vol. CXIX, col. 1014.

[4] 'De libris profanis, quos a Saracenis vos abstulisse, ac apud vos habere perhibetis, quid faciendum sit, inquiritis.' *Resp.* 103, loc. cit. col. 1014.

[5] There is, however, some evidence that the teaching of Islam was known in Bulgaria for several centuries after the introduction of Christianity. See I. Ivanov, Богомилски книги и легенди, p. 368. Tsukhlev's opinion that Moslem teachers came to Bulgaria from among the Bulgars of the Volga (История на българската църква, pp. 242–3) cannot be more than hypo-

To sum up the preceding considerations, it can be said that the introduction and consolidation of Christianity in Bulgaria in the third quarter of the ninth century was considerably hindered by the active proselytism of Christian heretics and teachers of non-Christian religions, who exploited the religious instability of the country in their efforts to secure the adhesion of the Bulgarians to their conflicting doctrines. The Bulgarian Church historian Tsukhlev has aptly compared Bulgaria at that time to a debating hall echoing with the heated contests of foreign missionaries.[1]

In one sense these religious struggles had a beneficial effect on the Bulgarian Orthodox Church. Faced with the necessity of fighting heresy from the very moment of its foundation, it was compelled to organize itself on a unified and centralized basis. Though canonically subject to the authority of the Oecumenical Patriarch, the archbishop of Bulgaria enjoyed full autonomy in all matters of administration and interior discipline.[2] The dioceses (or eparchies) were organized according to a strictly hierarchical principle, closely modelled on that of the Byzantine Church. While the aim of Boris and of the Bulgarian ecclesiastical authorities was to restore wherever possible the ancient Christian sees which had existed before the Great Invasions, a number of new dioceses were also created.[3] The majority of the sees were situated in the

thetical. They may have come from the country of the Khazars, where they were numerous at that time. (See V. N. Zlatarski, История, vol. I, pt 2, p. 67.) In the first half of the ninth century a large Moslem population was transplanted by the Emperor Theophilus to Macedonia. They were settled on the lower Vardar and became known as the 'Vardar Turks' (see K. Jireček, *Geschichte der Bulgaren*, p. 222, Zlatarski, op. cit. vol. I, pt 1, p. 341, n. 2). F. Rački ('Bogomili i Patareni', *Rad Jugoslavenske Akademije* (Zagreb, 1869), vol. VII, p. 98) thinks that they were responsible for spreading Islam in Bulgaria. Zlatarski, however, has shown that the 'Vardar Turks' indulged in no missionary activity and that they were baptized shortly after their forced emigration to Macedonia (op. cit. vol. I, pt 2, p. 67). For the 'Vardar Turks', see F. Tafel, *De Thessalonica*, pp. 70–86; S. Novaković, Охридска Архиепископија у почетку XI века, *Glas Srpske Kraljevske Akademije* (Belgrade, 1908), vol. LXXVI, p. 61.

[1] История на българската църква, p. 242.
[2] See Zlatarski, История, vol. I, pt 2, pp. 145, 203 et seq.
[3] See Zlatarski, ibid. pp. 208–14; Tsukhlev, op. cit. pp. 360–70; Runciman, op. cit. pp. 135–6. In the reign of Boris there is positive evidence for the existence of the following sees, with resident metropolitans: Ochrida, Bregalnitsa, Morava, Sardica, Philippopolis, Provadia, Dristra (residence of the archbishop of Bulgaria).

south-west, i.e. in Macedonia, that ancient cradle of Christianity, and in the north-east, which contained the cities of Pliska and Preslav and was thus the political and administrative centre of Bulgaria. Thus from the periphery of the land the work of enlightenment could spread to the wilder and more backward interior.

This centralized organization was all the more necessary as the Bulgarian Church, in its efforts to consolidate Christianity in the land, was faced with the necessity of a twofold struggle: it may be said that its war of defence consisted in preserving its flock from heresy, while its war of attack was waged against paganism.

The baptism of Boris had only dealt a superficial blow to paganism. Agelong traditions could not be uprooted by the spiritual and political act of the ruler. Evidence that pagan beliefs and customs survived the events of 864 is supplied by the *Responsa Nicolai*, which inform us that, in 866, many Bulgarians worshipped idols,[1] performed pagan rites before going to battle,[2] wore amulets round the neck to obtain recovery from illness,[3] chipped off pieces from a stone endowed with magical qualities,[4] and took solemn oaths on their swords.[5]

By declaring war on paganism the Church was pursuing a persistent and elusive foe. As in so many other countries, Christianity was forcibly imposed by the prince on his subjects, and the new teaching spread from the court to the more remote districts of Bulgaria. This process was perforce a very slow one and into some parts of the country, particularly the north, Christianity scarcely penetrated at first, owing to the slowness and difficulties of communications and to the comparative scarcity of available clergy.

[1] 'Qui Christianitatis bonum suscipere renuunt, et idolis immolant, vel genua curvant.' (*Resp.* 41, *P.L.* vol. cxix, col. 995.)

[2] 'Refertis quod soliti fueritis, quando in proelium progrediebamini, dies et horas observare, et incantationes, et joca, et carmina, et nonnulla auguria exercere.' (*Resp.* 35, ibid. col. 993.)

[3] 'Perhibentes quod moris sit apud vos infirmis ligaturam quamdam ob sanitatem recipiendam ferre pendentem sub gutture.' (*Resp.* 79, ibid. col. 1008.)

[4] 'Refertis quod lapis inventus sit apud vos..., de quo si quisquam ob aliquam infirmitatem quid accipit, soleat aliquoties remedium corpori suo praebere.' (*Resp.* 62, ibid. col. 1003.)

[5] 'Perhibetis vos consuetudinem habuisse, quotiescunque aliquem jurejurando pro qualibet re disponebatis obligare, spatham in medium afferre, et per eam juramentum agebatur.' (*Resp.* 67, ibid. col. 1005.)

An added obstacle was the language difficulty: originally the Bulgarian clergy, consisting of bishops, priests and deacons, was entirely Greek, and the liturgy and most of the preaching were performed in a language with which the vast majority of the people were unfamiliar.[1] Among the upper class there was, as it has been shown, strong opposition to Christianity and a resulting restlessness. The masses, on the other hand, though they had accepted baptism, freely or by force, could not easily abandon their ancient beliefs and rites, with which their life from the cradle to the grave was intimately linked. As a result of this situation, the two ways of life, the old and the new, after the first inevitable clash, gradually merged and produced that ambiguous state which existed in various forms and at different times in all Slavonic and indeed in all Christian countries, and which the Russian Churchmen called the 'dual faith' (двоевѣріе).[2] The assimilation of pagan gods to the Christian saints and the adaptation of pagan festivities to the feasts of the Church gradually softened the differences between the two conceptions of life, but they were never able to destroy them completely. The Orthodox Church, often freer in this respect than the Roman Catholic,[3] adapted some of the ancient rites to its own doctrines and has thus preserved to the present day certain customs which are pagan in origin.[4] On the whole, however, the Orthodox Church rigorously denounced all vestiges of paganism, and its sermons, prohibitions, instructions and hagiographies contain frequent references to the 'dual faith'.[5] Strict measures were taken, not only against idolatry, but also against any games or songs directly or indirectly connected with pagan practices. These, however, were never entirely successful, and, as against the incessant denunciations of the Churchmen, there was a mass of indistinct, ever-shifting con-

---

[1] V. N. Zlatarski, История, vol. I, pt 2, pp. 204-5.
[2] See L. Niederle, *Slovanské Starožitnosti — Život starých Slovanů*, pt II, vol. I, pp. 9–12, and *Manuel de l'Antiquité Slave*, vol. II, pp. 128, 168. Valuable information on the relation between paganism and Christianity in Russia may be found in E. V. Anichkov's Язычество и древняя Русь (St Petersburg, 1914), and in Mansikka's *Die Religion der Ostslaven*. Cf. A. N. Pypin, История русской литературы (3rd ed.; St Petersburg, 1907), vol. I, pp. 73–4.
[3] L. Niederle, *Slov. Starož.* loc. cit. pp. 273-4, and *Manuel*, vol. II, p. 168.
[4] See *Slov. Starož.* vol. I, pt 1, pp. 292, 294-5.
[5] Ibid. pt II, vol. I, pp. 27-8.

ceptions, gradually assimilating more and more Christian elements, but still retaining a measure of their original duality.

But the dangers of paganism in Bulgaria were not limited to this passive, and often unconscious, resistance of the masses. Already in 866 a section of the boyars had attempted by a *coup d'état* to extirpate Christianity from the country. They had been cowed by Boris's energetic counter-measures and were effectively silenced for the rest of his reign. But when Boris retired from the throne in 889 and entered a monastery they seized the opportunity to strike again. They secured the support of Prince Vladimir himself, Boris's eldest son and successor. Vladimir, who in contrast to Boris's noted austerity led a dissolute life, completely reversed his father's policy. Furthermore, he encouraged the revival of paganism,[1] ordered the destruction of churches and even started a persecution of the clergy.[2] In his monastic retreat, however, Boris became aware of this threat to the whole of his life-work. In 893 he suddenly appeared in Pliska, rallied those who had remained faithful to him, reassumed the position of ruler and had Vladimir blinded and imprisoned. He then summoned a general assembly of the land, which ratified the following decisions: Boris's third son, Symeon, became ruler, the capital was transferred from Pliska to Preslav and the Greek language was officially replaced by Slavonic in the whole country. His work accomplished, Boris returned to his monastery. The official pagan revival had lasted four years.[3] Of all the measures promulgated by the 'sbor' (council) of 893 the official recognition of Slavonic as the spoken and written language of Bulgaria was the most far-reaching in its effects. This decision was directly related to an event of paramount

[1] Regino, *Chronicon*, M.G.H. Ss. vol. I, p. 580: 'Interea filius eius quem regem constituerat, longe a paterna intentione et operatione recedens, praedas coepit exercere, ebrietatibus, comessationibus et libidinibus vacare, et omni conamine ad gentilitatis ritum populum noviter baptizatum revocare.' Sigebertus, *Chronicon*, M.G.H. Ss. vol. VI, p. 341; Annalista Saxo, ibid. p. 575: 'Sed cum filius eius iuveniliter agens, ad gentilitatis cultum vellet redire.'

[2] Cf. Zlatarski, ibid. pp. 246–9. The Archbishop Joseph was imprisoned.

[3] See Zlatarski, ibid. pp. 249–60, Runciman, op. cit. pp. 134–5. A proof of the organic unity of Boris's religious and political policy—and of Vladimir's wholesale opposition to it—lies in the fact that the latter concluded an alliance with King Arnulf of Germany, thus reversing his father's pro-Byzantine policy and reverting to that of Omortag.

importance in Bulgarian history which had occurred eight years previously: Clement, Naum and Angelarius, who, as the principal disciples of St Methodius, had been expelled from Moravia owing to the intrigues and persecution of the German clergy, arrived in Bulgaria in 885.[1] They brought with them the Slavonic version of the Scriptures, translated by St Methodius, and the Slavonic Liturgy hitherto used in Moravia and Pannonia.[2] The missionaries were cordially received by Boris, who must have immediately understood that in them he had found the means of achieving his old wish of founding a truly Slavonic Church in which the services, the preaching and the very hierarchy would be close to the people. The work of St Clement and St Naum directly resulted in a deep transformation of the Bulgarian Church and of the whole religious life of the country and hence, as it will be shown, indirectly affected the growth of heresy in Bulgaria.

Shortly after his arrival at Pliska, Clement was sent to Macedonia, where he took up his residence not far from Ochrida. There he laboured unceasingly among the Macedonian Slavs for seven years; he baptized those who were still pagan, preached the Gospel in Slavonic, built churches, founded the monastery of St Panteleimon in Ochrida and the first Bulgarian Slavonic school at Devol, and improved local agricultural conditions.[3]

Meanwhile Clement's companion and friend, Naum, remained at Pliska, where in direct collaboration with Boris and in permanent contact with the Macedonian school he built up a second Slavonic centre in north-eastern Bulgaria. He founded the monastery of

---

[1] See F. Dvorník, *Les Slaves, Byzance et Rome*, pp. 312–13; Spinka, op. cit. pp. 46–7.

[2] The whole question of the Slavonic liturgy in central Europe in connection with the mission of Constantine and Methodius is dealt with exhaustively in Dvorník's *Les Slaves, Byzance et Rome au IXe siècle* and *Les Légendes de Constantin et Méthode*.

[3] Devol remained the centre of St Clement's activity until his nomination in 893 to the bishopric of Debritsa and Velitsa. He taught Slavonic letters in person and prepared his pupils for the duties of readers, subdeacons, deacons and priests in the Bulgarian Church. In the course of seven years, 3500 passed through his hands. See Theophylact of Ochrida, *Vita S. Clementis Bulgarorum Archiepiscopi, P.G.* vol. CXXVI, cols. 1193 et seq.; cf. N. L. Tunitsky, Св. Климент, епископ словенский (Sergiev Posad, 1913); Zlatarski, op. cit. vol. I, pt 2, pp. 226–39. An extensive bibliography of works dealing with St Clement is given by F. Dvorník, *Les Slaves, Byzance et Rome*, pp. 313–16.

St Panteleimon near Preslav at the mouth of the river Ticha, which became a second Ochrida.[1]

In this manner the work of St Cyril and St Methodius, banished from Moravia and Pannonia, was saved for Bulgaria, where it bore abundant fruit for a time. As the result of the labours of St Clement and St Naum, the reign of Symeon became in many respects the golden age of Slavonic letters,[2] when in the space of some two decades an astonishing number of works in Old Slavonic was produced.

The school of Preslav was particularly noted for its Slavonic literary productions. The majority were translations and adaptations from the Byzantine Fathers,[3] but the rest were original.[4] At first Prince Symeon himself was the chief inspiration and moving force in the school of Preslav. As a young man he had studied Greek literature and philosophy in Constantinople, probably at the famous school of the Magnaura, where he became known as ἡμιάργος. At Preslav, his knowledge of both Slavonic and Greek were a valuable asset to the school.[5]

But this very dependence on the monarch, though beneficial at first, rapidly became a source of weakness for the Slavonic school at Preslav. Symeon, who began his reign in the best traditions of his father, became obsessed in later years by his desire to crush Byzantium. His ceaseless and bitter wars with the Empire and his quest for external glory and prestige occupied all his attention, and he was not likely to devote much interest or give support to a group of ecclesiastical writers who were making the Byzantine Fathers accessible to the Bulgarians.

Moreover, even in its most glorious period, the school of Preslav suffered from a dangerous defect: it remained mainly ecclesiastical, largely imitative and somewhat academic; its

[1] See Tunitsky, op. cit. pp. 251–5; Zlatarski, op. cit. pp. 239–43.
[2] See S. N. Palauzov, Век болгарского царя Симеона (St Petersburg, 1852).
[3] For the outstanding ecclesiastical writers of the school of Preslav: John the Exarch, Bishop Constantine, the monks Gregory, Khrabr and Duks, see Palauzov, op. cit. passim; K. Kalaidovich, Иоанн, Ексарх болгарский (Moscow, 1824); Tsukhlev, op. cit. pp. 419–46; Zlatarski, op. cit. pp. 169–70, 258–9, 347–50, 853–60.
[4] These included a Slavonic grammar by John the Exarch and an apology for Slavonic letters by the monk Khrabr.
[5] See Zlatarski, op. cit. pp. 279–81. Cf. Tunitsky, op. cit. pp. 251–6.

productions were invaluable to Bulgarian Churchmen, but difficult of access to the masses. This, no doubt, partly explains the fact that the Old-Bulgarian literature, while performing the invaluable service of transmitting to Russia and Serbia the Byzantine tradition through the medium of Cyrillic, failed to develop all its potentialities in its original home.[1]

It should be noted here that the failure of Bulgarian Orthodox literature to become firmly grounded on popular foundations partly explains the later growth of heterodox and heretical literature in Bulgaria.[2]

St Clement's school of Ochrida was less literary, more educative and nearer to the people. Moreover, for historical and geographical reasons, it remained apart from the stream of Byzantinism which, particularly after Symeon's death, inundated eastern Bulgaria. During the life of its founder it built the basis of a truly Slavonic Bulgarian Church, but after the death of St Clement in 916 its apostolic activity was greatly curtailed. Here again, Symeon's ceaseless wars with Byzantium seriously damaged the work of the disciples of St Cyril and St Methodius in Bulgaria. This work had the full approval and support of the highest ecclesiastical authorities in Constantinople, which was the criterion of its validity and an essential condition of its success.[3] St Clement well understood this. Guided by his Christian principles and Boris's will, he always maintained towards the mother Church of Byzantium a respectful and filial attitude.

Unhappily for the Church of Bulgaria, this wise policy was not followed by Symeon: in his pretensions to equality with the

---

[1] A. N. Pypin and V. D. Spasovich, История славянских литератур (2nd ed.; St Petersburg, 1879), vol. I, p. 67. It was not until the middle of the fourteenth century that Bulgarian literature enjoyed a new efflorescence.

[2] Cf. infra, pp. 154–5.

[3] The interest and sympathy with which the Byzantine government regarded the work of St Cyril and St Methodius is shown by the following episode: after the collapse of Methodius's work in Moravia, some of his disciples were sold as heretics to the Jews. They were discovered in Venice by the ambassador of Basil I, who bought them and brought them back to Constantinople. The emperor received them with honour and provided them with benefices. Some of them even went on to Bulgaria, probably on Basil's suggestion, and thus joined their comrades who had journeyed down the Danube from Moravia. See F. Dvorník, *Les Slaves, Byzance et Rome*, pp. 298–9.

Byzantine Basileus and to complete ecclesiastical independence from Constantinople, he arbitrarily raised the archbishop of Bulgaria to the rank of patriarch and had himself crowned 'Tsar and Autocrat of all Bulgarians'. This act of rebellion against the Byzantine Church probably took place in 918, two years after the death of St Clement, and it is safe to assert that the great apostle of Slavdom would never have agreed to it.[1] By placing the Bulgarian Church in the position of an outlaw and usurper towards the Oecumenical See, Symeon betrayed the work of St Clement. Moreover, in his desire to overthrow the Empire, he was forced to open his country to Byzantine influences, in an attempt to conquer Byzantium by her own weapons. Symeon was unsuccessful in his policy of violence, and in the reign of his son Peter Byzantine ideas and institutions overran most of Bulgaria and seriously crippled the Slavonic national development. In the widening gulf between the Hellenized Church and State and the masses, many of whom were ignorant, indifferent or hostile, the legacy of St Clement failed to play the part of cultural intermediary which it might have performed. The existence of this gulf between the Church and the people was, by its very nature, favourable to the spread of heresy. The Paulicians, in particular, benefited considerably from it, for the aim of their proselytism was precisely to detach the Bulgarian masses from the Orthodox Church by attacking the corruption and worldliness of the latter's representatives.

Symeon's disastrous policy threatened furthermore to obscure the most precious gift bestowed on Bulgaria by the disciples of St Methodius, namely the Slavonic Liturgy. Already in the reign of Boris, when, after 885, it was first introduced into Bulgaria, it roused opposition among certain members of the local Greek clergy, whose exclusive position in the country as teachers and administrators was threatened by the vernacular liturgy and the consequent rise in the numbers of Bulgarian priests. Some of them upheld the view against the Slavonic liturgy that it was only lawful to worship God in three languages, i.e. in Hebrew, Greek

---

[1] See Zlatarski, op. cit. vol. I, pt 2, pp. 399–401; Spinka, op. cit. pp. 52–3. Runciman, however, does not accept Zlatarski's chronology and places the establishment of the Bulgarian Patriarchate in 926 (op. cit. p. 174).

and Latin. The Bulgarian monk Khrabr denounced this opinion under the name of the 'three languages heresy'.[1]

At a time when there was harmony and understanding between the higher representatives of the Byzantine and Bulgarian Churches, the opposition to the Slavonic liturgy of some of the local Greek clergy could have no more than a transitory importance, particularly as the authorities in Constantinople were openly sympathetic to it.[2] But the mutual distrust and hatred between Bulgarians and Greeks, brought about by Symeon's wars and accentuated in the course of the tenth century, considerably hampered the growth of Slavonic Christianity. It encouraged the Greeks to regard the Slavonic liturgy as an obstacle to their domination over the Bulgarian Church, and the Bulgarians to use it as a weapon of religious nationalism against Byzantine imperialism. Thus the Orthodox principle of vernacular liturgies, connected with that of autocephalous Churches, was frequently obscured or misunderstood, and the ground was prepared for the growth of religious nationalism in Bulgaria.[3]

[1] In his apology for Slavonic letters, ѡ писменехъ чрноризьца Храбра, written, according to Zlatarski (op. cit. p. 860) between 887 and 894.

The notion of the three sacred languages arose fairly early in the West. It can be found in Isidore of Seville (*Etymologiarum* lib. IX, c. 1, *P.L.* vol. LXXXII, col. 326): 'tres autem sunt linguae sacrae: Hebraea, Graeca, Latina, quae toto orbe maxime excellunt. His namque tribus linguis super crucem Domini a Pilato fuit causa ejus scripta.'

In the polemics between the eastern and western Churches, the Latins were sometimes accused of the 'three languages heresy' (see J. Hergenröther, *Photius, Patriarch von Constantinopel* (Regensburg, 1867–9), vol. III, pp. 206–8). The question was raised in Moravia, in connection with the opposition of the German clergy to the Slavonic liturgy (see N. L. Tunitsky, Св. Климент, pp. 131–4). These considerations have led M. S. Drinov to the view that Khrabr, in attacking the adherents of the 'three languages heresy', was aiming at the Roman clergy (Исторически прѣгледъ на българската църква, Sofia, 1911, pp. 46–7). But, as far as is known, at the time when Khrabr wrote there were no Latin priests in Bulgaria; and his accusations of heresy can only have been directed against the extreme section of the Greek party in Bulgaria (see also Tsukhlev, op. cit. pp. 587–689; S. Stanimirov, История на българската църква, Sofia, 1925, pp. 92–3).

[2] See Tunitsky, op. cit. pp. 239–48. The attitude of the Byzantine Church to vernacular liturgies is discussed by F. Dvorník, *National Churches and the Church Universal* (London, 1944).

[3] The term 'nationalization of the Bulgarian Church' used by Zlatarski and other Bulgarian historians to describe that union of Orthodoxy and Slavdom, which Boris and St Clement had largely succeeded in achieving, is

THE RISE OF BALKAN DUALISM 93

This extreme nationalism, which is mainly due to the loss of the true understanding of the Oecumenical significance of the Church, is often not unrelated in a general sense to the growth of heresy. It is now necessary to consider the development of heresy in the reign of Symeon and its relation to the work of St Clement and St Naum.

In an immediate sense it cannot be doubted that the activity of the disciples of St Methodius dealt a considerable blow to heresy in Bulgaria. Before 885 the attacks of the heretics (in particular of the Paulicians) on the Orthodox Church were facilitated by the inevitable gulf between the Greek clergy and the Bulgarian people which was due to differences of language and nationality. After 885, and especially after 893, the success of Boris and of the 'Holy Seven'[1] in bringing the Bulgarian Church into closer touch with the people deprived the accusations of the heretics of much of their ground and enabled the Church to convert many hitherto obdurate pagans and to consolidate wavering Christians. Liturgical and hagiographical evidence mentions, among the exploits of the 'Holy Seven', the extirpation of heresy. The Greek canon composed in their honour glorifies them for 'completely destroying the heresy of those terrible wolves, the Massalians'.[2] The Greek life of St Vladimir of Dioclea refers to St Clement's extirpation of 'the heresy of the Bogomils and Massalians'.[3]

The reference to the Bogomils is anachronistic and no doubt due to a later addition in the Greek version of the Life of Saint Vladimir, itself a translation, in many places inaccurate, of a

unfortunate and misleading, but aptly describes the distortions to which this union was later subjected. The growth of such religious nationalism can be observed only too frequently in the history of the southern Slavs.

[1] The 'Holy Seven' (οἱ ἅγιοι ἑπτάριθμοι) is the name given by the Orthodox Church to the seven most prominent Slavonic apostles, beginning with St Cyril and St Methodius.

[2] τὴν αἵρεσιν λύκων τῶν δεινῶν, Μασσαλιανῶν, ὑμεῖς παντελῶς ἀπεσβέσατε (see B. Petranović, Богомили, Црьква Босаньска и крьстяни, Zara, 1867, p. 98).

[3] Symeon is described as συνδρομεὺς τοῦ μακαριωτάτου ἁγίου Κλήμεντος... εἰς ἀναίρεσιν τῆς αἱρέσεως τῶν Βογομίλων καὶ Μασσαλιανῶν. ('Ακολουθία τοῦ ἁγίου ἐνδόξου, βασιλέως, καὶ μεγαλομάρτυρος 'Ιωάννου τοῦ Βλαδιμήρου καὶ Θαυματουργοῦ, Venice, 1774. I have been unable to consult this source, and quote from V. Levitsky, Богомильство—болгарская ересь, *Khristianskoe Chtenie*, 1870, pt I, pp. 57–8.)

Slavonic original.[1] At the time of St Clement, the Bogomil sect had not yet arisen, but the mention of the Massalian heresy as an anti-Orthodox force in Bulgaria is very probably authentic and can be borne out by historical evidence.

This earliest reference to Massalianism in Bulgaria is particularly important in view of the fact that this heresy, together with Paulicianism, exercised a considerable influence on the Bogomil sect.

The penetration of Massalians from Asia Minor to Bulgaria is prima facie a plausible hypothesis. In view of the connections which very probably existed in Armenia and Asia Minor between the Massalian and the Paulician sects[2] it is extremely likely that among the 'Syrians' and 'Armenians' transplanted to Thrace in the eighth century some at least were Massalians. Some of the colonists settled there in 757 by Constantine Copronymus came from Melitene, which was an important Massalian centre at the end of the fourth century and probably also in later times. This hypothesis is confirmed by the evidence of Cedrenus, who asserts that those Massalians who had been driven out of Syria by Flavian found refuge in Pamphylia and thence spread in large numbers to the western part of the Empire,[3] which, in all probability, means Thrace.[4] The testimony of Cedrenus is corroborated by Michael Psellus, who describes the numerous Massalians in Thrace towards the middle of the eleventh century.[5]

From Thrace the Massalian heretics could easily penetrate to various parts of Bulgaria in the same manner as the Paulicians, and there can be little doubt that by the beginning of the tenth century, if not before, Massalianism existed there together with the Paulician heresy as a threat to the Orthodox Church. Although the statement in the life of St Vladimir regarding the extirpation of Massalianism by Saint Clement is certainly an exaggeration (since we possess unmistakable evidence of the prevalence of this heresy in Bulgaria in the latter part of the tenth century), the reference to the saint's fight against it is perfectly acceptable.

It is important to remark that although the Paulician and

---

[1] See Yu. Trifonov, Бесѣдата на Козма Пресвитера и нейниятъ авторъ, *S.B.A.N.* (1923), vol. XXIX, pp. 49–52.   [2] Cf. supra, p. 51.
[3] Cedrenus, *C.S.H.B.* vol. I, p. 516: εἰς δὲ τὴν Παμφυλίαν ἀνεχώρησαν καὶ ταύτην τῆς λώβης ἐπλήρωσαν, νῦν δὲ σχεδὸν εἰπεῖν καὶ τὴν πλείονα δύσιν.
[4] Such is the opinion of J. Gieseler, *Lehrbuch der Kirchengeschichte*, vol. II, pt 1, pp. 401–3. Cf. I. von Döllinger, *Beitr. Sektengesch. Mittelal.* vol. I, p. 34.
[5] Cf. infra, pp. 183–8.

Massalian heresies in Bulgaria had numerous points of contact both in their doctrines and in their common opposition to the Church, they always remained clearly distinct from one another. There were important differences in the manner of life of their adherents: thus we find no trace among the Paulicians of the extreme asceticism or of the extreme immorality characteristic of the Massalians. In contrast to the contemplative life of the latter, the former retained their active and warlike qualities, and whereas the Paulicians were noted for their aversion to monks,[1] the Massalians had a particular predilection for the monastic life.

Evidence of the proselytism of the heretics in Bulgaria in the reign of Symeon can be found in the attack of John the Exarch on the 'filthy Manichaeans and all pagan Slavs...who are not ashamed to call the Devil the eldest son [of God]'.[2] The reference to the pagan Slavs is significant, as it is the earliest direct indication of the alliance between heresy and paganism in Bulgaria. It is indeed very probable that the heretics, in their hostility to the Church, appealed to those elements which were the most refractory to its influence and which expressed their dissatisfaction with the new Christian order by falling back on their old pagan traditions.

The term 'Manichaean' is clearly used in a general sense by John the Exarch, who conforms to the common habit of Orthodox writers of using this epithet to designate a number of sects whose teaching was based to a greater or lesser degree on the dualism associated with the doctrines of Mani, and especially the Paulicians.[3] There is, however, no evidence that the Paulicians taught that the Devil was the eldest son of God. The precise origin of this doctrine is unknown, but from the second half of the tenth century it is frequently ascribed to the Bogomils.[4]

[1] Cf. supra, p. 42.
[2] Да се срамлѣютъ оубо вьси пошибенни и сквръннни манихеи и вси поганнии словѣне...то же не стыдетсе диявола глаголюще старѣиша сына (Ivanov, op. cit. p. 20).
[3] From the eighth century the Paulicians are described as 'Manichaeans' in Byzantine sources. See V. Grumel, *Regestes des Actes du Patriarcat de Constantinople* (Constantinople, 1932), vol. I, fasc. II, pp. 6, 27.
[4] See infra, pp. 122, n. 4, 184–6, 207. It may be of interest to observe that the cosmological myth of the two brothers, the elder, and evil, one, who has dominion over this world, and the younger, and good, one, who will inherit the Kingdom which is to come, occurs in Iranian Zarvanism, which, as it has been pointed out, had a marked influence on Manichaeism (cf. supra, pp. 14–15).

It is not unreasonable to claim that the growth of heresy in Bulgaria was indirectly facilitated by Symeon's aggressive policy towards Byzantium. A strong hostility towards the Greeks spread among the people and did not abate during the reign of his successor Peter. At the same time Peter's reign was characterized by extreme Byzantinization in the ecclesiastical, administrative and social spheres and by a policy of servility towards the Empire. Consequently, the heretics were now faced not only with the growing power of the Byzantine Church which they particularly disliked, but also with the increasing influence of the concomitant secular institutions on every aspect of Bulgarian life; it is therefore only natural that the Bulgarian heretics exploited the anti-Greek feeling in the country for their own aims. This explains the important fact that heresy in Bulgaria, from being essentially a religious phenomenon, assumed in the course of the tenth century a distinctly social aspect.[1]

For this reason the true causes and character of Bogomilism cannot be understood without relating its growth to the social and economic aspects of Bulgarian life in the tenth century. It is particularly necessary to consider the trend of Byzantine influence in Bulgaria, since Bogomilism developed in the tenth and eleventh centuries partly as a reaction against it.

The influence of Byzantium on Bulgarian institutions was already strongly felt under Symeon, particularly after the establishment of the Bulgarian patriarchate. In the ecclesiastical sphere, the organization of metropolitan and episcopal eparchies, begun under Boris, was completed.[2] The court of the patriarch was closely modelled on that of the Oecumenical See, with a patriarchal synod and a great number of ecclesiastical officials.[3] The clergy was established as a new class in the State, its legal powers being determined by Byzantine canon law and its subsistence assured by regular income.[4] Gradually, however, a gulf appeared between the higher clergy, metropolitans and bishops,

[1] See infra, pp. 136–8.   [2] See Tsukhlev, op. cit. pp. 458–80.
[3] Ibid. pp. 482–96. Cf. M. S. Drinov, Южные славяне и Византия въ X веке, pp. 71–2.
[4] Evidence of the material support received by the clergy at the time of Symeon can be obtained from the words of the monk Duks to John the Exarch: 'What other business have the priests, except to teach and to write books?' See Tsukhlev, op. cit. p. 500.

who enjoyed numerous privileges and whose interests were allied to those of the State, and the ordinary parish priests, whose social position by the middle of the tenth century was often not very different from that of the free peasants.[1] In the spheres of the court and government, many Byzantine institutions were adopted during the reign of Symeon. The supreme position of the autocrat in the State, the magnificence of the court of Preslav which held the provincial visitor speechless with wonder,[2] the titles of the ranks in the government and civil service, were directly borrowed from Byzantium.[3] It is significant that in spite of the wars between the two countries the relations between Byzantium and Bulgaria were not completely interrupted in the reign of Symeon.[4]

In the first year of the reign of Peter (927) the alliance of Bulgaria and Byzantium was cemented by a treaty, by which the emperor formally recognized Peter's title of βασιλεύς and the autocephality of the Bulgarian Church; moreover, through Peter's marriage with Maria Lecapena, granddaughter of the Emperor Romanus, Byzantine influence gained a stronghold at the Bulgarian court and the Empire a useful eye into the internal affairs of the State.[5]

The picture of the Byzantinization of Bulgaria in the reign of Peter (927–69) would be incomplete without some reference to the social and economic conditions, which were among the important causes of the growth of heresy.

The economic structure of the Byzantine Empire was undergoing a severe crisis in the tenth century. The power of the aristocracy (the δυνατοί) was growing. By their uncontrolled acquisition of land they were hastening the development of *latifundia*, and by buying up the free peasant and military holdings (so characteristic a feature of Byzantine agrarian economy in the seventh to ninth centuries) they were threatening to deprive the

[1] See S.S. Bobchev, Симеонова България отъ държавно-правно гледище, *G.S.U.* 1926–7, vol. XXII, pp. 82–4, 88. Cf. Tsukhlev, op. cit. pp. 499–500.
[2] Cf. Runciman, op. cit. pp. 141–2.
[3] See Bobchev, loc. cit. pp. 79, 124–8.
[4] This can be seen, for instance, in the correspondence between the Patriarch Nicholas Mysticus and Symeon. See Drinov, op. cit. pp. 11 et seq.; Zlatarski, op. cit. vol. I, pt 2, pp. 388 et seq.; S. Runciman, *The Emperor Romanus Lecapenus and his Reign* (Cambridge, 1929), pp. 81 et seq.
[5] See Zlatarski, op. cit. pp. 526–36.

State of the mainstay of its military power as well as of its best taxpayers, thus causing serious anxiety to the central authorities. Moreover, the small proprietors and the serfs, in view of the extent and pressure of taxation, frequently resorted to the practice, known in the West as *patrocinium* and in eastern Europe as *prostasia*, which consisted in seeking, in return for labour and economic dependence, the material support and protection of the 'powerful'. The Byzantine emperors, seeing the danger of the gradual feudalization of the State, took vigorous measures to protect the small freehold peasant against the encroaching tendencies of the magnates. Thus a violent struggle arose between the central authority and the δυνατοί, carried on during the tenth and the first quarter of the eleventh centuries. But this imperial policy had little effect, for it was opposed not only by the great landowners and the very officials who were responsible for its execution, but also by the peasants who, under the burden of taxation, could not resist the attractions of *patrocinium*. Thus, in spite of repressive measures taken by the government, the development of *latifundia* and *prostasia* continued unabated in tenth-century Byzantium.[1]

The question of the precise extent to which these social and economic conditions prevalent in Byzantium were also to be found in tenth-century Bulgaria, important for a proper understanding of the nature of Bogomilism, has never been studied in any detail.[2] Consequently, it is not always possible when examining particular

[1] For the social and economic background of tenth-century Byzantium, see G. Ostrogorsky, 'Agrarian conditions in the Byzantine Empire in the Middle Ages', *Cambridge Economic History*, vol. I, ch. v, pp. 194–223; A. A. Vasiliev, 'On the Question of Byzantine Feudalism', *B.* (1933), vol. VIII, pp. 584–604; Th. Uspensky, К истории землевладения в Византии, *Zh. M.N.P.* (February, 1883), vol. CCXXV, pp. 323 et seq.; C. Diehl, *Byzance, grandeur et décadence* (Paris, 1919), pp. 165–71; A. Andréadès, 'Deux livres récents sur les finances byzantines', *B.Z.* (1928), vol. XXVIII, pp. 287–323.

[2] Apart from Zlatarski's general history, S. Bobchev's article (Симеонова България отъ държавно-правно гледище, loc. cit.) gives much useful information, but does not sufficiently take into account the gradual evolution of Bulgarian institutions under the influence of Byzantium. The inscriptions unearthed at Pliska and interpreted by Th. Uspensky give some valuable indications (*Aboba-Pliska*, Материалы для болгарских древностей, *I.R.A.I.K.* Sofia, 1905, vol. X). Cf. F. Dvorník, 'Deux inscriptions gréco-bulgares de Philippes', *Bull. Corresp. Hellénique de l'École Franç. d'Athènes* (Paris, 1928), and V. Beshevliev, Първобългарски надписи, *G.S.U.*, ист. фил. фак. (1934), vol. XXXI, pt I.

Bulgarian institutions to decide whether they were directly borrowed from Byzantium, whether they arose independently but through similar conditions, or even whether in some cases their origin was much earlier. Generally speaking, however, those institutions which existed in the reign of Peter appear to have a twofold origin: some were remnants of the old Bulgar order, others appeared with or after the introduction of Christianity and clearly possess a Byzantine character. Those of the former which were not in contradiction with the Christian order were retained and sanctioned: thus the title of boyar existed in Bulgaria before the Baptism and was most common in the tenth century,[1] and the boyars, according to the Byzantine conception of authority, participated in some degree *de jure* in the divine nature of the tsar's power.[2] Under Symeon and Peter many of them held titles, the very names of which were borrowed from Byzantium.[3] The boyars appear to correspond exactly to the two classes of Byzantine δυνατοί, i.e. the imperial officials and the landed gentry.[4] But the precise social and economic position which they occupied in the Bulgarian State is not very clear.[5] It seems that they were obliged

[1] See *Aboba-Pliska*, loc. cit. pp. 201–3; Bobchev, loc. cit. pp. 77–81; Drinov, op. cit. p. 82.

[2] The priest Cosmas, writing in the late tenth century, emphasizes against the derogations of the Bogomil heretics the divine origin of the authority of the tsar and the boyars: яко цари и боляре Богомъ соутъ учинени (M. G. Popruzhenko, Козма Пресвитер, болгарский писатель X века, Sofia, 1936, p. 35.)

[3] See Bobchev, loc. cit. p. 80.

[4] The Bulgarian boyars were divided in the tenth century into three classes, the six 'Great Boyars', the 'Inner Boyars' and the 'Outer Boyars'; the first 'probably comprised the Khan's confidential Cabinet', the second 'were probably Court officers', the third 'provincial officers' (Runciman, op. cit. p. 284); cf. Drinov, op. cit. pp. 82–4.

[5] Uspensky, in his study of the inscriptions of Pliska (*Aboba-Pliska*, pp. 204–12), analyses the expression often found on monuments in honour of the dead: θρεπτὸς ἄνθρωπός μου ('μου' refers to the Khan). He suggests that this is a translation of a Bulgar technical term, serving to describe a man hired to fight in a subordinate capacity in return for sustenance, and refers to its probable connection with the German *comes* and similarity with the Byzantine *foederatus*. If Uspensky's hypotheses are correct, we should find in Bulgaria, by the eighth and ninth centuries, that personal relation between subject and ruler, based on the obligation of military service, which is characteristic of a pre-feudal state of society. Uspensky notes the frequent occurrence in Bulgarian sources of the term *comes*, in its Greek form (κόμης). In the ninth and tenth

to render military service to the sovereign,[1] but whether or not they paid him tribute is not known. Equally uncertain is their exact territorial relation to the tsar; apparently in some cases they were given territorial grants, though it is very doubtful whether in tenth-century Bulgaria there was anything similar to a regular system of *beneficia* (or πρόνοια).[2] It seems more likely that at that time military service based on a personal relationship with the ruler and the tenure of land existed as separate and largely uncoordinated institutions. The integrated relationship between the two within the feudal conception of πρόνοια was achieved in the following century and it is only then that it becomes possible to speak of 'feudalism' in Bulgaria.[3]

As in Byzantium, so in Bulgaria, these magnates, particularly the provincial lords or 'Outer Boyars', were frequently a menace to the central authority. In the eighth and ninth centuries, whenever they were strong enough, they strove to influence the Khan or even to control his election to the throne. Boris crippled their power for a time by his drastic repressions (866 and 893), but under Symeon and especially under Peter it rose again.[4] The wars with the Empire increased their prestige and influence as military commanders and purveyors of man-power and, on the other hand, by forcing the peasants to resort to them for pro-

centuries the term is often used in Bulgaria to describe a man of position and authority, usually in the military sense, and often a provincial administrator (see N. P. Blagoev, Произходъ и характеръ на царь Самуиловата държава, *G.S.U.* 1925, vol. xx, pp. 524–8, 558), particularly in western Bulgaria (*Aboba-Pliska*, p. 212). In Bulgarian chrysobulls these magnates are sometimes referred to as 'владящи' or 'владалци господствующи по царьство ми', a term suggestive of a considerable degree of independence, and distinct from the mere 'владалци царьства ми'. (See Bobchev, loc. cit. p. 79; Drinov, op. cit. pp. 84–5; A. Rambaud, *L'Empire Grec au Xme siècle*, Paris, 1870, pp. 318–23.)

[1] See Bobchev, loc. cit. p. 81.

[2] Bobchev (ibid.) assumes the generalization of the institution of πρόνοια already under Symeon. But this view seems untenable, since the regular existence of the πρόνοια cannot be certified in the Byzantine Empire before the second half of the eleventh century. (Cf. A. A. Vasiliev, 'On the Question of Byzantine Feudalism', *B.* 1933, vol. viii, p. 591.)

[3] See Th. Uspensky, Значение византийской и южнославянской пронии, *S.L.* pp. 3–4.

[4] Symeon, to curb the independence of the provincial magnates, was in the habit of appointing them to various posts in his capital (see Bobchev, loc. cit. p. 80).

tection against foreign attack, hastened the movement of feudalization. The growth of a new class of powerful landowners by the end of the ninth century coincided with a corresponding decline of small peasant holdings in Bulgaria and hastened the ruin of the agricultural commune.[1] The increased taxation and the economic misery resulting from the frequent wars, particularly in Thrace, the perpetual battlefield between Bulgaria and Byzantium, were among the factors which brought about a catastrophic decline in the productivity of the land and induced the peasants to resort to *prostasia*.[2] The terrible famine and plague which followed the exceptionally severe winter of 927–8[3] and several bad harvests caused the 'powerful' in many parts of the Byzantine Empire to buy up the land from the starving population at very low prices or in exchange for food.[4] It is known that the famine also ravaged Bulgaria at the same time,[5] and it is permissible to suppose that it gave a similar impetus to the movement of *prostasia* in Bulgaria.

These economic conditions were undoubtedly conducive to the spread of heretical teachings. Not only did the widespread misery which accelerated the development of *prostasia* provide excellent food for the proselytism of the heretics, but also the gradual feudalization, which, in a country where the rapid inrush of Byzantinism accumulated all the power and wealth in the hands of a privileged minority, tended to deprive the masses of all means of economic subsistence. There is evidence that this form of social inequality was opposed by the Bogomils, whose successful proselytism in Bulgaria was partly due to the fact that they appeared as defenders of the people against their oppressors.[6]

To complete the picture of Byzantine influence in tenth-century Bulgaria, it is now necessary to consider the development of monasticism; for it clearly reflects both the good and the bad

[1] See I. Klincharov, Попъ Богомилъ и неговото време (Sofia, 1927), pp. 108–15.
[2] For the development of *prostasia* in Bulgaria, see Bobchev, loc. cit. pp. 88 et seq.
[3] In Constantinople the ground was frozen for 120 days; see S. Runciman, *The Emperor Romanus Lecapenus*, pp. 226–7.
[4] *C.E.H.* vol. I, p. 205.
[5] The famine in Bulgaria was accompanied by an invasion of locusts. See Zlatarski, op. cit. p. 518.
[6] See infra, pp. 137–8, 172–3.

features of this influence. Furthermore, monasticism at that time was not unconnected with the growth of heresy. The relation between Bogomilism and certain aspects of Bulgarian monasticism, which will be pointed out below, was established not through the Paulicians, who were opposed to the monastic ideal, but through the Massalians, who were notorious for spreading their teachings in Orthodox monasteries. For this reason an examination of certain features of Bulgarian monasticism in the tenth century forms a necessary introduction to the study of Bogomilism.

The reign of the Tsar Peter has been called 'the monastic reign'.[1] It witnessed the foundation of an astonishing number of monasteries, particularly in southern and south-western Bulgaria. Macedonia contained a very large number, especially round Ochrida, Skoplje, Bitolj and Thessalonica; in the neighbourhood of Thessalonica alone there were in Peter's time more than twenty monasteries, and in the mountains to the north-east of the city there was a continuous chain of houses, occupied by large numbers of monks and nuns.[2] This region was known as 'the second Holy Mountain', or 'the little Byzantium'.[3] It is significant that

[1] See Tsukhlev, op. cit. pp. 510 et seq. The training of Bulgarians in the monastic life was instituted by Boris soon after his baptism. A letter of Photius tells us that a number of young Bulgarians had been sent to Constantinople to seek the monastic vocation and had been entrusted to the Higumen Arsenius (Photii Patriarchae *Epistola* xcv, *P.G.* vol. cii, cols. 904–5; cf. J. Hergenröther, *Photius*, vol. II, p. 221; Zlatarski, op. cit. pp. 218–19; Dvorník, op. cit. p. 300). Boris himself spent the last eighteen years of his life (excepting his brief return to power in 893) as a monk in the foundation of St Panteleimon near Preslav. Symeon, in his younger days, took the monastic vows in Constantinople, but renounced them in order to ascend the throne. Peter was a man of great piety with a strong inclination for the monastic life; he showed great zeal and generosity in founding and endowing monasteries (see Tsukhlev, op. cit. pp. 512 et seq.).

[2] Tsukhlev, op. cit. pp. 518–20.

[3] This comparison is motivated by the considerable growth of monasteries in the Byzantine Empire in the ninth and tenth centuries, where monasticism had become a very powerful force, particularly after the defeat of Iconoclasm. Its influence was felt among all classes of the population, from the emperor to the peasants, and the foundation and endowment of monasteries were a common practice. See I. Sokolov, Состояние монашества въ византийской церкви с половины IX до начала XIII века (Kazan, 1894), pp. 33 et seq.; J. M. Hussey, *Church and Learning in the Byzantine Empire* (Oxford, 1937), pp. 159 et seq.

the same province of Macedonia became the original centre of Bogomilism in the Balkans.[1]

The thirst for sanctity and the ascetic life was furthered by men of great spiritual power and popular appeal. They were the real leaders of the people in their hours of severe trial.[2] The greatest of them was St John of Rila (d. 946), who lived for many years first as a hermit in a hollow oak and then in a cave in the Rila Mountains and was destined to become the patron saint of Bulgaria.[3] At the places of their ascetic endeavour, generally in deserted spots or high up in the mountains, monasteries would arise, built and inhabited by their disciples and pilgrims from all over the country who gathered round the saints in search of guidance and wisdom.[4] In other cases, when the monasteries were founded or endowed by the tsar or other secular persons, they remained generally in greater contact with the outside world; it was there that Byzantine influence was the strongest, particularly through the different rules or *typica*, which were borrowed from those used in Byzantine monasteries with only slight modifications necessitated by local conditions.[5]

This search for holiness, which was one of the principal causes of the uncommonly rapid development of monasticism in tenth-century Bulgaria, was indirectly strengthened by the political, social and economic instability of the times. Both in Byzantium

[1] See infra, pp. 151 et seq.
[2] See Tsukhlev, op. cit. pp. 548-81.
[3] See I. Ivanov, Св. Иванъ Рилски и неговиятъ монастиръ (Sofia, 1917), and Сѣверна Македония (Sofia, 1906), pp. 85-90.
[4] Such was the origin of the celebrated Rila Monastery.
[5] The Orthodox *typica* are based on the rules of St Basil and St Pachomius, both of which were translated into Bulgarian very early (see Tsukhlev, op. cit. p. 532). The most common in Bulgaria was the Studite rule, also prevalent in Byzantium. The monastery of Studion had a great reputation throughout the Balkans and was often visited by the high dignitaries of the Bulgarian Church. The 'Jerusalem typicon' of St Sabbas was introduced into Bulgaria in the eleventh century. Besides these traditional *typica* there were others composed by founders of new monasteries, though generally in accordance with the principles formulated by St Basil. The most celebrated of these was the *typicon* of Gregory Pacurianus, founder of the Bulgarian monastery of Bachkovo, based on the Studite rule. See L. Petit, 'Typicon de Grégoire Pacurianos pour le monastère de Pétritzos (Bačkovo) en Bulgarie', *V.V.* (1904), vol. XI, Suppl. no. 1.

Both forms of Byzantine monasticism, the coenobitic and the idiorrhythmic, existed in Bulgaria at the time of Peter (Tsukhlev, op. cit. pp. 532-3).

and Bulgaria, the monasteries appeared as the only stable places of retreat and peace amid the surrounding confusion and misery. Suffering from the economic exhaustion which followed Symeon's wars with the Empire and from the periodical devastations wrought after 934 by invaders from the north, Magyars, Pechenegs and Russians, many Bulgarians looked to the monasteries as the only refuge from the evils of the world. Personal suffering in many cases undoubtedly brought about the realization of the monastic vocation; on the other hand, the great quantitative increase of monks was often prejudicial to the quality of the monasteries. It is in the numerous defects of the monastic life in tenth-century Bulgaria that we find the origin of heresy.

The best picture of monasticism in tenth-century Bulgaria is painted by the priest Cosmas, in his *Sermon against the Heretics*,[1] written soon after 972[2] and containing bitter attacks on the distortions of the monastic ideal at that time.[3] It is significant that this work is directed at once against the Bogomil heresy and the abuses of contemporary monasticism; between the two Cosmas traces a definite connection.[4]

He inveighs against those who enter monasteries without sufficient preparation or because they are unable to support their

---

[1] Св. Козмы Пресвитера Слово на Еретики (ed. Popruzhenko; Odessa, 1907). The following quotations from Cosmas's work are taken from the more recent edition of the *Sermon against the Heretics* by Popruzhenko: Козма Пресвитер, болгарский писатель X века (Sofia, 1936). The *Sermon* has been admirably translated into French by Vaillant and analysed in detail by Puech: H.-C. Puech and A. Vaillant, *Le traité contre les Bogomiles de Cosmas le prêtre* (Paris, 1945).

[2] See infra, Appendix I.

[3] See in particular the chapters entitled: ѡ мятущихъ ся чернцѣхъ (Cosmas, op. cit. pp. 42 et seq.), о хотящих ѡтити в черныя ризы (pp. 46 et seq.), о затворницѣх (pp. 55 et seq.). While attacking its abuses, Cosmas expounds with great force and insight the true purpose and significance of the monastic life.

[4] The precise functions exercised by Cosmas in the Bulgarian Church are not known. His title of *presviter* suggests a secular priest of somewhat high ecclesiastical standing. (The ordinary village priest was generally called *pop*, e.g. Bogomil himself.) Vaillant supposes that after the suppression of the Bulgarian Patriarchate in 972, Cosmas held a position corresponding to that of a vicar-general. The tone of authority which he adopts even towards the Bulgarian bishops certainly suggests that he occupied an influential position in the Church. See Puech and Vaillant, op. cit. pp. 29, 35.

families and who abandon their children to starvation.¹ While denouncing these individual weaknesses, Cosmas also points out a dangerous error, based on a perversion of the true meaning of monasticism and a distortion of the whole Orthodox view of life, and apparently fairly widespread among Bulgarian monks at that time: the opinion that those who live in the world cannot be saved and that possessions, family cares, worldly occupations and miseries are unsurmountable obstacles to sanctity.²

This view was tantamount to a rejection of marriage as a sinful capitulation to the world, and was held, according to the testimony of Cosmas, by the Bogomil heretics against whom his *Sermon* is directed. It was also, however, to be found among the Orthodox, and the denial of the sanctity of marriage is denounced by Cosmas as 'nothing but a heretical thought'.³

This important point of contact between heresy and monasticism shows that by the middle of the tenth century, if not earlier, heretical proselytism had been active and often successful in the monastic circles in Bulgaria. The condemnation of the world as an obstacle to salvation—and hence evil—was not, in practice, very different from the Paulician teaching regarding the creation of this world by the evil principle, especially as the Paulicians outwardly accepted all the doctrines of the Church and concealed their metaphysical dualism under the cloak of 'pure' Christian ethics. The condemnation of the world in the name of a false asceticism is, however, characteristic of the Massalians, whose direct influence one is tempted to see here, especially in view of their predilection for monasteries.

---

¹ Аще ли кто нищеты бѣжа ѿходит в манастырь и не могыи дѣтми пещи ся ѿтбѣгаетъ их, то оуже не любве Божья тамо ищетъ. (Cosmas, op. cit. p. 47.) И дѣти бо осиреныа имъ гладомь измирающе...и во мнози плачи кленуть и глаголюще въскую роди ны ѿтець нашь, и мати наша ѡстави ны. (ibid. p. 48.)

² И глаголеши нѣсть мощно в миру семь живуще спасти ся, понеж пещи ся есть женою дѣтьми силою. Ещеж и работы настоять владыкъ земьныхъ и ѿт дружины пакость всяка и насилья ѿт старѣишихъ. (Ibid. pp. 43–4.)

³ Слышим...и ѿт наших добрыя блазняща ся ѡ законнѣи женитвѣ, и не творяще достоины спасенья живущихъ въ твари сеи, рекши въ миру. (Ibid. pp. 42–3.) Аще ли скверну мня миръ сіи ѿходиши, и житіе съ женою ѡхуляеши, немощно творя спасти ся сице живущему, то ничимьж кромѣ еси мысли еретически. (Ibid. p. 58.)

It can thus be inferred that by the middle of the tenth century the Massalian heretics had succeeded in corrupting the orthodoxy of many Bulgarian monks. Their proselytism was undoubtedly facilitated by other serious defects of Bulgarian monasticism of the time.

Among the principal ones was the lack of stability of the foundations. Though we possess no detailed evidence of the inner organization of Bulgarian monasteries in the tenth century, it is probable, by analogy with the situation in Byzantium, that their very number, the rapidity with which they sprang up, and their frequent dependence on secular patrons and benefactors caused many of them, particularly the less important, to lapse after the death of their founder into disrepair, neglect and eventually ruin.[1] A further element of instability appeared after 934 with the frequent invasions of the Magyars, Pechenegs, Russians and Greeks, and the consequent devastations which the monasteries suffered. Cosmas admits that the destruction of monasteries through enemy warfare increased the number of homeless and vagabond monks who were such a scandal in his time and who were particularly receptive subjects for heresy.[2]

There is evidence that the bane of Byzantine monasticism, i.e. the frequently ephemeral nature of the monastic vows, spread also to Bulgaria.[3] Symeon exchanged the cowl for the throne.

---

[1] See J. M. Hussey, *Church and Learning in the Byzantine Empire*, pp. 165–6. In Byzantium at that time there was an irresistible tendency to build new monasteries, instead of endowing or repairing old ones, which reached the point of 'manifest disease', and even 'madness' (see Zachariae von Lingenthal, *Jus Graeco-Romanum*, Lipsiae, 1857, pars III, pp. 292, 295). I. Sokolov, Состояние монашества в византийской церкви, pp. 98–9; Tsukhlev, op. cit. p. 510. The Byzantine emperors tried to check this process, which was ruining the State by depriving it of military man-power and taxable population. The novel of Nicephorus Phocas, issued in 964, forbade the building of new houses and urged the necessity of repairing the older ones. But no measures could arrest the feverish growth of new monasteries. (See G. Schlumberger, *Un Empereur Byzantin au Xe siècle, Nicéphore Phocas*, Paris, 1890, pp. 387–92; Sokolov, op. cit. pp. 97–116.)

[2] Cosmas, op. cit. pp. 51: Аще ли ти случит ся расыпати ся мѣсту нашествіемь ратныхъ или иною виною.

[3] In Byzantium, men would enter monasteries when faced with defeat or failure in their public life and not uncommonly would resume their secular existence if fortune favoured them once more. (See Hussey, op. cit. pp. 162–3.)

Monasteries were often used as prisons, where the dangerous enemies of the tsar could be conveniently confined for life: thus Symeon, to assure the throne for his second son Peter, forced his eldest son Michael to enter a monastery. Peter dealt similarly with his younger brother John who had conspired against him.[1] Naturally enough, these so-called monks were usually only birds of passage and sought the first opportunity to escape from their monasteries and to resume the pursuit of their secular ambitions.

These inherent defects of Bulgarian monasticism in the tenth century explain the sorry picture of it painted by Cosmas. He devotes a large part of his *Sermon against the Heretics* to exposing particular defects and vices of the monks he observed. He attacks the hypocritical monks, the image of the Biblical Pharisees,[2] and deplores the inability of so many to shake off their worldly inclinations. Some, he complains, live unchastely, as 'an object of ridicule for men',[3] are 'slaves to their bellies and not to God',[4] indulge in idle gossip and, like the *gyrovagi* of western Europe, wander from house to house relating their adventures in foreign lands,[5] suffer from a restlessness which drives them to pilgrimages to Jerusalem and Rome, instead of remaining in their cells and obeying their higumen;[6] others, unable to endure the numerous prayers and rigorous fasts prescribed by the rule, return to the world.[7] Cosmas devotes a chapter to the pitfalls of the eremitical life and rebukes those monks who, wishing to avoid obedience to their superior or because they cannot live in peace with their brethren, leave their monasteries and become a law unto themselves; they lead a worldly life, engage in trade and business and,

[1] See V. N. Zlatarski, История, vol. I, pt 2, pp. 516, 536.
[2] Оупокритомь оуподобивше ся. (Cosmas, op. cit. p. 49.)
[3] Ѡви в них своя жены поемлют, смѣху суще человѣкомь. (Ibid. p. 43.)
[4] Чреву суще раби а не Богу. (Ibid. p. 47.)
[5] Ини...ипреходятѿдому в домы чюжая незатворяюще ѿ многорѣчья оусть своих повѣдающе и прилагающе сущая на инѣх землях. (Ibid. p. 43.)
[6] Ѡтходять въ Іерусалимъ, інниж в Римъ, и въ прочая грады и тако помятше ся възвращают ся в домы своя. (Ibid. p. 43.)
[7] Мнозѣ...ѿ ходящих в монастыря, не могоущих терпити сущих ту молитвъ и троудовъ прибѣгают и възвращают ся акы пси на своя блевотины. (Ibid. p. 46.)

puffed up with pride, try by every means to gain the reputation of holy men.¹

Thus the main characteristics of the Byzantine influence on the cultural life of Bulgaria in the tenth century are apparent in Bulgarian monasticism. From one point of view this influence was undeniably very beneficial, for it made accessible to the Bulgarian people the treasures of the Orthodox tradition, carefully preserved in Byzantium throughout the ages, and the civilizing power of the Empire in the intellectual, political and social spheres; in this respect Bulgaria became the eldest daughter of Byzantium, her treasured heir and the transmitter of her civilization to the other Slavonic peoples. But from another point of view Byzantine influence brought with it inherent defects from which the Empire at that time was suffering and many of which became accentuated in Bulgaria. This is particularly clear in the case of monasticism, which, in its new home, was not always able to resist heretical tendencies. Moreover, the Byzantinization of Bulgarian life was so violent and sudden that it met with strong resistance from many sides; in the reign of Peter this inner struggle created a dangerous social and economic rift in the country which, again, furthered the cause of heresy in Bulgaria.

The wholesale introduction of Byzantine customs and institutions was effected with the direct co-operation of the Tsar Peter, of his uncle the Regent Sursubul, and of that section of the boyars who gained titles and position owing to their collaboration with or subservience to Byzantium. But among the people as a whole there was strong opposition to the foreign influence and a violent dislike of the Greeks.² Moreover, those boyars who remained loyal to the policy of Symeon were now in opposition to Peter's pro-Byzantine government which, in their opinion, threatened to swamp Bulgaria in a sea of Hellenism. 'Symeon's nobles',³ for

---

¹ О затворницѣх. (Cosmas, op. cit. pp. 55 et seq.) The wandering monk or cleric (the *gyrovagus* or *clericus vagus*) was the bane of medieval monasticism in eastern and western Europe. Helen Waddell has collected the most important passages from the acts of the Church councils condemning those monks and clerics who break the rule of stability. (*The Wandering Scholars*, 7th ed., London, 1942, pp. 244–70.)

² See M. S. Drinov, Южные славяне и Византия въ X вѣке, pp. 70–1.

³ The Byzantine chroniclers call them οἱ μεγιστᾶνες τοῦ Συμεών (see Zlatarski, op. cit., vol. I, pt 2, p. 537).

this reason, fomented a series of revolts with the object, according to the old tradition of Bulgarian politics, of replacing Peter by their own candidate, first by his younger brother John (in 928), then by his elder brother Michael (in 930). The first rebellion was brutally crushed, the second—more serious and widespread—ended with the timely death of its ringleader. In this case a large district of Macedonia appears to have been in open revolt against the authority of the tsar.[1]

In this manner all outward opposition to the Byzantine influence in Bulgaria was successfully repressed in the reign of Peter. The consequent weakening of the Slavonic element in all national institutions was among the principal causes of the collapse of Bulgarian independence in the beginning of the eleventh century and of the establishment of Byzantine domination over the country for 168 years. During this period the resistance to Byzantine oppression was carried out from within by the Paulicians and the Bogomils.

From the preceding facts and considerations we may draw the following conclusions:

By the middle of the tenth century, the Orthodox Church of Bulgaria appears on the surface firmly established owing to the policy of centralization which was carried out according to Byzantine principles by Boris, Symeon and Peter. But in reality the situation of the Church was most critical. It was seriously affected by the religious, social and economic unrest which reigned in Bulgaria throughout the tenth century. The country was still rent by the ethnic and social dualism which had caused so much disorder in the past three centuries. The three Christian monarchs of the past century were unable to destroy this dualism owing to the lack of continuity in their policy. Boris's work of peacefully building a Christian State under the guidance of Byzantium was undone by Symeon, who, in his attempts to destroy the Empire, brought economic ruin on his country and almost entirely neglected the work of inner reconstruction. However, so long as he lived, the very strength and prestige of his personality kept the country together and ensured its political power. But after Symeon's death his son Peter broke away from his father's

[1] See Zlatarski, ibid. pp. 536–9; Runciman, op. cit. pp. 187–8.

policy and, incapable of resisting the inrush of Byzantine influence, provoked several revolutionary movements which weakened the country and laid it open, after Peter's death, to foreign invasions and inner anarchy. These inconsistencies and waverings strengthened the various centrifugal forces which were working against the centralizing policy of the monarchs and in opposition to the Church and to the State. The condition of the Bulgarian Church at that time was not such as to command unqualified obedience and respect. Its prelates in many cases had become Byzantinized and had lost that contact with the people which had been the strength of men like St Clement. The minor clergy, monks and parish priests, could not escape the accusation of intellectual and moral decadence levelled against them by Cosmas. These shortcomings of many sections of the clergy considerably strengthened paganism and heresy, the two principal enemies of the Church. The former was still by no means overcome in the tenth century, and the 'dual faith' continued to live among the people.[1] Heresy, sometimes connected with paganism, developed in Bulgaria as a result of two factors: on the one hand, a basis of Eastern dualistic doctrines, Paulician and Massalian, which penetrated to Bulgaria as a result of the colonizing policy of the Byzantine emperors, and on the other, pre-existing and contemporary conditions in Bulgaria exceptionally favourable to the spread of anti-ecclesiastical teachings. During the first part of the tenth century, however, as far as it is possible to judge, Bulgarian heresy remained somewhat indistinct and unformed. The boundaries between Paulicianism and Massalianism and between both these heresies and paganism are not yet clear. Heresy in Bulgaria was awaiting a leader who would unify the various teachings of the heretics and organize more effectively the struggle against the Church.

[1] According to the *Life of St Naum*, at the beginning of the tenth century some Macedonian Slavs still worshipped stones and trees. See L. Niederle, *Život starých Slovanů*, vol. II, pt 1, pp. 28–9. Paganism is attacked by John the Exarch (see supra, p. 95). Cosmas also complains of the sway that pagan beliefs and rites hold over the people: Мнози бѡ ѿт человѣкъ паче на игры текут, неже въ церкви, и кощуны и бляди любять пач[е] книгъ...да по истинѣ нѣсть лѣпо нарицати христіаны творящих таковая...аще со гусльми и плесканіемъ и пѣсньми бѣсовскыми вино піютъ и срящамъ и сномъ и всякому оученію сотонину вѣруютъ. (Popruzhenko, op. cit. p. 74.) In the eleventh century St George of Iberia baptized a pagan Slavonic tribe in Thrace (see Tsukhlev, op. cit. p. 170).

CHAPTER IV

BOGOMILISM IN THE FIRST BULGARIAN EMPIRE

I. *The beginnings of Bogomilism:* A letter of the patriarch of Constantinople to the tsar of Bulgaria. The 'ancient and newly appeared heresy'. Fusion of Paulician and Massalian doctrines. The priest Cosmas, the 'pop' Bogomil and the name Bogomils. How to recognize a Bogomil.

II. *The teaching of the Bogomils:* Their doctrines and their ethics. The organization and discipline of the Bogomil community. The Bogomils and contemporary society; their social anarchism. Two basic trends of Bogomilism: dualism and reformation of Christianity. Reasons for its success. Bogomilism and the other Bulgarian dualistic sects.

III. *The growth of Bogomilism in Macedonia:* Bogomilism after the death of the Tsar Peter. New transplantations of heretics to Thrace and Macedonia. Reasons for the growth of heresy in Macedonia in the late tenth century. The Tsar Samuel and Bogomilism. Macedonia as the cradle of Bogomilism.

Towards the middle of the tenth century a twofold transformation can be observed in the Bulgarian heretical sects: on the one hand, the teachings of the Paulicians and the Massalians, hitherto largely uncoordinated and distinct from one another, now coalesced; on the other hand, sectarianism ceased to be a predominantly foreign movement in Bulgaria and assumed specifically Slavonic characteristics. The outcome of this fusion of the two early dualistic heresies and of this Slavicization was Bogomilism.

Our earliest evidence of this transformation is contained in a letter written by Theophylact, patriarch of Constantinople, to Peter, tsar of Bulgaria.[1] Theophylact was the fourth son of the Emperor Romanus Lecapenus and the uncle of Peter's wife

[1] This document was first published from a photograph of the original in the Ambrosian Library in Milan by the Russian scholar N. M. Petrovsky, Письмо патриарха Константинопольского Феофилакта царю Болгарии Петру: *Izvestiya otdeleniya russkogo yazyka i slovesnosti Imperatorskoy Akademii Nauk* (1913), vol. xvIII, tom. 3, pp. 356–72. (A Russian translation is appended to the Greek text.) A Bulgarian translation together with a brief historical survey of the document can be found in V. N. Zlatarski's История, vol. I, pt 2, App. xI, pp. 840–5. A French synopsis of the letter is given by Grumel (*Regestes des Actes du Patriarcat de Constantinople*, vol. I, fasc. 2, pp. 223–4) and an abridged English translation by V. N. Sharenkoff (*A Study of Manichaeism in Bulgaria*, New York, 1927, pp. 63–5).

Maria-Irene.[1] An exact determination of the date of this letter would contribute in a large measure towards the solution of the problem of the origins of Bogomilism. Unfortunately, however, the letter can only be dated approximately. It appears from the text that Peter had appealed to the patriarch for guidance on the manner of dealing with a 'newly appeared' heresy in Bulgaria and that Theophylact had sent him a reply; Peter, however, wrote back, requesting a clearer and fuller explanation.[2] Theophylact's first letter has not come down to us; our document is his second reply, composed after a careful study of the new heresy from Bulgarian sources, and 'in plain letters', as the tsar had requested.[3] Hence it cannot have been written at the very beginning of the patriarchate of Theophylact, who occupied the Oecumenical See from 2 February 933 to 27 February 956,[4] particularly as he became patriarch at the age of sixteen and could scarcely have given the husband of his niece such fatherly advice while still in his teens. The letter can be dated with the greatest probability between 940 and 950.[5] Theophylact describes the heresy which confronted Peter as 'ancient' (παλαιᾶς) and at the same time as 'newly appeared' (νεοφανοῦς).[6] He defines it as *Manichaeism mixed with Paulicianism*.[7] The significance of this definition will become clear from the patriarch's exposition of the

---

[1] Petrovsky (loc. cit. p. 361, n. 2) erroneously states that Peter's wife was the sister of Theophylact. In reality, however, she was the daughter of Christopher Lecapenus, brother of Theophylact, and hence the niece of the patriarch. See S. Runciman, *The Emperor Romanus Lecapenus*, p. 97.

[2] 'Επεὶ δέ σοι καὶ ἤδη περὶ τῆς νεοφανοῦς ἀντεγράφη κατὰ τὰ ἐρωτηθέντα αἱρέσηος, καὶ νῦν τρανότερόν τε καὶ διεξωδικώτερον γράφομεν πάλιν, ὡς ἐπεζήτησας. (Ibid. p. 362.)

[3] ...τελεώτερον ἀναμαθόντες ἐξ ὑμῶν τοῦ δόγματος τὸ ἐξάγιστον. Γράφομεν δὲ σαφεῖ λόγῳ, γυμνὰ τιθέντες τὰ πράγματα, διὰ λιτῶν γραμμάτων, καθὼς ἠξίωσας. (Ibid.) In explaining the meaning of διὰ λιτῶν γραμμάτων, Grumel (op. cit. p. 223) adopts the interpretation of L. Petit ('Le Monastère de N.-D. de Pitié', *I.R.A.I.K.* vol. VI, Sofia, 1900, pp. 134–6), who takes plain letters to mean separate letters, i.e. uncial letters. Theophylact's first letter seems to have been very hard to read for Peter, who must have been unfamiliar with the Byzantine cursive.

[4] See Grumel, op. cit. p. 222.

[5] See Zlatarski, op. cit. vol. I, pt 2, p. 563, n. 1.

[6] Loc. cit. pp. 362, 365.

[7] Μανιχαϊσμός...ἐστι, παυλιανισμῷ συμμιγής, ἡ τούτων δυσσέβεια (loc. cit. p. 363).

doctrines of the new heresy which was causing Peter so much uneasiness. But it is obvious that he distinguished it from Paulicianism, while recognizing the connections between the old and the new heresies.

The teachings of the heretics are briefly set out by Theophylact in his list of anathemas to be used against them by the Bulgarian Church. All except one can be found in Peter of Sicily's *Historia Manichaeorum* and are hence undoubtedly Paulician.[1] These are the dualism between a Good and an Evil Principle, the one the creator of Light, the other the creator of Darkness, Matter and all the visible world;[2] the rejection of the Mosaic Law and the Prophets as originating from the Evil Principle;[3] the Docetic Christology, according to which the Incarnation, Crucifixion and Resurrection of our Lord were only κατὰ φαντασίαν καὶ δόκησιν, and not κατὰ ἀλήθειαν;[4] the denial of the Real Presence in the Eucharist and the figurative interpretation of the Words of Institution as referring not to the Body and Blood of Christ, but to the Gospels and the Acts of the Apostles;[5] the denial of the virginity of Our Lady and the assertion that she was the 'higher Jerusalem'.[6]

The influence of the *Historia Manichaeorum* is apparent in Theophylact's letter. Not only is the patriarch's formulation of the Paulician doctrines practically identical with that of Peter of Sicily, but the heresiarchs of the 'ancient and newly appeared heresy' to whom Theophylact devotes his four last anathemas

[1] Cf. supra, pp. 38–42.
[2] Ὁ δύο ἀρχὰς λέγων καὶ πιστεύων εἶναι, ἀγαθήν τε καὶ κακήν, καὶ ἄλλον φωτὸς ποιητὴν καὶ ἄλλον νυκτός,...ἀνάθεμα ἔστω.
Τοῖς τὸν πονηρὸν διάβολον ποιητὴν ὑπάρχειν καὶ ἄρχοντα τῆς ὕλης καὶ τοῦ ὁρωμένου τούτου κόσμου παντὸς καὶ τῶν σωμάτων ἡμῶν κενολογοῦσιν, ἀνάθεμα. (Loc. cit. p. 364.)
[3] Τοῖς τὸν μωσαϊκὸν νόμον κακολογοῦσι καὶ τοὺς προφήτας μὴ εἶναι λέγουσιν ἀπὸ τοῦ ἀγαθοῦ, ἀνάθεμα. (Ibid.)
[4] Τοῖς τὸν...Υἱὸν καὶ Λόγον τοῦ Θεοῦ...κατὰ φαντασίαν καὶ δόκησιν, ἀλλ' οὐ κατὰ ἀλήθειαν ἄνθρωπον χωρὶς ἁμαρτίας γεγονέναι βλασφημοῦσιν....τοῖς τὸν σταυρὸν καὶ τὸν θάνατον τοῦ Χριστοῦ καὶ τὴν ἀνάστασιν ὡς δόκησιν φαντασιοσκοποῦσιν, ἀνάθεμα. (Ibid. p. 365.)
[5] Τοῖς μὴ κατὰ ἀλήθειαν σῶμα Χριστοῦ καὶ αἷμα πιστεύουσιν, τὸ ὑπ' Αὐτοῦ ἐν τῷ 'λάβετε, φάγετε' τοῖς ἀποστόλοις ῥηθέν τε καὶ ἐπιδοθέν, ἀλλὰ τὸ Εὐαγγέλιον καὶ τὸν 'Ἀπόστολον τερατολογοῦσιν, ἀνάθεμα. (Ibid.)
[6] Τοῖς τὴν Ὑπεραγίαν Θεοτόκον μὴ τὴν Παρθένον Μαρίαν,...ἀλλὰ τὴν ἄνω 'Ἰερουσαλήμ,...ληρωδοῦσιν, ἀνάθεμα. (Ibid.)

are enumerated in the same order as in the *Historia Manichaeorum*.[1] Moreover, it seems that the Tsar Peter himself was not ignorant of Paulicianism: the rapid enumeration by the patriarch of the ancient heresiarchs with scarcely a word of comment suggests that Peter was already in some measure familiar with these personages. As it is unlikely that Theophylact's first letter to Peter contained any detailed reference to them, since we know that it was brief, it is legitimate to conclude that the tsar was probably acquainted with the treatise of Peter of Sicily, which must have reached Bulgaria at the end of the ninth century.[2]

However, one heretical doctrine mentioned by Theophylact is of a non-Paulician origin: the heretics, he writes, reject lawful marriage and maintain that the reproduction of the human species is a law of the demon.[3] This exaggerated and distorted asceticism, essentially characteristic of Bogomilism, is a logical consequence of metaphysical dualism, according to which Matter, the product of the Evil Principle, is a source of limitation and suffering for the divinely created soul; hence marriage, as the means of reproduction of Matter, is to be condemned and avoided. The Paulicians, however, somewhat illogically, did not apply their dualistic teaching to this particular sphere of ethics: their active and warlike mode of life no doubt prevented them from indulging in any extreme form of asceticism. Abstention from sexual intercourse was enforced on the 'elect' of the early Manichaean sect[4] and Theophylact clearly uses the term 'Manichaeism' to describe this particular teaching.[5] But there is no serious evidence to suggest that any real Manichaeans existed in the Balkans at that time and hence that Manichaeism could have exerted anything but an indirect influence on Bogomilism.[6] In Bulgaria,

[1] These are the Egyptian Scythianus, his disciple Terebinthus, Mani, Paul and John (the two sons of Callinice), Constantine, Symeon, Paul, Theodore, Gegnesius, Joseph, Zacharias, Baanes and Sergius. (Loc. cit. pp. 365–7.) Cf. Petrus Siculus, *Historia Manichaeorum*, *P.G.* vol. CIV, cols. 1257–1300.

[2] Cf. supra, p. 30.

[3] Τοῖς τὸν εὔνομον γάμον ἀθετοῦσι καὶ τοῦ δαίμονος εἶναι νομοθεσίαν τὴν αὔξησιν τοῦ γένους ἡμῶν καὶ διαμονὴν δυσφημοῦσιν, ἀνάθεμα. (Loc. cit. pp. 364–5.) [4] See supra, p. 6, n. 2.

[5] It is interesting to note that in contrast to nearly all medieval Byzantine writers Theophylact does not identify Paulicianism and Manichaeism.

[6] See Puech and Vaillant, *Le traité contre les Bogomiles de Cosmas le prêtre*, pp. 304–16.

as we have seen, the condemnation of marriage was taught by the Massalians and was implicit in certain distorted forms of Orthodox monasticism. Although precise evidence on this point is lacking, it seems probable that this teaching developed in the tenth century from the interaction of Massalian dualism and the unbalanced asceticism and acosmism of certain monastic circles, which found its justification in a dualistic metaphysic of matter.

The significance of Theophylact's definition of the new heresy as 'Manichaeism mixed with Paulicianism' now becomes clear: the first term refers to the teaching of the Massalians, and particularly to the condemnation of marriage, the second to the doctrines of the Paulicians as described in the *Historia Manichaeorum*. Moreover, this Bulgarian heresy was 'ancient' because its component parts, Paulicianism and Massalianism, were both old heresies and had existed in Bulgaria for probably more than a century; yet it was 'newly appeared' because a fusion had recently occurred between a number of teachings of both these sects, which resulted in the rise of a new heresy. This new heresy, which became the most important Bulgarian and indeed Balkan sectarian movement, was later given the name of Bogomilism.

The measures prescribed by the patriarch against the new Bulgarian heretics are particularly interesting, as they show that in spite of his correct analysis of the 'ancient and newly appeared heresy', Theophylact was ignorant of its real origin. He writes: 'their leaders and teachers of dogmas alien to the Church who reject and curse their own impiety are to be rebaptized, *according to the 19th canon of the* [first] *Council of Nicaea.... For their impiety is Manichaeism mixed with Paulicianism.*'[1] Now the 19th canon of the Council of Nicaea is concerned with the rebaptism not of Paulicians, who did not yet exist in the fourth century, but of 'Paulianists' or the followers of Paul of Samosata.[2] The theory that the Paulicians were descended from Paul of Samosata was held by a number of Byzantine theologians, but, as it has been shown, without the

[1] Loc. cit. pp. 362–3. [The italics are mine.]
[2] See G. Rhalles and M. Potles, Σύνταγμα τῶν θείων...κανόνων (Athens, 1852), vol. II, pp. 158–9; cf. Theodore Balsamon, *In Can. XIX Conc. Nicaen. I*, *P.G.* vol. cxxxvii, col. 308: Παυλιανισταί...εἰσιν οἱ ἀπὸ Παύλου τοῦ Σαμοσατέως καταγόμενοι.

slightest justification.¹ This confusion between the Paulicians and the adherents of the heretical bishop of Antioch, which seems to have been current in Byzantine circles at least as early as the ninth century, explains the fact that the patriarch, while rightly ascribing to the Bulgarian heretics doctrines which he derived from the *Historia Manichaeorum*, at the same time described their teaching not as Παυλικιανισμός, but as Παυλιανισμός, and ordered them to be treated according to the measures prescribed by the First Oecumenical Council with regard to the followers of Paul of Samosata.²

It can thus be supposed that many of the patriarchal injunctions appeared irrelevant to the Tsar Peter: what he wanted was not a pronouncement on the various degrees of validity of heretical baptism³ (since this sacrament was rejected by Paulicians, Massalians and Bogomils⁴ alike, and hence the problem of whether they were to be rebaptized or not could never arise), but precise instructions on the method of dealing with the 'newly appeared' Bulgarian heresy. The sole practical advice given by Theophylact concerned the application to the heretics of the Christian secular laws (οἱ πολιτικοὶ τῶν χριστιανῶν νόμοι): while remarking that the rightful punishment was death, especially when heresy spread like a disease, the patriarch nevertheless urged the tsar to avoid excessive severity and to strive continually (ἔτι καὶ ἔτι) for the conversion of the heretics by the force of persuasion.⁵

But in spite of the failure of Theophylact to understand the true

[1] Cf. supra, pp. 55–7, and G. Bardy (*Paul de Samosate*, pp. 43–4), who also discusses the question of the 19th canon of the Council of Nicaea. (Ibid. pp. 412–23.)

[2] The same confusion was made by Theodore Balsamon in his commentary on the 19th canon (loc. cit. col. 301): Παυλιανισταὶ λέγονται οἱ Παυλικιανοί.

[3] Theophylact issued the following prescriptions: those who abjure their heresy and return to the Church are to be classed into three groups: (1) the heretical teachers should be rebaptized and the orders of their priesthood declared null; (2) those who lapsed into heresy through simplicity or ignorance should not be rebaptized, but merely anointed with chrism; their priests should be received after abjuration; (3) those who, without accepting any false doctrine, were unsuspectingly led to listen to the teachers of heresy are to be treated as follows: the laymen should be received into the Church after an exclusion of four months from the Sacraments, the priests should retain their orders. As for those who persist in their heresy, the Church leaves them to perpetual condemnation. (Loc. cit. pp. 362–4.)

[4] See infra, pp. 129–30.

[5] Petrovsky, loc. cit. pp. 364, 367.

origin of the new Bulgarian heresy, his letter remains a document of considerable importance. Not only does it show that by the middle of the tenth century the Paulician sect was widespread in Bulgaria, but it is also the earliest source pointing to the amalgamation of Paulician and Massalian teachings which formed the basis of Bogomilism.

The evidence concerning the Bogomil heresy supplied by the letter of the Patriarch Theophylact is confirmed and at the same time considerably enriched by the *Sermon against the Heretics* of the priest Cosmas,[1] composed in the second half of the tenth century, probably soon after the death of the Tsar Peter in 969.[2] Though written in a polemical, and often heated tone, its description of Bogomilism is very concrete and reasonably objective. Cosmas's treatise is a vivid and detailed account of an eyewitness and occupies, among the sources concerned with Bogomilism, a position of unique importance.[3]

After a short introductory discourse on the significance of heresy and an enumeration of some ancient heresiarchs, Cosmas writes: 'And it came to pass that in the land of Bulgaria, in the days of the Orthodox Tsar Peter, there appeared a priest (*pop*) by the name of Bogomil, but in truth "not beloved of God".[4] He was the first who began to preach in Bulgaria a heresy, of whose vagaries we shall speak below.'[5]

---

[1] Слово святаго Козмы презвитера на еретики препрѣніе и поученіе ѿт божественныхъ книгъ. For editions of this work, see supra, p. 104, n. 1. The hotly debated problem of the original form of Cosmas's work seems to have been satisfactorily solved by Zlatarski (Сколько бесед написал Козма Пресвитер? *Sbornik statey v chest'* M.S. Drinova, Kharkov, 1904) and by Popruzhenko (Козма Пресвитер, *I.R.A.I.K.*, Sofia, 1911, vol. xv), who have shown: (1) that the *Sermon* was originally written and not spoken; (2) that it was written as one whole, but was later divided into chapters.

[2] The problems of chronology connected with Cosmas are discussed in Appendix I.

[3] This position is admirably defined by Puech (op. cit. pp. 129–45).

[4] This is a pun on the name of Bogomil, the Slavonic translation of Θεόφιλος, i.e. 'the beloved of God'. Богу не милъ means 'not beloved of God'.

[5] Якож случис въ болгарьстіи земли в лѣта правовѣрнааго царя Петра быс[ть] попъ именемъ Богумилъ, а по истинѣ рещи Богу не милъ, иже начя первое оучити ересь в земли болгарстѣ, юже блядь на прежде поидуще скажемь. (Popruzhenko, op. cit. p. 2). The usual and correct form of the heresiarch's name is Богомилъ. Богумилъ is used here by Cosmas to introduce the words Богу не милъ. Cf. Puech, op. cit. p. 54, n. 3.

That this new heresy was first taught in Bulgaria in the reign of Peter by the priest Bogomil is confirmed by a thirteenth-century Bulgarian document, the *Synodicon of the Tsar Boril*.[1]

Apart from the evidence of these sources, which show that the *pop* Bogomil was a contemporary of the Tsar Peter (927–69), we know next to nothing about the person of the greatest heresiarch of the southern Slavs.[2] A Russian sixteenth-century document mentions him as a writer of heretical books in Bulgaria.[3] It is probable that Bogomil taught before the composition of Theophylact's letter to Peter, i.e. in the late thirties or the early forties of the tenth century.[4] The fact that Theophylact does not mention his name cannot be taken as proof that Bogomil only began to spread his teaching after the composition of the patriarch's second letter to Peter.[5] In fact, the 'mixture' of Paulicianism and Massalianism which we find in Theophylact's analysis of the new heresy is eminently characteristic of the teaching of Bogomil,[6]

---

[1] Cf. infra, p. 238.

[2] I. Klincharov (Попъ Богомилъ и неговото време, pp. 22–30) has no difficulty in showing that the doubts sometimes cast on the historicity of Bogomil are without foundation. But Klincharov's own highly imaginative and idealized portrait of Bogomil (ibid. pp. 32–3) bears little or no relation to the evidence of the sources. In particular, there seems to be no serious reason for believing that Bogomil belonged to 'a Slavonic noble family' (p. 32). However, Klincharov's view that Bogomil lived in Macedonia is, in view of our present knowledge of the origins of the Bogomil sect, quite acceptable.

[3] I. Ivanov, Богомилски книги и легенди, p. 50.

[4] Puech's assertion (op. cit. p. 289) that Bogomilism appeared in the first quarter of the tenth century, perhaps even as early as 915, does not seem to me to be based on conclusive evidence. The fact that John the Exarch before 927 mentions the heretical belief that the Devil is the eldest son of God (cf. supra, p. 95)—a doctrine adopted by the Bogomils—is scarcely proof that the sect already existed in Bulgaria at that time. It is, perhaps, safer to accept Cosmas's statement that the actual founder of the sect was Bogomil, a contemporary of the Tsar Peter.

[5] This argument is put forward by Spinka (*A History of Christianity in the Balkans*, p. 63), but is not conclusive. The patriarch may well have not known of Bogomil even though he had familiarized himself with his teaching. The heresiarch's name was probably less widely known at the time of Theophylact than it was in the days of Cosmas. Moreover, Theophylact was solely preoccupied with expounding the doctrinal errors of the heretics and instructing Peter how to treat them according to the law of the Church. Cosmas, on the other hand, whose knowledge of the heresy was direct and much fuller and who was writing for a wider public, naturally emphasized the origins of the heresy.

[6] Cf. infra, pp. 198, 206–7.

BOGOMILISM IN THE FIRST BULGARIAN EMPIRE 119

and, moreover, both Bulgarian sources, the *Sermon against the Heretics* and the *Synodicon of the Tsar Boril*, expressly state that this 'mixture' was first taught in Bulgaria by Bogomil himself. It is thus possible to place the teaching of the *pop* Bogomil in Bulgaria at the beginning of the reign of the Tsar Peter.[1]

Cosmas mentions the name of the *pop* Bogomil only once, at the very beginning of his *Sermon*, and his followers are, throughout the work, not referred to by any other name except that of 'heretics'. The names Paulicians or Manichaeans do not occur at all in this source. The term Bogomils (Βογομίλοι,[2] Богомили, Bogomili), under which the followers of the *pop* Bogomil became known in history, is of a later date and appears for the first time, as far as can be ascertained, in its Greek form in a letter of the Byzantine monk Euthymius, written *c*. 1050.[3] The name became famous, however, owing to the learned Byzantine theologian Euthymius Zigabenus who, at the beginning of the twelfth century, entitled one of the chapters of his *Panoplia Dogmatica*, Κατὰ Βογομίλων.[4] Since Cosmas seems unacquainted with this name and as Zigabenus recognizes its Bulgarian origin, we can infer that the name of the *pop* Bogomil became a generic term serving to designate his followers in Bulgaria either at the end of the tenth century or at the beginning of the eleventh.

Zigabenus, however, gave a false etymology of the word Βογομίλοι, deriving it from the Slavonic 'Bog' (God) and 'mil' which, he asserted, means 'have mercy'; he concluded that Bogomil signifies 'a man who implores the mercy of God'.[5] He overlooked the fact that the root *mil* in Slavonic has not only the meaning of 'mercy' (as in помилуй = ἐλέησον), but also 'dear,

---

[1] Cf. J. A. Ilić, *Die Bogomilen in ihrer geschichtlichen Entwicklung* (Sr. Karlovci, 1923), p. 18. The MS. of the *Sermon against the Heretics* published by Popruzhenko contains in its title the significant words: 'The newly appeared heresy of Bogomil'. (На новоявившую ся ересь Богомилоу.) Новоявившую ся is the exact equivalent of the term νεοφανοῦς used by Theophylact.

[2] For the other Greek forms of the name, such as Βογόμυλοι, Πογόμιλοι, Πογόμηλοι, see G. Ficker, *Die Phundagiagiten* (Leipzig, 1908), index, p. 278.

[3] See infra, p. 177.

[4] See infra, p. 207.

[5] *Panoplia Dogmatica*, tit. XXVII, *P.G.* vol. CXXX, col. 1289: Βόγον μὲν γὰρ ἡ τῶν Βουλγάρων γλῶσσα καλεῖ τὸν Θεόν, Μίλον δὲ τὸ ἐλέησον. Εἴη δ' ἂν Βογόμιλος κατ' αὐτοὺς ὁ τοῦ Θεοῦ τὸν ἔλεον ἐπισπώμενος.

beloved' (as in милъ). The latter meaning is the correct one here.¹ The true meaning of Bogomil is therefore 'beloved of God', and the name is the Slavonic translation of the Greek Θεόφιλος. Ivanov has shown that the Christian name Bogomil (Theophilus) was prevalent in Bulgaria even before the time of the Tsar Peter.² It can scarcely be doubted to-day that the term Bogomil is derived from the name of the heresiarch Bogomil-Theophilus.³

[1] See L. Léger, 'L'Hérésie des Bogomiles', *Revue des Questions Historiques* (1870), vol. VIII, p. 486. [2] Op. cit. p. 22, n. 3.
[3] The Slavonic equivalent of οἱ τοῦ Θεοῦ τὸν ἔλεον ἐπισπώμενοι would be *Bogomoli*, which corresponds to the Greek εὐχίται, another name for the Massalians. As the Massalians were, from the thirteenth century, generally identified with the Bogomils, the etymological confusion between 'Bogomili' and 'Bogomoli' is understandable, though originally both terms were distinct.
The false etymology of Zigabenus was adopted by a number of non-Slav historians. See B. de Montfaucon, *Palaeographia Graeca*, p. 333; C. Du Cange, *Glossarium ad scriptores mediae et infimae Graecitatis*, p. 207. Even in recent days some of the opinions expressed concerning the origin of the Bogomils are vitiated by Euthymius's error. Thus M. Gaster, in his article on the Bogomils (*Encyclopaedia Britannica*, 11th ed.), still asserts that 'the word [Bogomil] is a direct translation into Slavonic of *Massaliani*, the Syrian name of the sect corresponding to the Greek Euchites'. Here again we see the confusion between 'Bogomili' and 'Bogomoli'.
According to this theory, the Bogomils, by reason of their name, were supposed to pray frequently with the words κύριε ἐλέησον (see C. Schmidt, *Histoire et doctrine de la secte des Cathares ou Albigeois*, Paris, 1849, vol. II, p. 284). In reality, however, there is not the slightest evidence that they used this particular prayer; the only one they recognized was the Lord's Prayer.
Gieseler and Kopitar have shown the falsity of Euthymius's etymology; the latter wrote that his interpretation 'cum slavicae linguae indole conciliari nequit....Nomen illud...cum precatione κύριε ἐλέησον nihil commune habet, quam Slavi partim Gospodine pomiluj, partim Hospodine smiluj se vertunt.' (See F. Rački, 'Bogomili i Patareni', *Rad*, VII, pp. 94–5.) For this and other reasons, I find it difficult to accept the categorical statement of A. Vaillant, which is echoed by Puech, that the name Bogomil is a pseudonym, whose meaning is 'que Dieu prend en pitié' or else 'qui supplie Dieu' (see Puech, op. cit. pp. 27, 282–3).
Another group of scholars has proposed a solution more in accordance with Slavonic etymology by taking the term Bogomils to mean 'beloved of God'. See G. Arnold, *Kirchen- und Ketzer-Historie*, pt 1, vol. IV, ch. 8, § 66, p. 211: 'Bogomilos...auf der Bulgarischen Sprache — von Gott geliebte.' Cf. J. L. Oeder, *Dissertatio...prodromum historiae Bogomilorum criticae exhibens* (Gottingae, 1743), pp. 9–10. These scholars, however, did not know of the *pop* Bogomil.
But even in recent times the direct relation between the name of the sect and that of its founder has been denied. Thus V. Jagić maintained that the name Bogomils comes not from the name of the heresiarch but from the mode of life

The *Sermon against the Heretics*, apart from shedding some light on the origins of Bogomilism, is also the most complete account we possess of the doctrines and the behaviour of the Bogomil heretics. In his desire to save his compatriots from falling a prey to their insidious teachings, Cosmas uses his personal experience of the heretics to describe their outward appearance and thus permit their identification:

'The heretics in appearance are lamb-like, gentle, modest and silent, and pale from hypocritical fasting. They do not talk idly, nor laugh loudly, nor show any curiosity. They keep away from the sight of men, and outwardly they do everything so as not to be distinguished from righteous Christians, but inwardly they are ravening wolves.... The people, on seeing their great humility, think that they are Orthodox and able to show them the path of salvation; they approach and ask them how to save their souls. Like a wolf that wants to seize a lamb, they first cast their eyes downwards, sigh and answer with humility. Wherever they meet any simple or uneducated man, they sow the tares of their teaching, blaspheming the traditions and rules of Holy Church.'[1]

The strength of the heretics lay in their tenacious attachment to their errors; according to Cosmas, they were incapable of being converted:

'You will more easily bring a beast to reason than a heretic; for just as a swine passes by a pearl and collects dirt, so do the heretics swallow their own filth. And, just as an arrow which, aimed against a slab of marble, not only cannot pierce it, but rebounds and strikes whoever stands behind [the one who shoots], so will a man who tries to instruct a heretic not only fail to teach him, but will also pervert one weaker of mind.'[2]

of his followers (История сербско-хорватской литературы: *Uchenye zapiski imperatorskogo Kazanskogo Universiteta*, 1871, p. 101).

But this opinion is against the judgement of the two most eminent authorities on Bogomilism, Rački and Ivanov, who derive the name of the sect from that of the *pop* Bogomil: 'Bogomili i Patareni', *Rad*, VII, p. 94; Богомилски книги и легенди, p. 22.

However, it is quite probable that in later times the name Bogomils was used by the heretics themselves in a moral sense and represented their pretensions to the pure life and true understanding of the Gospels. V. Levitsky has pointed out that this name typifies the strivings and the claims of the Bogomils to the title of true Christians κατ' ἐξοχήν (Богомильство—болгарская ересь: *Khristianskoe Chtenie*, 1870, pt I, p. 371). The name Bogomils had undoubtedly a strong moral appeal, like that of Cathars in Western Europe.

[1] Op. cit. p. 3.   [2] Ibid. p. 5.

The doctrines of the heretics are expounded in far greater detail by Cosmas than by Theophylact. The *Sermon* lays great emphasis on their fundamental teaching, i.e. the cosmological dualism, according to which the Devil is the creator of the visible world: 'They say that everything belongs to the Devil: the sky, the sun, the stars, the air, man, churches, crosses; all that comes from God they ascribe to the Devil; in general, they consider all that is on earth, animate and inanimate, to be of the Devil.'[1] The Bogomils attempted to support this view by Scriptural references, in particular by the Parable of the Prodigal Son (Luke xv. 11–32): 'Having heard what our Lord says in the Gospel in the parable of the two sons, they claim that Christ is the elder and think that the younger, who deceived his father, is the Devil; they call the latter Mammon and assert that he is the creator and author of earthly things.'[2] The belief that the Devil is the son of God and the brother of Christ was already ascribed at the beginning of the tenth century to the Bulgarian heretics by John the Exarch,[3] with the difference, however, that, according to Popruzhenko's text of Cosmas, the Devil is presented as the younger brother,[4] while in the words of John the Exarch he is held to be the elder brother. We do not know whether the Bogomils really differed from one another on this point, since Cosmas is the only one to mention the doctrine that the Devil is the younger son of God; the general belief among the Bogomils was that he was the elder; this was taught in particular by the Byzantine Bogomils of the eleventh and twelfth centuries.[5] The position of the Devil in Bogomil exegesis is defined by Cosmas as follows: 'They call the Devil

[1] Cosmas, loc. cit. p. 26.
[2] Слышаще бо въ евангеліи Господа рекша притчю ѡ двою сыну Христа оубо творять старѣишаго сына, меншааго же еже есть заблудилъ ѡтца діавола мѣнять, и сами и мамоноу прозваша и того творца нарицають и строителя земныимь вещем. (Ibid. p. 26.)
[3] Cf. supra, p. 95.
[4] However, a variant quoted by Popruzhenko from a sixteenth-century MS. of the *Sermon against the Heretics* states that the heretics believed that Christ was the younger son of God (Творяще Господа нашего сына меншаго, ibid. p. 26, n. 10).

Cf. infra, pp. 207 et seq. As it will be shown, the doctrines of the Byzantine Bogomils of the eleventh and twelfth centuries, compared to those of the Bulgarian heretics of the tenth century, are more developed and complex, but true to the original teaching of the sect. This identity in essence and evolution

BOGOMILISM IN THE FIRST BULGARIAN EMPIRE 123

the creator of man and of all God's creatures; and because of their extreme ignorance, some of them call him a fallen angel and others consider him to be the unjust steward.'¹ The name of 'unjust steward' (οἰκονόμος, икономъ) is taken from the parable in St Luke xvi. 1–9, which the Bogomils interpreted as referring to the Devil.²

This conception of the Devil differs notably from the Paulician dualism as described by Peter of Sicily and the Patriarch Theophylact. Whereas these writers emphasize the belief in two principles (ἀρχαί), parallel and independent of one another,³ the

in form can be seen particularly clearly in the Bogomil teaching on the Devil.

An ingenious explanation of this discrepancy is put forward by Puech (op. cit. pp. 190–2), who compares the Bogomil interpretation of the Parable of the Prodigal Son, described by Cosmas, with the later accounts of the Bogomil teaching on God the Father and His two Sons, given by Psellus and Euthymius Zigabenus. According to Psellus (see infra, p. 185), the Bogomils taught that the elder son, creator and ruler of the visible world, incurred the hostility of his younger brother, the prince of the heavens, who 'is jealous of him,... envies him his good arrangement of the earth, and, smouldering with envy, sends down earthquakes, hailstorms and plagues'. Puech believes that, in the Bogomil interpretation of St Luke's parable, the prodigal son, the younger of the two brothers, represented the Devil, 'who deceived his father', and that consequently Christ was logically regarded as the elder brother. He points, moreover, to the similarity between the anger which, in the parable, the *elder* son showed at the return of his *younger* brother, and the envy which, in Psellus's account, the *younger* son manifests towards his *elder* brother. In both cases, it is Christ who shows anger, only the respective seniority of the two brothers is reversed. Puech explains this reversal with reference to Zigabenus's account of the Bogomil teaching on the rebellion and fall of the Devil, as the result of which his heavenly throne and his seniority passed over to his brother Christ (cf. infra, p. 207, n. 8). From that time onwards Christ became the elder, and the Devil the younger, brother.

This interpretation would thus seem to overcome the apparent discrepancy between Cosmas and the other sources. The seniority of Christ would correspond not to the initial phase of the Bogomil cosmology (since all the sources concerned with this phase state quite plainly that the Devil at the beginning was Christ's elder brother), but to a later stage in the history of the universe, when the position of the two brothers is reversed.

¹ Діавола творца нарицающе человѣкѡм и всеи твари божіи и ѡт многыа грубости ихъ, иниж аггела ѡтпадша наричють и, друзииж оуконома неправеднааго творять и. (Op. cit. p. 22.)
² This identification of the 'unjust steward' with the Devil is a typically Bogomil feature. Cf. infra, p. 227.
³ Cf. supra, pp. 38, 113.

one good, the other evil, the dualism attributed to the Bogomils by Cosmas is not bitheistic, but is based on the recognition of the inferiority of the Devil and of his ultimate dependence on God. This inferiority is clearly expressed in the designations of 'fallen angel' and 'steward', applied to the Devil by the Bogomils. Moreover, the very terms 'devil' (διάβολος, діаволъ) and 'fallen angel' show that there were points of contact between the Bogomil cosmology and the Christian teaching on the fall of Satan. From the combined evidence of all the sources it can be asserted that the Bulgarian Bogomils never believed in the existence of two Principles or Gods. Their dualism consisted in rejecting the unity between God and His creation by interposing an intermediary endowed with demiurgical and creative powers, who was, in their belief, the author and Lord of the material world, described as ὁ ἄρχων τοῦ κόσμου τούτου (John xii. 31). These two forms of dualism, the Paulician and the Bogomil, are sometimes defined respectively as 'absolute' and 'moderate'.[1]

According to some scholars, the original teaching of the *pop* Bogomil was 'absolute' dualism, but at the end of the tenth century, at the time when the *Sermon against the Heretics* was composed, this dualism was 'mitigated' by the introduction of Christian influences.[2] According to them, both forms of dualism can be found in Cosmas's exposition, the 'absolute' dualism being represented by the words: 'They call the Devil the creator of man and of all God's creatures', and the 'moderate' dualism by the references to the 'fallen angel' and the 'unjust steward'. But this radical transformation of Bogomilism can be substantiated by no historical evidence.[3] Moreover, Cosmas does not in fact allude to 'absolute' dualism: the belief that the material world is the creation of the Devil, far from being exclusively characteristic of this form,

---

[1] See Rački, op. cit., *Rad*, x, pp. 163-4; Ivanov, op. cit. pp. 20-2; cf. infra, pp. 161-2.

[2] B. Petranović, Црьква Босаньска и крьстяне, p. 46; Rački, loc. cit. p. 164; M. S. Drinov, Исторически прѣгледъ на българската църква, p. 50.

[3] Runciman believes that a schism between representatives of extreme and moderate dualism in the Bogomil sect took place after the tenth century (op. cit. p. 69). I venture to disagree with him and think that these two trends which existed among the medieval Balkan sectarians correspond to Paulicianism and Bogomilism respectively, rather than to a division within the Bogomil sect. Cf. infra, pp. 161-2.

was, in fact, held by all Balkan dualists, whether Paulicians, Massalians or Bogomils, as N. Filipov has rightly pointed out.[1]

It can thus be affirmed that the teaching of the *pop* Bogomil, from the very moment of its appearance, retained the cosmology of the early dualistic heresies, by attributing the creation of the material world to the Devil, considered as an intermediate spirit, secondary to God, but renounced the bitheistic doctrine of the Paulicians.

The following difficulty, however, remains to be explained: the letter of Theophylact to the Tsar Peter, which describes the Bogomil heresy, refers to its cosmological dualism in terms which are purely Paulician and essentially different from those which were later used by Cosmas. The solution of this problem seems to lie in the point of view from which Theophylact regarded the Bulgarian heresy and in the methods of investigation to which he resorted. The patriarch's information on the heresy was indirect and probably derived from the hierarchs of the Bulgarian Church; it lacked, for this reason, the advantage of personal observation, so characteristic a feature of the *Sermon against the Heretics*. Moreover, Theophylact as a pastor and theologian was concerned above all with analysing the new heresy and treating its component parts according to the law of the Church;[2] the results of his analysis led him to conclude that it was 'Manichaeism mixed with Paulicianism'. The best account and refutation of the Paulician heresy which the patriarch possessed was no doubt that of Peter of Sicily, and he had all the more reason for relying on the *Historia Manichaeorum* as it had been composed with special reference to a Paulician mission in Bulgaria.[3] It is thus understandable that Theophylact should have accused the Bogomils of believing in two principles, since Peter of Sicily ascribed this doctrine to the Paulicians, although in reality it was neither taught by the *pop* Bogomil nor held by his followers in the tenth century.[4]

---

[1] Произходъ и сѫщность на богомилството, *Bŭlgarska Istoricheska Biblioteka* (Sofia, 1929), vol. III, pp. 46–8.

[2] Such an analysis, on the contrary, is totally lacking in the work of Cosmas.

[3] Cf. supra, p. 30.

[4] Ivanov's view that the heresy described by Theophylact corresponds to 'an extreme wing' of Bogomilism, 'very near to Paulicianism' (op. cit. p. 21), does not seem convincing, since the appearance of two dualistic 'Churches', to which he refers, took place much later, probably in the twelfth century (see infra, pp. 161–2).

Judging from Cosmas's exposition of the different Bogomil views on the Devil, there appears to have been a certain lack of doctrinal unity among members of the sect, for which he ridicules them: 'Their words are ridiculous for those who possess intelligence, for they do not agree with one another, and fall apart like a piece of rotten cloth.'[1] Contradictions and inconsistencies in matters of doctrine are not surprising in the case of a sect which laid the greatest emphasis not on dogma, but on the pursuit of moral purity and the evangelical life. This predominance in primitive Bogomilism of the ethical point of view can be seen from the fact that the exposition of the purely doctrinal errors of the heretics forms a comparatively small part of the *Sermon* of Cosmas, which is concerned above all with the moral and social aspects of the heresy.[2]

The doctrines ascribed to the Bogomils in the *Sermon against the Heretics*, with the important exception of their views on the Devil, are already to be found in the *Letter of Theophylact* and in the *Historia Manichaeorum*, so a brief enumeration of them will suffice.

The Docetic Christology, however, emphasized by Theophylact, is only hinted at by Cosmas in his anathema against those 'who do not love our Lord Jesus Christ'.[3] That this vague expression is in fact an allusion to Docetism is shown by the *Synodicon of the Tsar Boril*, which expressly states that Docetic Christology was taught by the *pop* Bogomil.[4] The false doctrines concerning Our Lady, described by Peter of Sicily and Theophylact, are alluded to by Cosmas, though they are not specified: he merely states that the heretics 'do not venerate the Most

---

[1] Op. cit. p. 23.

[2] In this respect the *Sermon against the Heretics* differs from the *Letter of Theophylact to the Tsar Peter*, which emphasizes above all the doctrines of the heretics and mentions their ethical applications (such as the rejection of marriage) only when they are very glaring. This difference between the two documents is, no doubt, partly due to the difference between the points of view from which they were composed. Theophylact wrote mainly as a theologian, to instruct the members of the Bulgarian hierarchy, while Cosmas, as a priest, was essentially concerned with exposing to the Bulgarian people those aspects of the Bogomil heresy which were more immediately accessible to them. In any case, the picture painted by Cosmas is the result of his personal observation and must be regarded, with its stress on the ethical side of the heresy, as more accurate than the letter of Theophylact.

[3] Op. cit. p. 62. [4] See infra, p. 238.

Glorious and Pure Mother of our Lord Jesus Christ, but talk much nonsense concerning Her; their insolent words cannot be written in this book'.[1] The attitude of the Bogomils towards the canon of the Scriptures was very similar to that of the Paulicians. Cosmas tells us that they rejected the Mosaic Law as contrary to the teaching of the Apostles and reviled the Old Testament Prophets.[2] Like the Paulicians, they based themselves exclusively on the New Testament, and, more particularly, on the Gospels and the Acts of the Apostles, which they interpreted not in conformity with the tradition of the Church, but in an individualistic manner: 'Although they carry the Holy Gospel in their hands, they interpret it falsely and thus seduce men...with the intention of destroying all Christian charity and faith....The Holy Gospel is in their hands...'as a jewel of gold in a swine's snout".'[3]

On the other hand Cosmas lays great emphasis on the ethics of the Bogomils. As the *Sermon against the Heretics* contains not only the earliest, but also the fullest account of the moral teaching of Bogomilism, it is necessary to examine his evidence in some detail.

The fundamental ethical teaching of the Bogomils, like that of the Manichaeans, was deduced from their cosmological dualism: if the visible world is the creation and realm of the Evil One, it naturally follows that, in order to escape from his domination and to be united with God, all contact with Matter and the flesh, which are the Devil's best instruments for gaining mastery over the souls of men, should be avoided. Hence the Bogomils condemned those functions of man which bring him into close contact with the world of the flesh, in particular marriage, the eating of meat and the drinking of wine. 'They say that he [i.e. the Devil] has ordered men to take wives, to eat meat and to drink wine. Briefly, in blaspheming all that is ours they claim to be the inhabitants of heaven and call those who marry and live in the world the servants of Mammon.'[4] Cosmas emphasizes that the heretics avoid marriage, meat and wine not from abstinence or Christian asceticism, but because 'they consider them abominable', as part of the natural law which they rejected as

[1] Op. cit. p. 17.   [2] Ibid. p. 16.
[3] Ibid. p. 25. The quotation is from Prov. xi. 22.
[4] Ibid. p. 26.

coming from the Devil.[1] The condemnation of marriage as an obstacle to holiness and a capitulation to the flesh was common in Bulgaria by the middle of the tenth century,[2] and developed as a result of mutual interaction between Massalianism and certain exaggerated and decadent forms of Orthodox monasticism. Here again the Bogomils utilized a pre-existing tendency to heresy for the successful promotion of their own doctrines. Although the abstention from meat for heretical reasons is not mentioned in any Bulgarian source prior to the *Sermon against the Heretics*, it was probably also preached, together with the rejection of marriage, in heterodox monastic circles, for in the Eastern Church the monastic rule prohibits the eating of meat. As for the condemnation of wine, its precise origin in Bulgaria is uncertain.[3] In later times, probably from the eleventh or twelfth centuries, this tenet became firmly fixed in the written tradition of the Bogomils and gave rise to the belief, recorded in particular in certain apocrypha, modified and used by the Bogomils for spreading their own doctrines,[4] and partly in a Bogomil legend,[5]

---

[1] Самиж всего того гнушающе ся не пріемлють, не въздержанія ради якож и мы, нъ скврьнаво творяще. (Ibid.)
The fundamental difference between abstention from marriage, meat and wine for the sake of discipline and abstinence and their rejection out of 'disgust' was made by Cosmas in accordance with the tradition of the Church concerning this matter, formulated in the 51st Apostolical Canon: εἴ τις...γάμου καὶ κρεῶν καὶ οἴνου οὐ δι' ἄσκησιν, ἀλλὰ διὰ βδελυρίαν ἀπέχεται, ἐπιλαθόμενος ὅτι πάντα καλὰ λίαν, καὶ ὅτι ἄρσεν καὶ θῆλυ ἐποίησεν ὁ Θεὸς τὸν ἄνθρωπον, ἀλλὰ βλασφημῶν διαβάλλει τὴν δημιουργίαν, ἢ διορθούσθω, ἢ καθαιρείσθω, καὶ τῆς Ἐκκλησίας ἀποβαλλέσθω. Theodore Balsamon, *In Canonem* 51 *Sanctorum Apostolorum*, P.G. vol. cxxxvii, col. 141. Balsamon in his commentary on this Canon refers in particular to the Bogomils. [2] See supra, p. 105.

[3] Pogodin has advanced the suggestion that the Bogomil aversion to wine may have been influenced by Krum's law ordering the extirpation of all the vines in Bulgaria (cf. supra, p. 64, n. 1) through the 'legal tradition' which this measure is supposed to have initiated. (История Болгарии, p. 13.) This somewhat far-fetched theory can be substantiated by no evidence: it can scarcely be doubted that the effect of Krum's law was purely temporary.

[4] The *Apocalypse of Baruch*. See I. Ivanov, Богомилски книги и легенди, pp. 196–7, 207: Сатанаиль же оусади лозоу...и рече ми ангель: слыши, Варохь, пръвоє дрѣво ѥсть лоза, въторое же дрѣво похоть грѣховна, иже излия Сатанаиль на Єоугоу и Адама; и сего ради проклель бѣ Господь лозоу, зане бѣ ю Сатанаиль оусади и тою прѣльсти пръвозданьнаго Адама и Єоугоу. Cf. infra, p. 154 n.

[5] *The Sea of Tiberias*. See Ivanov, op. cit. pp. 297, 324.

that the vine was planted in Paradise by Satanael (the Devil) and that it was that very Tree of the Knowledge of Good and Evil the tasting of whose fruits caused man's downfall. Thus the dualistic asceticism of the Bogomils was historically an outcome of the gradual fusion between Massalianism and perverted monasticism, and, logically, the application to the realm of ethics of the cosmological dualism of the Paulicians.[1] It is doubtful, however, that the rigid forbearance from sexual intercourse was equally enforced on all members of the Bogomil sect. The considerable proportions assumed by the sect in the course of its history are difficult to explain without the recognition that some of its members were perhaps permitted to have children. Although evidence is lacking on this point, it seems probable that the tenth-century Bogomils were divided into two distinct groups, the ordinary 'believers', who were not bound to rigorous asceticism, with regard either to sexual intercourse or to food, and the 'perfect' who were. This distinction, characteristic of the Manichaean sect, is attributed to the Bulgarian Bogomils by Rački and Ivanov[2] and existed among the Byzantine Bogomils of the eleventh and twelfth centuries.[3]

From their dualistic cosmology, the Bogomils were naturally led to deny the Christian view of Matter as a vehicle for Grace, and itself capable of sanctification, and to adopt the anti-sacramental view of the Paulicians. According to the evidence of Cosmas, they rejected the validity of Baptism and held John the Baptist to be the forerunner of Antichrist.[4] Their dislike of Baptism was apparently carried to curious extremes: according

---

[1] Asceticism, as it has been pointed out, was not practised by the Bulgarian Paulicians, who were permitted to marry, to eat meat and to drink wine. In this respect the Bogomils, by unifying their cosmology and their ethics, were more consistent.

[2] 'Bogomili i Patareni, '*Rad*, x, p. 177; Богомилски книги и легенди, p. 27.

[3] Cf. infra, pp. 214–17. An important difference can be observed here between the ethics of the Bogomils and those of the Massalians: for the former, as for the Manichaeans, sexual intercourse, if and when it was allowed, was regarded as an inevitable evil and a capitulation to human weakness. The Massalians, on the contrary, held that strict asceticism was necessary for the ordinary 'believers', while free indulgence in sexual intercourse was a prerogative of those who had succeeded in driving out the demon from within them and who were thus 'perfect'. (Cf. supra, pp. 49–50.)

[4] Иѡана же предтечю и зарю великааго солнца безчьствують, антихристова предтечю наричющеи. (Op. cit. p. 17.)

to Cosmas, they 'felt an aversion to baptized children' and, whenever they encountered a child, they would 'turn away and spit'.[1]

Like the Paulicians, the Bogomils rejected the sacrament of the Eucharist, spurned the liturgy, denied the Real Presence and interpreted the Words of Institution allegorically, as referring to the Gospels and the Acts of the Apostles.[2]

Likewise they spurned all the material objects used by the Orthodox as vehicles for Grace and supports for prayer, principally the Cross, which they hated as the symbol of Christ's suffering: 'About the Cross of our Lord...they say: how can we bow to it, for on it the Jews crucified the Son of God? The Cross is an enemy of God. For this reason they instruct their followers to hate it and not to venerate it, saying: if some one were to kill the son of a King with a piece of wood, is it possible that this piece of wood could be dear to the King? So is the Cross to God.'[3] According to Cosmas, in their hatred of the Cross the heretics are worse than demons, 'for the demons are afraid of the Cross of Christ, but the heretics cut down the crosses and make their tools out of them'.[4]

Churches were, for the Bogomils, material creations of man, and hence the abode of the Devil; they called them распутья (probably 'dispersio gentium': Jn vii, 35).[5] For the same reason they condemned the use of icons and the veneration of relics:

---

[1] Видите ли братье колми есть поразилъ діаволъ да святое крещеніе ѿмещуть гнушающе ся крестимыхъ младенець, аще бо ся имъ случить видѣти дѣтищь млад, то акы смрада зла гнушают ся, ѿвращающе ся плюють...сами смрадъ суще аггеломъ и человѣкомъ. (Op. cit. p. 31.) It is probable that these words must be taken to mean not only that the Bogomils condemned those children who had received baptism, but also, as Puech suggests (op. cit. pp. 266–7), that they regarded all children as participating, at least in some degree, in the impure and devilish character of the sexual act that procreated them.

[2] Что бо глаголють ѡ святѣмь комканіи, яко нѣсть божіемь повелѣніемь творимо комканіе...но акы...все и простое брашно. (Ibid. p. 8.) Кто бо вы оуказа...яко нѣсть то речено ѡ томь святѣмь хлѣбѣ и ѡ чаши, акож то вы иеретици блазняще ся бесѣдуете яко ѡ тетровангѣ то есть речено и ѡ праксѣ апостолѣ, а не ѡ святѣмь комканіи. (Ibid. p. 10.) Како ли...глаголете не соуть апостоли литургіа предали ни комканіа, но Іѡань Златоустыи. (Ibid. p. 11.)

[3] Ibid. pp. 6–7.   [4] Ibid. p. 5.

[5] Ibid. p. 34; Rački, op. cit., *Rad*, x, p. 189.

'the heretics do not reverence icons, but call them idols...the heretics mock [the relics of the saints] and laugh at us when we reverence them and beg help from them.'[1] 'They read St Paul who says, about idols, that we must not obey gold and silver created by man's device. They think, the accursed, that this is said about the icons, and, finding their justification in these words, they do not reverence the icons.'[2] The miracles performed through the relics of saints they ascribed to the Devil: 'They say that the miracles are not wrought by the will of God, but that the Devil performs them to deceive men.'[3] They rejected the cult of saints.[4] They recognized the miracles performed by Christ, but interpreted them in a non-material sense, falling back, as in their explanation of the Eucharist, on the use of allegory:

'They do not confess that Christ performed miracles. On reading the evangelists who... wrote about the miracles of Our Lord, they distort their meaning, to their own ruin, saying: Christ neither gave sight to the blind, nor healed the lame, nor raised the dead, but these are only legends and delusions, which the uneducated evangelists understood wrongly. They do not believe that the multitude in the desert was fed with five loaves of bread; they say it was not loaves of bread, but the four Gospels and the Acts of the Apostles.'[5]

---

[1] Cosmas, op. cit. p. 5.
[2] Ibid. pp. 18-19. The origin of the iconoclastic tendencies in Bogomilism is uncertain. Popruzhenko supposes a direct influence of Iconoclasm on Bogomilism through the Paulicians transplanted into Thrace by Constantine Copronymus (Синодик царя Бориса, *I.R.A.I.K.* (1900), vol. v, Suppl. pp. 121-6). The role of the Paulicians as spreaders of Iconoclasm is emphasized by G. Ostrogorsky (*Studien zur Geschichte des byzantinischen Bilderstreites*, Breslau, 1929, p. 27, n. 1). E. J. Martin, however, while recognizing the existence of some common elements in Paulicianism and Iconoclasm, denies that there was interdependence between them (*A History of the Iconoclastic Controversy*, pp. 275-8). Cf. supra, p. 41.
It is true that the Byzantine Bogomils honoured the memory of the Iconoclastic emperors, particularly of Constantine Copronymus (cf. infra, p. 214, n. 9). This, however, may be sufficiently explained by the similarities in the teachings of the Bogomils and the Iconoclasts regarding in particular the veneration of images and the cult of Our Lady, which made the Bogomils look to Copronymus as to an early advocate of their faith. The identification of the Bogomils and the Iconoclasts occurs only at the beginning of the fifteenth century, when Bogomilism had practically disappeared. (Cf. infra, p. 166.)
[3] Ibid. p. 5.   [4] Ibid.   [5] Ibid. p. 32.

In view of their rejection of most of the Orthodox tradition, it is not surprising to find that the Bogomils were as hostile to the instituted Church as the Paulicians. The ecclesiastical hierarchy, as the mainstay of the Church and purveyors of Christian law, naturally became the visible object of attack for the heretics. Their inherent dislike of the hierarchy was strengthened by the decadent state of sections of the secular and monastic clergy, so vehemently denounced by Cosmas himself,[1] and which supplied the heretics with potent material for their attacks on the Church. According to Cosmas, the priests were accused by the heretics of laziness, hypocrisy and immorality: 'But what do the heretics say?—We pray to God more than you do; we watch and pray and do not live in idleness as you do.'[2] 'Why do you abuse the priests... calling the Orthodox priests blind pharisees?'[3] 'The heretics reply: the priests are given to drinking and robbing.'[4] The truth of these accusations is admitted by Cosmas. He even places on the clergy the main responsibility for the spread of heresy: 'whence arise these wolves, these wicked dogs, these heretical teachings? Is it not from the laziness and ignorance of the pastors?'[5]

The final chapter of the *Sermon against the Heretics* is an exhortation to bishops and priests to guard their flocks and an inherent criticism of their negligence and indifference to heresy.[6] Thus the direct relation between the appearance and spread of heresy and the contemporary decadence in the Bulgarian Church and society in the tenth century is confirmed by one of the most outstanding Churchmen of the time.

But, even though Cosmas admits the truth of the contentions of the heretics, he refuses to acknowledge their validity, for no heretic has the right to criticize a priest, heresy itself being the greatest of sins: 'although the Orthodox priests live a lazy life as you say, blaming them, they do not, however, blaspheme God as you do.'[7] In accusing the priests, the heretics are guilty of pharisaic self-righteousness.[8] This could be tolerated all the less,

[1] Cf. supra, pp. 104 et seq. [2] Op. cit. p. 4.
[3] Ibid. p. 12. [4] Ibid. p. 13.
[5] Ѡткуду бо исходять волци сии злїи пси еретическаꙗ оученїа, не ѡт лѣности ли и грубости пастушьскы. (Ibid. p. 75.)
[6] О епископѣх и ѡ попѣх. (Ibid. pp. 74 et seq.)
[7] Ibid. p. 12.
[8] Ѡ ле подобнаꙗ рѣчь къ фарисею. (Ibid. p. 4.)

as the heretics rejected the very Order of Priesthood and the sanctity of the Apostolic Succession: 'Why do you heretics blaspheme against the sacred orders that are transmitted to us by the Holy Apostles and the divinely inspired Fathers?'[1] 'Whosoever does not believe that the ecclesiastical orders are established by Our Lord and the Apostles, may he be cursed!'[2]

The rejection of the Order of Priesthood and of the Apostolic Succession in the Christian Church was common to the Bogomils and the Paulicians. The latter applied the conception of the Church to their own communities which were governed by elders who were not, it appears, invested with any special hieratic dignity.[3] It is, however, not possible to conclude with certainty from the account of Cosmas whether the Bogomils possessed any similar organization in the tenth century. We know from his evidence that the Bogomils, while rejecting the sacrament of penance, confessed their sins to one another,[4] which suggests that they recognized no essential distinction between the priesthood and the laity. This fact, together with the pronounced anti-sacerdotalism with which they are taxed by Cosmas, makes it impossible to attribute anything but a very democratic organization to the early Bogomil communities. It is probable that the Bogomils in the tenth century possessed leaders or elders who held the primacy of teaching. The Byzantine Bogomils of the eleventh and twelfth centuries and the Bulgarian ones of the twelfth and thirteenth called their leaders 'apostles',[5] and it is possible that the same term was originally applied to the immediate disciples of the *pop* Bogomil in Bulgaria.[6] There is, however, no foundation for the claim that Bogomil himself instituted a regular sectarian

[1] Ibid. p. 11. [2] Ibid. p. 63.
[3] Cf. supra, pp. 41–2.
[4] Еретицы...сами в себе исповѣдь творять. (Ibid. p. 39.)
[5] Cf. infra, pp. 199, 238.
[6] V. Levitsky (Богомильство—болгарская ересь, *Kh. Ch.* 1870, vol. I, pp. 372–3) and R. Karolev (За Богомилството, *Periodichesko Spisanie*, Braila, 1871, vol. III, p. 105) have drawn attention to Cosmas's designation of the Bogomils as новіи апостоли и предтеча антихристовы (op. cit. p. 35) and conclude that the name 'apostles' may have been given to the elders of the Bogomil communities in the tenth century. As the existence of this term is certified in later periods, this interpretation of Cosmas's words is not impossible. Yet the argument is not decisive, since the expression 'apostles of Antichrist' may well be a general derogatory term referring to all the heretics.

hierarchy or a 'Bogomil Church'.¹ The existence of such a hierarchy is not attested by historical evidence until the second half of the twelfth century.²

Together with the hierarchy, orders, sacraments and liturgy of the Orthodox Church, the Bogomils rejected its discipline, which they replaced by their own, of a rigorous nature. Cosmas accuses them of keeping Sunday as a day of fasting and work and of not celebrating the Orthodox feasts of our Lord and of the martyrs.³ They also spurned all the prayers used by the Orthodox Church, which they considered, together with the liturgy, to be 'babblings' (многоглаголанья),⁴ with the solitary exception of

---

¹ This claim has been put forward by A. Gilferding (Собрание сочинений, vol. I: История сербов и болгар, p. 132), Rački (op. cit., Rad, VII, pp. 103-4) and Ivanov (op. cit. pp. 27-30). The arguments adduced in support of this view are: (1) the existence of a semblance of a hierarchy among the Byzantine Bogomils; (2) the analogy with the Italian and French Cathars who possessed a highly developed hierarchy and who, moreover, in the twelfth century regarded Bulgaria as the original home of their doctrines; (3) the evidence of Latin twelfth- and thirteenth-century sources which expressly mention several ranks in the hierarchy of the Bulgarian Bogomils (cf. infra, pp. 242 et seq.) But it is most unsafe to argue from later and foreign sources, since Bogomilism, even in Bulgaria, underwent a process of continual evolution. As it will be shown, the inner discipline and organization of the Bogomil sect was undoubtedly much influenced by its contact with Byzantine civilization in the late eleventh and early twelfth centuries. Only from that period is it possible to speak of any definite hierarchy among the Bulgarian Bogomils.

The failure to take into account this historical evolution vitiates the best studies of Bogomilism such as Rački's and Ivanov's. Considering Bogomilism as a static phenomenon, they transpose into tenth-century Bulgaria notions and institutions which are found in the thirteenth century among the dualistic heretics of Bosnia, Italy and even southern France. Thus Rački (ibid.) attributes to Bogomil himself the foundation of a 'Bulgarian Church' (crkva bugarska) by analogy with the 'Bosnian (Patarene) Church' (crkva bosanska).

² See infra, pp. 244-5.

³ Кто...вы указа въ день въскресенія господня постити ся, и кланяти и ручная дѣла творити. Да глаголете человѣцы то суть оуставили, а не пишет того въ евангеліи, и вся господьскыа празникы, и память святых мученикъ и wтець не чтете. (Ibid. p. 33.)

The practice of fasting on Sundays existed already among the Manichaeans (cf. supra, p. 22). The rejection of Sunday as the day of the Resurrection was probably connected with the denial of the Christian dogma of the Resurrection of the Body, which, though not mentioned by Cosmas, is a natural consequence of the views of the Bogomils on matter. The Bogomils are accused of denying the Resurrection of the Dead in an eleventh-century source. (Cf. infra, p. 181.)

⁴ Ibid. p. 34.

BOGOMILISM IN THE FIRST BULGARIAN EMPIRE   135

the Lord's Prayer; this they recited day and night at regular intervals and with appointed prostrations: 'shutting themselves up in their houses, they pray four times a day and four times a night, and they open all the five doors which, as it is ordered, should be closed.[1] Bowing they say "Our Father which art in Heaven", but for this they must be greatly condemned, because only in words do they call the creator of Heaven and earth Father, and in reality they ascribe the creation to the Devil. When they bow they do not make the sign of the Cross.'[2] These prayers and also the fact that the Bogomils confessed their sins to one another and gave each other absolution would seem to show that their communities recognized some form of ritual, but it cannot be admitted that it was anything but very rudimentary in the tenth century.[3] These rites of confession were also performed by women.[4]

It is obvious that the Christian conception of the Church, both in its divine and human aspects, as the mystical Body of Christ and as a hierarchical institution on earth, was profoundly alien to the teaching of the Bogomils. It seems that they avoided using the very term Church.[5] The Orthodox Church, by stressing the

[1] This somewhat obscure simile of the 'five doors' is explained by Trifonov (Бесѣдата на Козма Пресвитера и нейниятъ авторъ (1923), *S.B.A.N.* vol. XXIX, pp. 29–30) as referring to the five senses. The meaning of this passage is that, although the heretics when praying close the doors of their houses, they leave those of their senses open; thus instead of achieving concentration in prayer, they allow their senses and their imagination to receive outside impressions.

[2] Ibid. p. 32.

[3] Rački (op. cit., *Rad*, x, pp. 189–206), Ivanov (op. cit. pp. 29, 113 et seq.) and Klincharov (op. cit. pp. 59–62) ascribe to the Bulgarian Bogomils of the tenth century rites which existed among the Byzantine sectarians in the late eleventh century (such as βάπτισμα and τελείωσις which correspond to two different degrees of initiation into the sect). Here again, however, this transposition is historically unjustifiable. Bogomil ritual developed gradually and the comparatively complex character it later assumed in Byzantium is undoubtedly due to the influence of Orthodox ritual. The ceremonies of initiation practised by the Byzantine Bogomils in particular (cf. infra, pp. 215–16) very probably evolved under the direct influence of the rites of ordination of monks, priests and bishops as performed in the Eastern Church. See B. Petranović, Богомили, pp. 65 et seq.

[4] Еретицы...сами в себе исповѣдь творять...не точью мужи...но и жены, еж ругу достоино есть. (Ibid. p. 39.) Cf. supra, p. 50.

[5] In this they doubtless differed from the Paulicians, who applied the term 'Catholic Church' to their own communities (cf. supra, pp. 40, 42).

material as well as the spiritual aspect of life, had, in their opinion, capitulated to Mammon and was incapable of guiding men to salvation. True Christianity, according to the Bogomils, could only be found in their own communities; hence they claimed for themselves the exclusive right to the name of Christians:[1] they alone lived 'according to the Spirit'[2] and were 'the inhabitants of heaven'.[3]

A teaching so fundamentally opposed to Orthodoxy in the spheres of doctrine, ethics and ritual could not fail to have important repercussions on every branch of Bulgarian social life. In particular, at a time when the interests of Church and State were closely linked, the rejection of Orthodoxy was inevitably also a rebellion against the secular laws and a challenge to the whole contemporary society. It is thus not surprising to find that already in the tenth century the Bogomils attracted the serious attention of the State authorities. The letter of the Patriarch Theophylact testifies to the grave concern with which the Tsar Peter viewed the growth of heresy in Bulgaria. Some of the accusations of Cosmas against the Bogomils point to the social significance of the heresy in the second half of the tenth century. The heretics are presented as idlers with no fixed abode, as parasites on society: Cosmas cites 'their other words, with which they entrap the souls of ignorant people, saying that it is unbecoming for a man to labour and to do earthly work. As the Lord said: "Take no thought, saying, what shall we eat, or what shall we drink, or wherewithal shall we be clothed, for after all these things do the Gentiles seek", they do not want to do anything with their hands, but wander from house to house, devouring the property of the people they deceive.'[4] The condemnation of manual labour,

[1] Аще же и хотять лгати по своему юбычаю глаголюще яко христіани есмь, не имѣте имъ вѣры. (Op. cit. p. 31.) This assumption of the name of Christians κατ' ἐξοχήν by the Bogomils is among the principal reasons for the accusations of imposture and hypocrisy frequently levelled at them by the Orthodox. The Manichaeans and the Paulicians incurred the same accusations for similar reasons.

Cf. the title of 'good Christians', assumed by the Bogomils at the beginning of the thirteenth century: Χριστιανοὺς καλοὺς ἑαυτοὺς ὀνομάζουσιν (Germanos II: *In exaltationem venerandae crucis et contra Bogomilos*, P.G. vol. cxl, col. 637).

[2] то по плоти живуть явѣ, а не по духу якож и мы. (Ibid. pp. 13–14.)
[3] Небесніи жителіе. (Ibid. p. 26.) Cf. infra, p. 179, n. 1.
[4] Ibid. p. 35.

together with voluntary poverty, which produced the type of wandering monk denounced so vigorously in the *Sermon against the Heretics*, is characteristic of the Massalians, whose direct influence on the Bogomils is evident here.[1]

Cosmas puts forward an even graver accusation against the Bogomils—that of preaching civil disobedience: 'They teach their own people not to obey their masters, they revile the wealthy, hate the tsar, ridicule the elders, condemn the boyars, regard as vile in the sight of God those who serve the tsar and forbid every serf to work for his lord.'[2] To this social anarchism Cosmas opposes the teaching of the Church, which recognizes the sanctification of the temporal power: the tsars and the boyars are established by God. Unfortunately, Cosmas limits himself to these scanty details and the doctrine and practice of civil disobedience which he ascribes to the Bogomils are not conclusively confirmed by any other source.[3] Caution therefore should be observed when interpreting this passage. There is no reason to doubt these words of Cosmas: his evidence in general is very trustworthy and his position as parish priest qualified him for direct and constant observation of the Bogomils. Nevertheless, one should beware of attributing too much importance to the social anarchism of the Bogomils or of seeing in them Slavonic communists of the Middle Ages.[4] It would seem that their doctrine of social equality was

[1] Cf. supra, p. 50. Puech (op. cit. pp. 276–7) rightly points out that only the 'perfect' were obliged to eschew manual labour.

[2] Оучать же своя си не повиноватися властелем своим, хуляще богатыа, царь ненавидять, ругаються старѣишинам, оукаряют боляры, мерзъкы Богу мнять работающаа царю и всякому рабу не велять работати господиноу своему. (Ibid. p. 35.)

[3] These words of Cosmas are sometimes coupled with the passage in the *Synodicon of the Tsar Boril*, cursing 'those who assist thieves, murderers, robbers and other such people'. But it is doubtful whether these words do in fact refer to the Bogomils. (Cf. infra, pp. 247–8.)

The accusation of preaching civil disobedience was levelled in the twelfth century against the Byzantine heretic Constantine Chrysomalus. But Constantine's affiliation to Bogomilism has not been proved. (See infra, p. 220.)

[4] There have, of course, been attempts, as biased as they are unhistorical, to over-emphasize the social and political role played by the Bogomils in Bulgaria. See, in particular, M. Popowitsch, 'Bogomilen und Patarener. Ein Beitrag zur Geschichte des Sozialismus', *Die Neue Zeit*, 24. Jahrg., Bd 1, Stuttgart, 1905, pp. 348–60. An extreme exponent of this point of view is I. Klincharov (Попъ Богомилъ и неговото време). He goes so far as to describe the Bogo-

deduced from their pursuit of spiritual poverty and moral purity, and their declared war against the powerful of this world was a transposition on to the social plane of the cosmic struggle between Good and Evil. In this sense alone can the Bogomils be said to have opposed the growth of Byzantine feudalism in Bulgaria, which was based only too often on the oppression of the weak by the powerful. But Bogomilism was not essentially a social and still less a political movement. Gabriel Millet has rightly pointed out that the Bogomils remained above all religious preachers, indifferent to secular affairs.[1]

From this analysis of the *Sermon against the Heretics* some basic features of Bogomilism can be brought out, which will permit a clearer understanding of the relation of the sect to the Bulgarian Church and State and of the reasons for its successful spread in the country. This in its turn should explain many aspects of the future history of Bogomilism in Bulgaria. Moreover, at least three features of Bogomilism—the doctrine of the two sons of God, the Devil and Christ, the teaching on the introduction of the soul into the body of Adam, the first man,[2] and the exclusive use of the Lord's Prayer—cannot be explained by any outside influence and are probably original.[3]

Bogomilism, which arose under the double influence of Paulicianism and Massalianism, was not an uncoordinated mixture of these earlier heresies. Many of their doctrines were reshaped and woven into a unified whole in the tenth century, probably by the

mil sect, already in Peter's reign, as 'the strongest and best organized party in Bulgaria' (p. 30)—a party both 'religious and political' (p. 72). The 'programme' he attributes to the Bogomils is more reminiscent of the twentieth century than of the tenth. The aim of the Bogomils, if we are to believe Klincharov, was 'the foundation of independent political communes' (p. 121) and 'the re-establishment of small agricultural property' (p. 126). To achieve their purpose, these 'first agrarians of the Balkan peninsula' (p. 116) (!) aimed at 'seizing political power' by 'concrete political means' (p. 120)(?). This theory is a particularly deplorable example of the tendency to interpret the past in terms of present-day conceptions and events. Klincharov's notions of the political role played by the Bulgarian Bogomils are warranted neither by any serious documentary evidence nor by our knowledge of the true character of the Bogomil movement.

[1] 'La Religion Orthodoxe et les hérésies chez les Yougoslaves', *Revue de l'histoire des religions* (1917), vol. LXXV, p. 292.
[2] See infra, pp. 180, 208.
[3] See Puech, op. cit. pp. 336–40.

pop Bogomil himself. Finally, Bogomilism was strongly influenced by Christianity;[1] from this point of view it can be regarded as an attempt to bring the dualism of the Paulicians into greater harmony with the teaching of Christ. Thus an analysis of Bogomilism reveals the presence in it of two basic trends, the one doctrinal, the other ethical: the first is a dualistic cosmology of foreign origin, imported into Bulgaria from the Near East; the other, largely autochthonous, is a revolutionary attempt to reform the Christian Church, based on the dissatisfaction with its existing state and a desire to return to the purity and simplicity of the apostolic age. These two trends together produced Bogomilism.[2] These Balkan reformers, while accepting the dualistic doctrines and indeed applying them to all branches of life, nevertheless laid the greatest stress on ethics, which were derived exclusively from

[1] From the historical point of view the question of what teachings, apart from Paulicianism, Massalianism and Christianity, influenced Bogomilism must be considered at present insoluble. Attempts have been made to find doctrinal similarities between Bogomilism and earlier religions and sects. Thus Filipov has tried to prove the influence on Bogomilism of Gnosticism and Marcionism (Произходъ и сжщность на богомилството, loc. cit. pp. 33 et seq. and p. 55). J. Lavrin sees in Bogomilism 'certain Gnostic principles' and 'a sprinkling of Buddhism' ('The Bogomils and Bogomilism', *The Slavonic and East European Review* (London, 1929), vol. VIII, p. 270). There certainly exist important points of resemblance between Bogomilism on the one hand and Gnosticism and Marcionism on the other, particularly as regards the dualistic cosmology, the Docetic Christology, the rejection of parts of the Old Testament and the condemnation of marriage. See W. Bousset, *Hauptprobleme der Gnosis* (Göttingen, 1907), pp. 91 et seq.; E. de Faye, *Gnostiques et Gnosticisme* (Paris, 1913), pp. 431–45 and passim; G. Bareille, 'Gnosticisme', *D.T.C.* vol. VI, cols. 1456 et seq.

It is possible that Gnostic elements may have penetrated into Bogomilism from Syria or Asia Minor through the Paulician or Massalian sects. But the influence of Gnosticism on these sects has not been proved. Until such a historical connection has been established, the link between Gnosticism and Bogomilism must remain largely hypothetical. Cf. Puech, op. cit. pp. 337–9.

The existence of Buddhist elements in Bogomilism is highly questionable and cannot be substantiated. In particular, there is no evidence for Lavrin's assertion that the Bogomils believed in reincarnation (loc. cit. p. 227).

[2] Both these trends, no doubt, existed already in Paulicianism; but the Paulicians remained essentially foreigners in Bulgaria. The movement of reformation, to gain sufficient power, had to spring out of pre-existing local conditions in the Church and in the whole of society and to assume a specifically Slavonic temper. This could be achieved only by a national movement like Bogomilism.

the New Testament. This somewhat paradoxical union of anti-Christian dualism with Christian morality was made possible by a rationalistic and individualistic interpretation of the Scriptures. Such an attitude to the Holy Writ, together with a strong anti-ritualistic and anti-sacerdotal tendency, explains two important features of Bogomilism which are also to be found in later movements of the Reformation: the general priesthood of the laity and the view of the Holy Scripture as the unique source of revealed faith.

It need hardly be emphasized that Bogomilism from its inception and in its essence was in complete opposition to the Orthodox Church. Its dualistic cosmology explicitly denied the dogma of the unity of God and implicitly rejected the Incarnation of Christ, the sanctity of His Mother, the sanctification of all Matter by means of the sacraments and, generally, the whole Orthodox tradition. In these circumstances, no possible agreement or compromise could exist between the Bogomil sect and the Orthodox Church. The former considered that the latter had irrevocably betrayed Christ. The Church could have only one policy towards Bogomilism—that of never-ceasing war, aimed at the complete extermination of her enemy. Bogomilism can scarcely even be called a heresy in the strict sense of the word; for it represented, not a deviation from Orthodoxy on certain particular points of doctrine or ethics, but a wholesale denial of the Church as such. It can safely be said that after the final defeat of Iconoclasm in the ninth century, Bogomilism was in the Balkans the most dangerous enemy of the Orthodox Church in the whole of the Middle Ages. But it was not only the Church which was menaced by the Bulgarian sect: the whole social structure was in peril. A teaching which resolutely condemned married life as sinful threatened to undermine the foundations of the family, the community and the State. These foundations, as it has been shown, were already sufficiently shaken in Bulgaria by the middle of the tenth century. Moreover, in their opposition to established authority, temporal as well as spiritual, which in their eyes was the social reflection of the evil inherent in all created things, the Bogomils preached a crusade against the great and powerful of this world, the rich, the elders, the boyars, the tsar himself. In this they undoubtedly profited by the social oppression, the ruinous wars, the economic decline and the restlessness caused among the people by the wholesale Byzantinization of Bulgarian life in the tenth century. By espousing the cause of the serfs against their masters, of the oppressed against the oppressors, the Bogomils appealed directly to the peasant masses who regarded them as liberators and were often led to accept their doctrines.

It is thus understandable that the struggle against Bogomilism was an urgent necessity for the Bulgarian State authorities as well as for the Orthodox Church. Unfortunately, our knowledge of this struggle in the early period of the sect's history is scanty.[1] Our sole evidence of the persecution of the Bogomils in the tenth century is a hint dropped by Cosmas: he laments the blindness of those many who 'do not know what their heresy is, and imagine that they suffer for truth and wish to receive reward from God for their chains and imprisonment; but how can they be pleasing to God, even if they suffer in vast numbers, when they call the Devil the creator of men and of all God's creation?'[2]

This halo of martyrdom which surrounded the Bogomils and which was recognized by their fiercest opponent was doubtless due to their great moral prestige as new spiritual leaders of the Bulgarian people. In contrast to the intellectual and moral decadence of the clergy, who only too often left their flock without adequate support or instruction, the Bogomils, owing to their saintly appearance, intimate knowledge of the Gospel, strict asceticism, ardent proselytism and courage in persecution, must have appeared to many Bulgarians as the bearers of true Christianity. Their clever simulation of Orthodoxy, which considerably facilitated their task of avoiding detection, was both a powerful weapon of proselytism and a protection against systematic persecution.[3] Thus Cosmas relates that in spite of their rejection of all the

---

[1] Klincharov (op. cit. pp. 52–4) tries to defend the Bogomils from the accusation that they were opposed to marriage and family life. His attempt is most unconvincing. It cannot be denied that the dualistic cosmology of the Bogomils led them to condemn on principle all forms of sexual intercourse, whether in wedlock or outside it, whatever concessions they may have made in practice to the 'weakness of the flesh'.

[2] Cosmas, op. cit. p. 22.

[3] The Bogomil practice of dissimulation, which seems to have been partly an outcome of the esoteric character of their teaching and partly a matter of tactics, is discussed by Puech (op. cit. pp. 145–61).

liturgical and sacramental life of the Church, 'out of fear of men they enter the church, and kiss the Crucifix and the icons, as we are informed by those of them who have returned to our true faith; they say "we do all this because of the people, and not according to our heart; we hold to our own faith secretly".'[1]

A further reason for the success of Bogomil proselytism in Bulgaria lies in its essentially popular and Slavonic character. From the moment of its appearance and throughout its entire history of four centuries in Bulgaria, Bogomilism was linked with the religious and social aspirations of a large dissatisfied section of the people, which explains its considerable appeal, particularly among the peasant class.[2] For this reason it was the strongest and most vital of all the sectarian movements in medieval Bulgaria.

Finally, the following psychological factor was favourable to Bogomilism. At a time when misery and suffering were so widespread in Bulgaria, the minds of the people were often not unnaturally preoccupied with the problem of the origin of evil: whence come wars, devastations, plagues, oppression of the poor by the rich? The Church taught that everything, visible and invisible, is created by God; but how could God, who is the Supreme Good, be the cause of suffering and Evil? There is evidence that the problem of *unde malum, et quare*, a source of anxious speculation in all times, preoccupied many Bulgarians at the time of Cosmas: a passage of the *Sermon* tells us that many Orthodox were seeking an answer to the question, 'why does God permit the Devil to work against men?'[3] Though Cosmas dismisses it as a product of a childish and unhealthy mind,[4] many contemporary Orthodox priests who, as we know, were fairly ignorant in matters of doctrine, must have been unable to reply satisfactorily to this question. The heretics, on the other hand, provided a remarkably convincing explanation of all calamities: suffering and Evil are inherent in this world, because this world is the creation of the Evil One.

The great strength of Bogomilism lay, as we have seen, in its

---

[1] *Sermon*, p. 19.
[2] See Rački, op. cit., *Rad*, VII, p. 103.
[3] Многы...слышимъ ѿт нашихъ бесѣдующа почто Богъ діаволу попоущаеть на человѣкъ. (Ibid. p. 24.)
[4] Но та словеса дѣтскыихъ суть и несъдравыихъ оумомъ. (Ibid.)

inner coherence and in its ability to unify the pre-existing Paulician and Massalian teachings. But dualistic heresy in tenth-century Bulgaria was not, it seems, exhausted by Bogomilism. There is some evidence that the Paulician and Massalian sects, while contributing to the formation of the new Bulgarian heresy, retained their individual existence.

Indirect evidence of the tenth-century sources suggests that a certain distinction existed at that time between the Bogomil and Paulician sects. The *Letter of Theophylact to the Tsar Peter*, which undoubtedly refers to Bogomilism, nevertheless presents some of the heretical doctrines in a Paulician form, thus testifying to the existence of the Paulicians in Bulgaria towards the middle of the tenth century. Traces of this distinction between the two heresies can also be found in the *Sermon against the Heretics*: the contradictions and inconsistencies in the heretical teachings, to which Cosmas alludes, may be significant in this respect; moreover, a study of his work reveals certain divergencies in the behaviour of the heretics: for instance, Cosmas accuses them of rebuking the Orthodox priests for leading an idle life and at the same time of despising manual labour.[1] Furthermore, the heretics are described as holding arguments with the Orthodox, mocking them and openly reviling their priests;[2] on the other hand, as simulating Orthodoxy out of fear and openly protesting their innocence of heresy whenever pressed by their enemies.[3] Although these differences may be accidental and due to local reasons, such as the presence or absence of persecution, it seems more likely that in each case Cosmas is referring to a separate group of heretics. These differences, moreover, are most significant: it should be remembered that the Paulicians were celebrated for their active and even warlike qualities,[4] while inactivity and the condemnation of manual labour were characteristic of the Bogomils who, in this respect, were influenced by Massalianism. Finally, the Paulicians were open and courageous proselytizers,[5] while insidious and hypocritical behaviour was associated with the Bogomils and the Massalians.[6]

As this distinction between the Bogomils and the Paulicians,

[1] Cf. supra, pp. 132, 136.
[2] Cf. supra, pp. 131, 132.
[3] Cf. supra, pp. 141–2.
[4] Cf. supra, pp. 29, 37–8.
[5] Cf. supra, p. 42.
[6] Cf. supra, p. 121.

implicit in the *Sermon against the Heretics*, is fully confirmed by the evidence of later sources, an indication of its most important features is appropriate here.[1]

The essential difference between the two sects lies in the fact that the Bogomil ideal was essentially contemplative (in this it was influenced by Massalianism), while the life of the Paulicians was primarily directed towards action. This explains the difference in the external features of the two sects: the Paulicians always appear in history as restless and troublesome, born soldiers with a great propensity for fighting, the Bogomils, on the contrary, as meek, humble and ascetic. The strict asceticism of the Bogomils was clearly unsuited to the mode of life of the Paulicians; hence marriage, the eating of meat and the drinking of wine, condemned by the Bogomils, were not forbidden among the Paulicians.

A final distinction is due to the different origins of the two sects. The Paulicians were predominantly foreigners in Bulgaria, they remained as self-contained ethnical and social units, organized in communities, living apart from the Orthodox, attacking them or attempting to convert them from the outside.[2] The Bogomils, on the contrary, grew from within the Bulgarian people and remained in continual contact with them. By proselytizing within the Bulgarian communities, they were able to bring the foreign dualistic ideas into harmony with the life of the people, who were still in many cases semi-pagan, and with their religious and social aspirations. Therein lies, in a large measure, the reason for the successful spread of Bogomilism in Bulgaria and in the other Balkan countries.

[1] It is particularly important to recognize the points of difference between the Bogomils and Paulicians, as both sects remained clearly distinct in Bulgaria until the disappearance of Bogomilism after the fourteenth century. In the following chapters the different roles played by both sects in Bulgarian history will be stressed. The best authorities on Bogomilism, such as Rački and Ivanov, fail to make this distinction sufficiently clear. For the doctrinal differences between the Bogomils and the Paulicians, see Puech, op. cit. pp. 319–25.

[2] This self-contained and isolated character of the medieval Paulician communities is illustrated by the fact that, in contrast to the Bogomil sect, they survived the Turkish invasion in the fourteenth century. The majority were converted to Roman Catholicism in the late sixteenth and early seventeenth centuries but retained many of their ethnical peculiarities. Their descendants living in Philippopolis and in a few surrounding villages call themselves Paulicians to the present day. (Cf. infra, p. 266.)

It is more difficult to establish a distinction between Bogomilism and Massalianism in the tenth century, as the direct influence exerted by the latter on the morals and the behaviour of the Bogomils does not permit any differentiation between the two sects on the sole evidence of the *Sermon against the Heretics*. However, from the doctrinal point of view, for all the resemblances in the teachings of the two sects (a non-material interpretation of Christianity, an emphasis on prayer to the exclusion of the sacraments, which both sects regarded as unnecessary, a dislike of the instituted Church and a cult of asceticism), the Massalians and the Bogomils differed in some important respects. The basic difference lay in the absence in Massalianism of any truly dualistic cosmology: we find no trace among the Massalians of the opposition between God, the ruler of the heavens, and the Devil, creator of the visible world and of man, a doctrine so fundamental to Bogomilism. The most that can be said of Massalianism in this respect is that, in so far as it emphasized the opposition between the Spirit and the demon *in the heart of man*, it led to a kind of 'anthropological dualism'.[1] Moreover, historically, it seems undeniable that the Massalian sect still existed in Bulgaria without entirely merging with Bogomilism, at least until the twelfth century. There is evidence which suggests that in the middle of the eleventh century both sects existed in Thrace separate from one another.[2] Only after the twelfth century, as it will be shown, does the notable increase of Massalian elements in Bogomilism point to a gradual fusion between them.[3]

The *Sermon against the Heretics* is the last direct evidence we possess of the Bogomil heresy in the tenth century. The *Synodicon of the Tsar Boril* mentions a certain Michael, disciple of the *pop* Bogomil, and gives a list of further disciples of the Bulgarian heresiarch—Theodore, Dobry, Stephen, Basil and Peter.[4] The

---

[1] See Puech, op. cit. pp. 325–36.
[2] Cf. infra, pp. 183–8.
[3] Cf. infra, p. 251.
[4] Тръклятаго Богомила и Михаила оученика его и Феѡдора и Добрѣ и Стефана и Василїа и Петра и прочяя еговы оученикы и единомудрьникы...анафема. (*Synodicon*, ed. by Popruzhenko; Sofia, 1928, p. 82.) The name Dobry is clearly Slavonic.

text of the document suggests that Michael was an immediate disciple and contemporary of Bogomil, but it is not certain at what time the others lived.[1] However, indirect information on Bogomilism in the late tenth and early eleventh centuries can be obtained from some contemporary events of Byzantine and Bulgarian history.

In the tenth century, with the sole exception of the Patriarch Theophylact, the Byzantine authorities seem to have paid no attention to the growth of heresy in Bulgaria. From 967 to 1018 the emperors were constantly concerned with Bulgaria, but only from the military point of view, since practically this entire period was occupied by wars on the northern frontier of Byzantium against the Russians and the Bulgarians. This exclusive predominance of military and strategic considerations is mirrored in an act which indirectly resulted in the strengthening of dualistic heresy in Bulgaria: the Emperor John Tzimisces (969–76) transplanted new colonies of warlike and ferocious Paulicians from Armenia and the land of the Chalybes (to the north-west of Armenia along the Black Sea coast) to Thrace, where he settled them around the town of Philippopolis.[2] His motives were identical with those which had prompted Constantine Copronymus and Leo the Khazar to transplant Syrian and Armenian heretics to the same province in the eighth century.[3] According to Anna Comnena 'this he did firstly to drive them [i.e. the Paulicians] out of their strong cities and forts which they held as despotic rulers, and secondly to post them as trustworthy guards against the inroads of the Scythians by which the country of Thrace was

[1] V. Levitsky (Богомильство—болгарская ересь, loc. cit. p. 372) and R. Karolev (За Богомилството, loc. cit. p. 128) identify Basil, the disciple of Bogomil, with Basil who was the celebrated leader of the Byzantine Bogomils from c. 1070 to c. 1110 (cf. infra, p. 200). This identification is, however, not very probable, as the Byzantine heresiarch Basil is the object of a separate paragraph in the *Synodicon*. (Cf. infra, p. 240.)

[2] Anna Comnena, *Alexiad*, lib. xiv, C.S.H.B. vol. ii, p. 298: ἀλλὰ τούτους δὴ τοὺς ἀπὸ Μάνεντος καὶ Παύλου καὶ Ἰωάννου, τῶν τῆς Καλλινίκης, ἀγριωτέρους ὄντας τὰς γνώμας καὶ ὠμοὺς καὶ μέχρις αἵματος διακινδυνεύοντας ὁ ἐν βασιλεῦσιν ἐκεῖνος θαυμάσιος Ἰωάννης ὁ Τζιμισκῆς πολέμῳ νικήσας ἐξανδραποδισάμενος ἐκ τῆς Ἀσίας ἐκεῖθεν ἀπὸ τῶν Χαλύβων καὶ τῶν Ἀρμενιακῶν τόπων εἰς τὴν Θρᾴκην μετήνεγκε καὶ τὰ περὶ τὴν Φιλιππούπολιν αὐλίζεσθαι κατηνάγκασεν. Cf. Michael Glycas, *Annales*, C.S.H.B. p. 623.

[3] Cf. supra, pp. 60–1.

BOGOMILISM IN THE FIRST BULGARIAN EMPIRE 147

often oppressed; for the barbarians crossed the passes of the Haemus and overran the plains below'.[1] Zonaras[2] and Cedrenus[3] affirm that the transportation of the heretics was effected by the express demand of Theodore, patriarch of Antioch, who no doubt wished to rid his patriarchate of these unruly and corrupting elements. In the tenth century the Paulicians were undoubtedly a lesser danger for the Empire in Thrace, which was then, according to Cedrenus, a 'desolate borderland'.[4] But they infused new life into the local heretical communities which had lived there for over two centuries and thus indirectly strengthened the Paulicians and Bogomils of Bulgaria.

Together with Thrace, Macedonia was likewise laid open in the late tenth century to penetration by a new wave of Eastern immigrants. In 988-9, according to the Armenian historian Asoghic, the Emperor Basil II transported a large number of Armenians into Macedonia and settled them on the borders of the Empire, to guard against Bulgarian attacks; the colonists, however, dissatisfied with the rule of their Byzantine masters, rebelled and passed over to the Bulgarians.[5] It can be supposed that some of them were Paulicians and that they united with their co-religionists, who had found their way into Macedonia together with the Armenian merchants in the ninth century.[6] As it will be shown, Macedonia in the tenth century was the centre of opposition to the Bulgarian State and the refuge of all malcontents against the government, and thus a particularly fertile ground for all anti-ecclesiastical movements.

In the late tenth and early eleventh centuries the internal situation in Bulgaria and more particularly in Macedonia was exceptionally favourable to the growth of heresy. The war of 969-72, fought with great ferocity between Greeks and Russians

[1] Anna Comnena, *The Alexiad* (tr. by Elizabeth Dawes; London, 1928), p. 385.
[2] *Epitome Historiarum*, C.S.H.B. vol. III, pp. 521-2.
[3] *Historiarum Compendium*, C.S.H.B. vol. II, p. 382.
[4] Ibid.
[5] See F. Dulaurier, 'Chronique de Matthieu d'Édesse', *Bibliothèque historique arménienne* (Paris, 1858), p. 389; H. Gelzer and A. Burckhardt, *Des Stephanos von Taron armenische Geschichte* (Leipzig, 1907), p. 186. According to Asoghic, the future Bulgarian Tsar Samuel was among these Armenian colonists.
[6] Cf. supra, p. 82.

over the stricken corpse of Bulgaria, resulted in the complete triumph of the armies of John Tzimisces, the establishment of Byzantine domination over the whole of eastern Bulgaria, which became a mere province of the Empire, and the abolition of the independence of the Bulgarian Church. Western Bulgaria, however, escaped this catastrophe, owing to the independent attitude of a local provincial governor Nicholas, who, together with his sons, the 'Comitopuli', cut himself off from the jurisdiction of Preslav. One of these sons, Samuel, became sole ruler in 987 and tsar in 997 or 998,[1] and rapidly built up a new Bulgarian Empire in Macedonia. For some years, Samuel's Empire enjoyed great external power. He re-established the Bulgarian patriarchate, the seat of which, after several changes, was finally fixed in Ochrida. Practically his entire reign was spent in bitter wars against Byzantium and its brilliant Emperor Basil II, who earned the sinister title of 'Bulgaroctonus', the Bulgar-slayer. The struggle ended with Samuel's death in 1014, with the final defeat of the Bulgarian armies and Basil's systematic conquest of a devastated Macedonia. In 1018 Basil entered Ochrida, Samuel's capital, and the independence of Bulgaria was destroyed for 168 years.[2]

Both in eastern and western Bulgaria these military disasters resulted in a decline and demoralization in all spheres of human life. Everything was crumbling in Bulgaria at that time, the Church, the State, the monasteries. The ceaseless wars for half a century, with the resulting social instability and economic misery of which the *Sermon against the Heretics* paints such an eloquent picture, accentuated the state of inner unrest already prevalent in the reign of Peter. Samuel, for all his greatness as a military leader, had probably neither time nor opportunity for inner reform and his Empire collapsed from inner weakness as rapidly as it had risen.[3] The direction of all the energies of the State into an exhausting military struggle naturally weakened its

---

[1] The older view that Samuel came to the throne *c*. 980 (see Runciman, *A History of the First Bulgarian Empire*, p. 219) is refuted by N. Adontz, 'Samuel l'Arménien, roi des Bulgares', *Mém. Acad. Belg.* (cl. des lettres) (1938), t. XXXIX, pp. 5–35.

[2] See Zlatarski, ibid. pp. 600–790; Runciman, ibid. pp. 205–52.

[3] See N. P. Blagoev, Произходъ и характеръ на царь Самуиловата държава, *G.S.U.* (1925), vol. XX, p. 578.

power of resistance to the dissident forces within Bulgaria, the strongest of which was Bogomilism. Moreover, by establishing an independent Bulgarian patriarchate Samuel rebelled against the Byzantine Church; thus the national Church of Bulgaria, severed from and not recognized by Constantinople, was deprived of much strength and guidance necessary for the struggle against heresy.[1]

Although we possess no direct contemporary evidence of the development of Bogomilism in Samuel's Empire, it cannot be doubted that his reign witnessed a considerable growth of the sect in western Bulgaria.[2] For this Samuel himself is often held responsible. The lack of evidence of any measures taken against the heretics in his reign is sometimes considered as a proof of his toleration of the Bogomils.[3] Some scholars have even maintained that Samuel was sympathetic to Bogomilism or even under its influence.[4] It is not possible to decide on this matter with any

---

[1] It should be noted that the very same factors conducive to the growth of heresy existed in Bulgaria at the time of Symeon. (Cf. supra, pp. 90 et seq.)

[2] See M. Drinov, Исторически прѣгледъ на българската църква, p. 52; Levitsky, Богомильство—болгарская ересь, loc. cit. p. 391.

[3] See G. Schlumberger, *L'Épopée Byzantine à la fin du Xe siècle* (Paris, 1896), vol. I, p. 615; D. Mishew, *The Bulgarians in the Past* (Lausanne, 1919), p. 135.

[4] In particular Gilferding (op. cit. vol. I, pp. 195, 235), Levitsky (ibid.), Karolev (loc. cit. p. 121). They adduce the following arguments:

(1) Had Samuel shown any Orthodox zeal in his treatment of the Bogomils, Cosmas would not have passed him over in complete silence.

(2) Asoghic asserts that Samuel was of Armenian origin (cf. supra, p. 147, n. 5); hence he may have had connections with the Paulicians of Philippopolis.

(3) The Greek version of the *Life of Saint Vladimir of Dioclea*, who, according to contemporary sources, was married to Samuel's daughter Kosara and was later murdered by order of Samuel's nephew the Tsar John-Vladislav, states that John-Vladislav and his wife were Bogomils: οἱ ὁποῖοι, ὡς αἱρετικοί, βασ-τῶντες ταῖς ῥίζαις τῆς ἰοβόλου αἱρέσεως τῶν Βογομίλων καὶ Μασσαλιανῶν, ὅπου εἰκόνας δὲν ἤθελαν νὰ προσκυνοῦν, ἀλλὰ ἦτον εἰκονομάχοι καὶ ἐχθροὶ τοῦ σταυροῦ. Having been unable to consult this source, I quote the passage as printed in Gilferding, op. cit. p. 235, n. 1; cf. Rački, op. cit., *Rad*, VII, p. 109. It should be noted that, as the result of an error in the Greek version of the *Life of Saint Vladimir* which states that John-Vladislav was Kosara's brother when he was in fact her first cousin, Rački mistook John-Vladislav for Samuel's son Gabriel-Radomir. This led him to the false conclusion that Samuel's son and daughter-in-law belonged to the Bogomil sect. Rački's error is repeated by Klincharov (op. cit. p. 73) and by Lavrin (op. cit., *S.R.* 1929, vol. VIII, p. 278). In any case this passage of the *Life of Saint Vladimir* cannot refer to Gabriel-Radomir and

certainty, as the internal history of Samuel's reign is almost completely unknown. The existing historical evidence, however, seems to contradict the hypothesis of Samuel's sympathy for the heretics. The Byzantine chroniclers would hardly have omitted to record any suspicion of heresy against this dangerous enemy of the Empire, had any such suspicion existed. Instead, Cedrenus recounts that in 986, after his capture of Larissa, Samuel transferred the city's holy treasure, the relics of St Achilleus, to his capital at Prespa.[1] A Bulgarian monument tells us that in 993 he built a church in a Macedonian village in memory of his father, his mother and his brother, from which an inscription has been preserved, engraved with several crosses and an invocation to the Blessed Trinity.[2] These facts in themselves refute the

his wife, as the saint was murdered by order of John-Vladislav. (See Zlatarski, op. cit. pp. 760–5.)

However, none of these arguments are conclusive:

(1) If Cosmas was a contemporary of Samuel, his silence, nevertheless, is not sufficient proof that the tsar had leanings towards Bogomilism. Moreover, it is very probable that the *Sermon against the Heretics* was composed before Samuel's accession to power, which took place about 997. (See Appendix I.)

(2) The epithet Bogomil, applied by the *Life of Saint Vladimir* to John-Vladislav and his wife, is rejected by Zlatarski (ibid. p. 765, n. 4) as fictitious and unhistorical. Trifonov has shown (Бесѣдата на Козма пресвитера, loc. cit. pp. 49–52): (*a*) that the Greek *Life of Saint Vladimir* is based on a confusion between this saint and a much older one, many of whose characteristics have been falsely applied to the martyr prince of Dioclea; (*b*) that the Greek version is a rather inaccurate translation of a Slavonic original. This original, which also confuses the two saints, mentions neither Bogomils nor Massalians, but states that the saint (who, in this case, is obviously not St Vladimir) was murdered by Novatians.

(3) The very relationship between Samuel and St Vladimir is of doubtful historicity. According to Adontz (*Samuel l'Arménien*, loc. cit. pp. 51–63) the marriage between Vladimir and Samuel's younger daughter (whose name, it seems, was not Kosara but Theodora) is a pious invention, based on the marriage which did take place between Samuel's eldest daughter Miroslava and the Armenian Prince Ashot.

The assertion that Bogomilism penetrated into the family of the Tsar Samuel is thus incorrect.

Samuel's Armenian origin is denied by some Bulgarian historians (see Blagoev, loc. cit. pp. 521–9), but is upheld by Adontz (loc. cit. pp. 36–50).

[1] Cedrenus, *Historiarum Compendium*, vol. II, p. 436.
[2] See I. Ivanov, Български старини изъ Македония (2nd ed.; Sofia, 1931), pp. 23–5; Adontz, *Samuel l'Arménien*, loc. cit. pp. 40–1.

hypothesis of Samuel's alleged 'indifference to Orthodoxy',[1] for the Bogomils rejected relics, crosses and churches and held unorthodox views on the Trinity. Thus Samuel's Orthodoxy seems established beyond doubt.

And yet in spite of this undeniable evidence, there remains attached to the name of Samuel a lingering suspicion of heterodoxy. It is significant that whereas the great Orthodox Tsars Boris, Symeon and Peter are frequently glorified in monuments of Bulgarian literature and have become the object of a national cult, Samuel's name is almost entirely absent from Bulgarian Orthodox literature and is always surrounded by a veil of reserve.[2] The explanation of this fact may well lie in the position of the Bulgarian Church in Samuel's Macedonian Empire. By refusing to recognize the abolition of the old Bulgarian patriarchate, decreed by John Tzimisces, and by setting up a patriarch of his own, Samuel had severed all relations with the Oecumenical See. In the eyes of the Byzantine Church and State, he always remained a rebel. Within his own Empire Samuel was obliged to pursue a policy which was essentially nationalistic, both ecclesiastically and politically; to be successful this policy required the collaboration of all parties and groups in Bulgaria in the pursuit of one aim, namely the destruction of the Byzantine power and of its domination over the Balkan Slavs. This collaboration, in its turn, presupposed a state of inner equilibrium, and it is understandable that Samuel could not afford to alienate the Bogomils, who at that time must have represented a notable proportion of his subjects. This probable toleration of the heretics for political motives may well have given rise to a popular legend associating the Tsar Samuel with Bogomilism.

The tenth-century sources do not explicitly show which region or regions of Bulgaria can be considered as the original home of the Bogomil sect. Nevertheless, from the combined evidence of geographical factors, of indirect historical data and of later sources, which must now be examined, it is possible to prove that the cradle and subsequent stronghold of Bogomilism in the Balkans was Macedonia.

The very geography of Macedonia made the country a most

[1] Levitsky, ibid.
[2] See Gilferding, op. cit. p. 236.

favourable ground for the spread and consolidation of heresy. A wild land of high lakes and valleys, dotted in the Middle Ages with a number of well-defended fortresses, it is surrounded on three sides by high mountains. Farther to the east, the high range of the Rhodope Mountains, impassable save for a few narrow defiles,[1] forms a second barrier between Macedonia and central and eastern Bulgaria. Thus, at the time of the appearance of Bogomilism in the reign of Peter, Macedonia, forming the furthermost western province of the Bulgarian Empire, was of very difficult access for the central ecclesiastical authorities who resided in Preslav, in the extreme east of the country. Hence the possibility of any large-scale repression of heresy in Macedonia at that time was remote, which lends some justification to Zlatarski's statement that Macedonia was 'for many centuries the principal centre and nursery of all heresy in the Balkan peninsula'.[2] The geographical isolation of Macedonia from the rest of Bulgaria made this region the centre of political opposition to Peter's government in Preslav.[3] After Peter's death the separatist movement led by the Comitopuli, which resulted in the creation of an independent Empire of western Bulgaria, also originated in Macedonia.

Moreover, the Rhodope Mountains separated Macedonia not only from the ecclesiastical authorities of Preslav, but also from the Paulicians of Thrace. Had Bogomilism developed in immediate and permanent contact with Paulicianism, it would probably have rapidly merged into the latter heresy, in view of the close similarities which existed between the doctrines of both sects. But if the origin of Bogomilism is placed in Macedonia, it becomes understandable why the Bogomils, separated from the Paulicians by a geographical barrier which could not easily be overcome, while being undoubtedly influenced by the doctrines of the latter, nevertheless developed some of their teachings and practices in different and sometimes opposite directions. The possibility of some Paulician missionaries working their way from the plains of Thrace across the high and wild mountain ranges

---

[1] The most famous of these was the pass of Cimbalongus, on the western side of the Rhodope range, where Samuel's army was trapped and routed by Basil II in 1014.

[2] История, vol. 1, pt 2, p. 65.   [3] See supra, p. 109.

into the distant heart of Macedonia cannot, of course, be excluded, especially as they were notorious for their courageous and enterprising proselytism. But it seems more likely that the Paulician influence on Bogomilism was due not so much to the proselytism of the Thracian heretics as to the presence in Macedonia of missionaries, merchants and colonists who came directly from Armenia, probably already in the ninth century, if not earlier still.[1] Geographical conditions undoubtedly facilitated their penetration into Macedonia; for this province, though encompassed by mountains on three sides, is easily accessible from the south-east. From Thessalonica, a great Armenian centre in the Middle Ages, the Paulicians could take either of the two routes leading into Macedonia: the ancient trade-route from Thessalonica up the Vardar to Niš and Belgrade, or the famous Roman *Via Egnatia* connecting Constantinople with Rome through Thessalonica, the Macedonian towns of Vodena, Bitolj and Ochrida, and Dyrrhachium (present-day Durazzo) on the Adriatic.[2]

Apart from these geographical factors, the rise of Bogomilism in Macedonia was furthered by historical circumstances. After the conquest of eastern Bulgaria by the Byzantines (972), Bulgarian national life became centred in Samuel's Empire of Macedonia. The religious and political malcontents, who had taken refuge in Macedonia from Peter's government, now found themselves in proximity to the Bulgarian central authorities. It has already been shown that Samuel's nationalistic policy, which required the conciliation of these elements, favoured the growth of the Bogomil sect.[3] The absence of State persecution goes far to explain the fact that Bogomilism developed into a powerful force in Macedonia in the second half of the tenth century.[4]

[1] Cf. supra, pp. 82, 147.
[2] For the *Via Egnatia*, see G. L. F. Tafel, *De Via Romanorum militari Egnatia, qua Illyricum, Macedonia et Thracia iungebantur* (Tübingen, 1837); K. Miller, *Itineraria Romana: Römische Reisewege an der Hand der Tabula Peutingeriana* (Stuttgart, 1916), cols. 516–27; M. P. Charlesworth, *Trade-Routes and Commerce of the Roman Empire*, 2nd ed. (Cambridge, 1926), p. 115.   [3] Cf. supra, p. 151.
[4] Trifonov's opinion that Cosmas conducted his polemical activities against the Bogomils in Macedonia (Бесѣдата на Козма Пресвитера, loc. cit. pp. 44–7) has received a serious setback by Vaillant's proofs of the east Bulgarian features displayed by Cosmas's language (Puech, op. cit. pp. 37 et seq.).

154   THE BOGOMILS

Finally, certain internal features of Bogomilism confirm the hypothesis that its cradle in the tenth century was Macedonia. Macedonia, as it has been pointed out, was the centre of the apostolic activity of St Clement. Although in an immediate sense this activity resulted in a temporary weakening of heresy at the beginning of the century, yet after the death of St Clement the failure of his Macedonian school to lay the foundations of a lasting Slavonic literature and the growth of religious nationalism due to a misuse of his legacy sowed the seeds of heresy in Bulgaria.[1] There is in some respects a curious similarity between the school of St Clement and Bogomilism: both were Slavonic and popular movements which were drawing at the same source, namely the Slavonic vernacular; Bogomilism, moreover, largely succeeded in achieving that which St Clement's school could have accomplished within the framework of Orthodoxy: it produced a popular religious literature answering to the interests and requirements of the masses.[2] This heretical literature, in which dualistic doctrines

[1] Cf. supra, pp. 90–2.

[2] The most complete study of Bogomil literature is that of Ivanov (Богомилски книги и легенди). An allusion to Bogomil stories can already be found in the *Sermon against the Heretics*: Cosmas accuses the heretics of 'babbling certain fables' (бающе никакы басни) (p. 22). Ivanov has shown that the so-called 'Bogomil books' are of two types: (1) those which are distinctly Bogomil in character and which contain doctrines held by the Bogomils—these formed the sectarian canon of Scriptures; (2) certain apocryphal writings, generally of Christian origin, but either interpreted or modified by the Bogomils in accordance with the views they professed and used for the purpose of proselytism (op. cit. pp. 54–9). The exact time when these Bogomil books appeared is unknown, but, according to Ivanov, the majority were compiled in the eleventh and twelfth centuries, or possibly earlier still.

To the first category belong the *Secret Book* or *Liber Sancti Johannis*, which was brought from Bulgaria to the Italian Patarenes in the twelfth century, and the legend of *The Sea of Tiberias*; to the second category belong the Old-Bulgarian versions of the *Vision of Isaiah*, the *Book of Enoch*, the *Apocalypse of Baruch*, the *Elucidarium*, the *Story of Adam and Eve*, the *Gospel of St Thomas*, etc. But, as Puech has pointed out (op. cit. pp. 129–31), most of these writings are either of non-Bogomil origin or are mixed with popular cosmogonical legends and tales of recent date, and cannot hence be properly regarded as primary sources for a study of Bogomilism. The only one which provides an authentic and reliable guide to some of the Bogomil doctrines is the *Liber Sancti Johannis*, of which an analysis will be given below (pp. 226–8).

The common factor underlying all Bogomil books is their dualistic cosmology: in one form or other they all contain the belief that the visible world is the creation of the evil principle, that the Universe witnesses a struggle between

are interwoven with apocryphal Old and New Testament stories and folk-lore, enjoyed a considerable vogue in the Middle Ages. This was particularly true of its cosmological and eschatological elements, which, despite the frequent prohibitions of the Church, were often regarded by the less educated as illustrations of the Orthodox canon of the Scriptures. The Bogomil preachers, whose appeal was primarily to the masses, adapted many of these written and oral productions to their own teaching and greatly contributed to the spread of these 'Bulgarian fables' in the Balkans and beyond. The fact that the Bogomils proselytized so successfully among the people largely explains the fact that many of these 'legends' have found their way into Bulgarian folk-songs and popular legends.[1] This analogy between Bogomilism and the Slavonic school of St Clement has given rise to the hypothesis that the former grew out of the latter.[2] It is indeed not improbable that there were points of contact between the two. Bogomilism may well have recruited many of its adherents from among those pupils of St Clement who were insufficiently grounded in Christianity and who fell away from Orthodoxy and imbibed dualistic

the Divine and the Satanic, which is destined to end in the victory of God, and that man, as a microcosm expressing this dualism, must pursue this struggle within himself. They afford a good illustration of the attempt of the *pop* Bogomil and of his followers to bring a dualistic creed into harmony with Christianity. The success of the Bogomil literature in Bulgaria is shown by its intimate connection with south Slavonic folk-lore; traces of this connection remain to the present day.

Several references to the most important of these books which reflect some basic features of Bogomilism are made in the course of this work.

An account of the Bogomil literature can also be found in the following works: F. Rački, 'Bogomili i Patareni,' *Rad*, x, pp. 230 et seq.; V. Jagić, История сербско-хорватской литературы, pp. 95 et seq.; A. N. Pypin and V. D. Spasovich, История славянских литератур, pp. 75–81; M. Gaster, *Ilchester Lectures on Greeko-Slavonic Literature* (London, 1887), pp. 17 et seq.; I. Broz, *Crtice iz hrvatske književnosti* (Zagreb, 1888), vol. II, pp. 153 et seq.; M. Murko, *Geschichte der älteren südslawischen Literaturen* (Leipzig, 1908), pp. 82 et seq.; D. Tsukhlev, История на българската църква, pp. 708 et seq.; D. Prohaska, *Das kroatisch-serbische Schrifttum in Bosnien und der Herzegowina* (Zagreb, 1911), pp. 37 et seq.; cf. P. Kemp, *Healing Ritual*, pp. 159 et seq.

[1] See Ivanov, op. cit. pp. 327–82. The term 'Bulgarian fables' is of Russian origin: the Balkan Bogomils were probably responsible for transmitting Bulgarian dualistic legends to Russia, where they were very popular in the Middle Ages. Cf. infra, pp. 281–2.

[2] B. Petranović, Богомили, p. 44, n. 1.

doctrines.[1] It is perhaps significant that the centre of St Clement's activity in Macedonia, the district of Debritsa to the north-east of Ochrida, was also the stronghold of Bogomilism in the tenth century.[2]

Macedonia was, moreover, the great centre of monasteries in the tenth century;[3] that it was also the cradle of Bogomilism is suggested by the contacts between Bogomilism and monasticism, indicated by Cosmas[4] and by later sources.

These general geographical and historical considerations are confirmed by the evidence of Latin sources of the twelfth and thirteenth centuries. Although these sources are of a later date, they unanimously testify to the predominance of Bogomilism in Macedonia and can hence be also used as an indirect proof that this province was the original home of the sect.

In 1167, the leader of the dualistic heretics of Constantinople, the celebrated 'bishop' Nicetas, presiding over a council of the Cathars at Saint-Félix de Caraman near Toulouse, mentioned among the heretical communities of the Balkans the 'Ecclesia Melenguiae'.[5] It is generally considered that this name is the

[1] The same process occurred in the development of heretical monasticism in tenth-century Bulgaria (cf. supra, p. 105).

[2] See G. Balaschev, Климентъ епископъ словѣнски и службата му по старъ словѣнски прѣводъ (Sofia, 1898), p. xxxiv. Ivanov has shown, moreover, that a large number of Old-Bulgarian apocrypha, including the *Vision of Isaiah* which was used by the Bogomils, originated in north-eastern Macedonia, i.e. not far from the centre of St Clement's activity. (Op. cit. pp. 163-4.)

Ivanov has also brought to light an interesting piece of evidence showing the later presence of Bogomilism in Ochrida, the centre of St Clement's diocese. He describes an icon he saw in the cathedral church of St Clement in Ochrida, on which are depicted the miracles wrought by St Naum, who succeeded to St Clement as bishop of Ochrida: St Naum, he writes, 'is represented in a mountainous region, pursued by Bogomils—διώκεται ὑπὸ τῶν Βογομηλον (sic)'. (Ibid. p. 34, n. 1.) This episode is clearly anachronistic and apocryphal, as the Bogomil sect had not yet arisen at the time of St Naum, but it shows that Bogomilism penetrated into the region of Ochrida at a later date. Unfortunately, Ivanov does not attempt to determine the date of this icon, beyond describing it as old.

[3] Cf. supra, p. 102. [4] Cf. supra, p. 104.

[5] See M. Bouquet, *Recueil des historiens des Gaules et de la France* (Paris, 1806), vol. XIV, p. 449. In this document the heretical bishop is called Niquinta; but N. Vignier (*Recueil de l'histoire de l'Égliſe*, Leyden, 1601, p. 268) refers to him as Nicetas. As Schmidt has pointed out (*Histoire et doctrine de la secte des Cathares ou Albigeois*, vol. I, p. 57, n. 3) there is no doubt that 'Niquinta' is a corrupted form of the Greek name Νικήτας.

Latin form of the Slavonic Melnik, the name of a town in eastern Macedonia in the Struma valley, on the western slopes of the Rhodope Mountains, and that the 'Ecclesia Melenguiae' refers to the Bogomil community of Melnik.[1]

The Dominican inquisitor Reinerius Sacchoni, who had himself for seventeen years been a heretical Patarene teacher in Lombardy, in his *Summa de Catharis et Leonistis*, written about 1250,[2] gives a list of sixteen heretical 'Churches' or communities of the Cathars, at least five of which were situated in the Balkans. The last two 'Churches' in Reinerius's list, the 'Ecclesia Bulgariae' and the 'Ecclesia Dugunthiae', were considered in his time to be the original source of all the others:[3]

'*Quot sunt ecclesiae Catharorum.* Sunt autem XVI omnes ecclesiae Catharorum; nec imputes mihi lector quod eas nominavi ecclesias, sed potius eis, quia ita se vocant. Ecclesia Albanensium vel de Donnezacho, E. de Concorrezo, E. Bajolensium, sive de Bajolo, E. Vincentina, sive de Marchia, E. Florentina, E. de Valle Spoletana, E. Franciae, E. Tolosana, E. Carcassonensis, E. Albigensis, E. Sclavoniae, E. Latinorum de Constantinopoli, E. Graecorum ibidem, E. Philadelphiae in Romania,[4] E. Burgaliae,[5] E. Dugunthiae,[6] et omnes habuerunt originem de duabus ultimis.'

The location of the last six of these heretical communities directly concerns the problem of the origin of Bogomilism. Reinerius himself gives no indication on this subject and limits himself to the vague statement: 'sunt omnis gentis.' More recent scholars, however, have attempted with some success to determine the whereabouts of these heretical 'Churches'.

[1] See J. K. L. Gieseler, 'Über die Verbreitung christlich-dualistischer Lehrbegriffe unter den Slaven', *Theologische Studien und Kritiken* (1837), p. 365; Schmidt, ibid, n. 5; Rački, op. cit., *Rad*, VII, p. 118. The significance of the term 'Ecclesia', applied to the Bogomil communities by Latin writers in the twelfth and thirteenth centuries, is discussed below. (See infra, pp. 243-4.)

[2] Further references will be made to this work of Reinerius Sacchoni, which provides valuable information on the organization of the Bogomil communities. (See infra, pp. 242 et seq.)

[3] For the various editions of the *Summa de Catharis et Leonistis*, see Schmidt, op. cit. vol. II, pp. 310-11. The above text is quoted from E. Martène et U. Durand, *Thesaurus novus anecdotorum* (Paris, 1717), vol. v, col. 1767; the variants in the following notes from the *Maxima Bibliotheca veterum Patrum* (ed. M. de La Bigne; Lugduni, 1677), vol. xxv, p. 269. Cf. the new edition in A. Dondaine, *Un traité néo-manichéen du XIIIe siècle*, Rome, 1939, pp. 67-78.

[4] 'Philadelphiae Romaniolae': variant.
[5] 'Bulgariae': variant.
[6] 'Dugranicae': variant.

Schmidt and especially Rački have shown conclusively that the 'Ecclesia Sclavoniae' was situated in Bosnia.[1] The 'Ecclesia Latinorum de Constantinopoli', which must have arisen as a result of the Fourth Crusade and the establishment of the Latin Empire of Constantinople (1204), was doubtless founded by those Cathars who had come to Byzantium with the crusading army.[2] The Greek heretical 'Church' of Constantinople, to which Nicetas himself belonged, was probably, as it will be shown,[3] the Byzantine Paulician sect.

The exact whereabouts of the last three communities mentioned by Reinerius, 'E. Philadelphiae in Romania', 'E. Bulgariae'[4] and 'E. Dugunthiae' (Dugranicae), particularly important for the problem of the origin of Bulgarian Bogomilism, have presented a long-standing puzzle to Slavonic historians. The first satisfactory explanation of the name 'E. Dugunthiae' (Dugranicae)[5] was offered by P. J. Šafařík, who related it to Δραγοβιτία, the Greek name for the country inhabited by the Balkan Slavonic tribe of Dragovichi or Dregovichi; this tribe lived in two different places: in Macedonia, to the north-west and west of Thessalonica, and in Thrace, along the River Dragovitsa, not far from Philippopolis.[6] Both these branches are mentioned in Byzantine sources.[7] Šafařík's solution

---

[1] *Histoire et doctrine de la secte des Cathares*, vol. I, p. 57; 'Bogomili i Patareni,' *Rad*, VII, pp. 162-3.

[2] Rački, loc. cit. p. 162.   [3] Cf. infra, pp. 161-2.

[4] 'Bulgariae' and not 'Burgaliae' is clearly the correct form. See E. Golubinsky, Краткий очерк истории православных церквей, p. 707.

[5] In Latin sources this name appears under different forms, including 'Drogometia' and 'Druguria'. See Schmidt, op. cit. vol. I, pp. 15-16, n.

[6] *Slovanské Starožitnosti*, pp. 619, 623; *Památky hlaholského písemnictví* (Prague, 1853), p. lx, přímének II: 'Bylit pak dvoji: jedni v Macedonii s městem Velici, sousedé Sakulatův, Berzitův, Runchinův...; druzí v Thracii na říčce Dragovici nedaleko Tatar-Bazarčíku a Filipopole, kdež Byzantinci Pavlikiany uměstují.' Cf. L. Niederle, *Slovanské Starožitnosti*, pt II, vol. I, pp. 424-5; G. L. F. Tafel, *De Thessalonica*, pp. lxxvii, 59, 252; F. Dvorník, *Les Slaves, Byzance et Rome*, p. 14.

[7] The bishop of the Macedonian Dragovichi, subordinate to the metropolitan of Thessalonica, took part in the deliberations of the Council of Constantinople in 879. In its records he is given the name of ὁ Δρουγουβιτείας. See M. Le Quien, *Oriens Christianus* (Paris, 1740), vol. II, cols. 95-6. Cf. P. J. Šafařík, *Slov. Star.* p. 623; *Pam. hlah. písem.* ibid.; Rački, loc. cit. p. 104. Moreover, the metropolitan of Philippopolis held in the Middle Ages the title of Exarch Θράκης Δραγουβιτίας. See G. Rhalles and M. Potles, Σύνταγμα τῶν θείων κανόνων, vol. v, p. 516; *Pam. hlah. písem.* ibid.

is generally accepted to-day, and Slavonic scholars have followed his principle that the origin of the Bogomil sect is to be sought in the regions inhabited by the Dragovichi, in Macedonia and Thrace.[1]

Šafařík, however, was not specifically concerned with the problem of the origin of Bogomilism and did not raise the question whether the 'Ecclesia Dugunthiae' was centred in Thrace or in Macedonia. This has been done by more recent Slavonic scholars. The results of some of their investigations must be examined here, since they permit of a more accurate delimitation of the regions occupied by the 'Ecclesia Dugunthiae', the 'Ecclesia Bulgariae' and the 'Ecclesia Philadelphiae in Romania', and thereby confirm the theory that the original cradle of Bogomilism was Macedonia.

According to Rački, the term 'Romania' was used by the medieval writers of western Europe to designate the province of Old Thrace, which was later called 'Rumili' by the Turks. Consequently Philadelphia was a town in Old Thrace. But since there is no evidence of a town of that name in Thrace, Rački supposed that 'Philadelphia' stood for Philippopolis and that the name had been borrowed from the Paulicians by the local Bogomil community. Accepting Šafařík's conclusions regarding the derivation of 'Dugunthia', and denying that the 'Ecclesia Dugunthiae' was situated in Thrace (since Thrace was, according to him, the centre of the 'Ecclesia Philadelphiae in Romania'), Rački was led to place the 'Ecclesia Dugunthiae' in the second region in-

[1] 'Kolébka Katharův čili Patarenův a Bohomilův tu v Dragovičích, v Macedonii a Thracii, nikdež jinde, pokládati se musí' (*Pam. hlah. písem.* ibid.). Schmidt, however, who was not conversant with Slavonic languages, derives the name Dugunthia, in its other form Druguria, from the Latin Tragurium, of which the Slavonic equivalent is Trogir, the name of a town on the Dalmatian coast (op. cit. pp. 15–16, 57–8). Most Slavonic scholars disagree with Schmidt on this point. Šafařík himself, whom Schmidt quotes as an authority, criticizes his interpretation (*Pam. hlah. písem.* ibid.). Petranović (Богомили, p. 95) strongly attacks it; according to him, there was never any community of dualistic heretics in Dalmatia and the cases of heresy were individual and scattered. However, Petranović appears to be mistaken on this last point, as the heretical Bishop Nicetas mentions in 1167 an 'Ecclesia Dalmatiae' (see Bouquet, ibid.). Nevertheless, Schmidt's etymology of Trogir seems unacceptable for philological and historical reasons. (See Gilferding, op. cit. vol. I, p. 133, n. 2.)

habited by the Dragovichi, i.e. in Macedonia. As for the 'Ecclesia Bulgariae', it could only be situated, according to this theory, in eastern Bulgaria.[1]

The weakness of Rački's argument, which lies in the arbitrary identification of Philadelphia with Philippopolis, was exposed by N. Filipov, who, at the same time, offered a fairly convincing solution of the problem.[2] Although the name 'Romania' was applied from early times to Thrace, particularly by the Slavs,[3] Filipov rightly points out that between 1204 and 1261 it served to designate the whole of the Latin Empire of Constantinople. Reinerius Sacchoni, who wrote in the period between these two dates, undoubtedly used the term 'Romania' in this wider sense. Filipov places 'Philadelphia' in Asia Minor, where two towns of that name existed in the Middle Ages, without, however, attempting any precise determination of its situation.[4] He shows, furthermore, that the 'Ecclesia Bulgariae' cannot rightly be placed in eastern Bulgaria: the term Bulgaria was not always used in the Middle Ages in the same sense as it is to-day. The 'Ecclesia Bulgariae' must have been organized at the very latest in the beginning of the twelfth century (and probably earlier still), since it is mentioned about the middle of that century in Latin sources. Now in the eleventh and twelfth centuries, the name Bulgaria was applied by the Byzantines to the regions which before 1018 had formed the Empire of Samuel, and after that date constituted, according to the administrative system introduced by Basil II, the Theme of Bulgaria, i.e. approximately present-day Macedonia, with Skoplje as the capital.[5] North-eastern Bulgaria formed the other Theme, of Paristrium. So ingrained was this terminology that when eastern Bulgaria revolted against the Empire in 1186, the Byzantine historians refer to this region not as Bulgaria, but as Moesia or

[1] Loc. cit. pp. 104–5.
[2] Произходъ и същность на богомилството, *B.I.B.* vol. III (1929), pp. 48–50.   [3] See K. Jireček, *Geschichte der Bulgaren*, p. 378.
[4] Tafel mentions two towns of Philadelphia in Asia Minor, the one in Isauria, the other in Lydia ('Symbolarum criticarum, geographiam Byzantinam spectantium, partes duae', *Abh. bayer. Akad. Wiss.* (hist. Kl.) (München, 1849), vol. V, Abt. 2, pars I, pp. 101–2).
[5] See Zlatarski, op. cit. p. 643, n. 1. The name Macedonia, on the other hand, was generally applied by the Byzantines to present-day Thrace. See M. S. Drinov, Южные славяне и Византия, pp. 102–3.

Wallachia.[1] However, in the thirteenth century, after the establishment of the Bulgarian Empire of Trnovo, the name Bulgaria was once more applied to the eastern provinces and thus regained the significance it had in the days of Boris, Symeon and Peter. But in western Europe it still remained traditionally attached to the western provinces of Bulgaria, and particularly to Macedonia. It is thus most likely that Reinerius Sacchoni, who wrote in the middle of the thirteenth century, used the term 'Ecclesia Bulgariae' to designate the Bogomil communities of Macedonia. As for the 'Ecclesia Dugunthiae', Filipov, contrary to Rački, places it in Thrace, the earliest centre of dualistic heresy in the Balkans.

Filipov's location of the 'Ecclesia Dugunthiae' in Thrace and the 'Ecclesia Bulgariae' in Macedonia, though not accepted by all Slavonic scholars,[2] seems the most satisfactory interpretation of Reinerius's text, as well as the most consonant with Slavonic and Latin sources. The very name 'Ecclesia Bulgariae', suggestive of a specifically Bulgaro-Slavonic community, points to the Bogomil sect; for Bulgarian Bogomilism was essentially a Slavonic movement, in contrast to Paulicianism, which retained some foreign elements.[3] Moreover, the location of the 'Ecclesia Dugunthiae' in Thrace is all the more convincing as Reinerius testifies to its antiquity, and as we know, on the other hand, that Thrace was the home of the earliest Balkan dualists, the Paulicians. It is hence legitimate to suppose that the 'Ecclesia Bulgariae' refers to the Bogomil sect in Macedonia and the 'Ecclesia Dugunthiae' to the Paulician communities of Thrace, and perhaps also to a later, and local, evolution of Bogomilism under the influence of Paulicianism. The latter possibility is not improbable, in view of the testimony of Anna Comnena that at the beginning of the twelfth century the Bogomils lived in Philippopolis alongside the Paulicians.[4]

Filipov's hypothesis is confirmed by the fact, which emerges from Latin sources, that in the twelfth and thirteenth centuries the 'Ecclesia Dugunthiae' and the 'Ecclesia Bulgariae' were

---

[1] See V. N. Zlatarski, *Geschichte der Bulgaren*, vol. I, p. 94.
[2] Golubinsky, it is true, considers that the 'Ecclesia Dugunthiae' was the heretical community of Philippopolis (op. cit. p. 707). But Gilferding (loc. cit. p. 133, n. 2) locates it in Macedonia, while Ivanov (op. cit. pp. 21–2) places the 'Ecclesia Bulgariae' 'mainly in northern Bulgaria'.
[3] Cf. supra, p. 144.   [4] Cf. infra, p. 189.

divided on a fundamental question of doctrine: the former was said to adhere to 'absolute dualism', while the latter adopted a 'mitigated dualism'; thus there appeared an 'Ordo de Dugrutia' and an 'Ordo de Bulgaria'.[1] The same schism occurred among the dualistic heretics of western Europe, the Italian Patarenes and the French Cathars: those of them who adhered to the doctrines of 'mitigated dualism' looked to Bulgaria as the source of their teaching, while the 'absolute' dualists were confirmed in their faith by Nicetas of Constantinople.[2] It is significant that the tenth-century Slavonic and Byzantine sources show the same difference between the teachings of the Bogomils and the Paulicians in Bulgaria. Without explicitly mentioning any schism between the dualistic sects in tenth-century Bulgaria, these sources reveal the fact that the Paulicians formulated their dualism in terms of two principles, the one good, the other evil, whereas the Bogomils never held a cosmological dualism in this 'absolute' sense, but 'mitigated' it by teaching that the creator of this world, the Devil or Satanael, generally considered to be a fallen angel, was dependent on and ultimately subordinate to God.[3] Thus the doctrinal distinction between the 'Ordo de Dugrutia' and the 'Ordo de Bulgaria' corresponds to the difference between the teachings of the Paulicians and the Bogomils in Bulgaria.[4]

[1] 'Haeretici, qui habent ordinem suum de Dugrutia..., credunt et praedicant...duos Dominos esse sine principio, et sine fine': Bonacursus, *Contra Catharos*, in S. Baluzius, *Miscellanea* (ed. Mansi), vol. II, p. 581; cf. Vignier, op. cit. p. 268; Schmidt, op. cit. vol. I, p. 58; Rački, loc. cit. pp. 120 et seq.
[2] See Schmidt, op. cit. vol. I, pp. 59, 61–2, 73–4; Rački, ibid.
[3] Cf. supra, pp. 123–4.
[4] A further proof of the truth of Filipov's theory is supplied by the following facts: (1) It is known that the teaching of the 'Ecclesia Dugunthiae' penetrated in the twelfth century into Constantinople where Nicetas was its most celebrated exponent. This influence is more likely to have originated from neighbouring Thrace than from more distant Macedonia. (2) On the other hand, Bogomilism, in its Bulgarian form, penetrated from Macedonia into Serbia in the twelfth century. (See Appendix IV.)

Runciman thinks that the 'Dragovitsan' and the 'Bulgarian' churches corresponded respectively to the original, completely dualist, and later, less rigidly dualist, trends within the Bogomil sect, and that a 'great schism of the Bogomils' occurred during the period of Byzantine rule in Bulgaria (op. cit. p. 69). But it is surely more satisfactory to explain the distinction between the 'absolute' and 'mitigated' forms of dualism with reference to the well-known differences between the Paulician and the Bogomil doctrines, rather

BOGOMILISM IN THE FIRST BULGARIAN EMPIRE 163

Attempts to ascribe a Slavonic origin to the other heretical 'Churches' mentioned by Reinerius Sacchoni have not yielded convincing results.[1]

Besides these Latin sources,[2] several Slavonic documents show

than to rely on the gratuitous hypothesis of a schism within the Bogomil community. Moreover, I cannot see that there is any real evidence to support Runciman's assertion that 'Bogomil founded the Dragovitsan church, while the later Bogomil church was founded by Jeremiah' (op. cit. p. 91). The identity and role of Jeremiah are discussed in Appendix II.

[1] Petranović made strenuous efforts to discover Slavonic roots in some of the names in Reinerius's list (cf. supra, p. 157) with a view to finding other centres of Bogomilism in the Balkans (op. cit. pp. 79–82). In his attempt to prove his case, he resorted to somewhat forced etymology. Thus in the 'Concorezenses' he saw the inhabitants of the village of Goritsa, in southern Thessaly. 'Bagnaroli' (a variant of 'Bajolenses') is, according to him, a literal translation of the Slavonic 'Polivaki', the name of a Bulgarian tribe to the north-west of Vodena, in south-eastern Macedonia. The name 'Runcarii', interpolated into the MS. of Reinerius (see Schmidt, op. cit. vol. II, pp. 283–4) and which belonged to a thirteenth-century German sect, corresponds, in his opinion, to the Slavonic 'Renjdane' (Greek 'Ρεντίνοι); this Slavonic tribe has left its name to the gulf of Rendina, to the north of Mount Athos. These conclusions have led Petranović to place the cradle of Bogomilism in southern Macedonia, in the region limited by the Lake of Ochrida, Janina, the Plain of Thessaly and Thessalonica. Petranović's ingenious hypotheses are not only philologically unproved, but are unacceptable from the historical point of view. As Levitsky has pointed out, there is no evidence for maintaining that Bogomilism was spread so far south in the early period of its existence (Богомильство—болгарская ересь, loc. cit. p. 371, n.). Moreover, even a superficial study of Reinerius's list shows that he begins his enumeration with the Italian 'Churches', continues with the French and ends with the Slavonic and Byzantine ones. Cf. Rački, loc. cit. pp. 161–2, n.

[2] It is generally thought that the anonymous author of the *Gesta Francorum* referred to the Bogomils of Macedonia in his account of the First Crusade. In 1096, he tells us, the Normans of Bohemond of Taranto, coming from Kastoria in southern Macedonia, encountered in the region of Pelagonia (round Bitolj and Prilep) a fortified town inhabited by heretics ('quoddam hereticorum castrum'). They took the city by storm, and proceeded to burn it together with its inhabitants: *Anonymi Gesta Francorum et aliorum Hierosolymitanorum* (ed. B. A. Lees; Oxford, 1924), p. 8. (See notes in *Histoire anonyme de la Première Croisade*, éd. et trad. par L. Bréhier, Paris, 1924, pp. 22–3.) The usual view is that these heretics were Bogomils (see Rački, op. cit. vol. VII, p. 118; Trifonov, loc. cit. pp. 52–3), and since Pelagonia was situated in the home of Bogomilism (cf. infra, pp. 164–6) this interpretation would appear legitimate. And yet the behaviour of these heretics, who defended their city against military aggression, is characteristic not of the Bogomils, to whom any recourse to arms was abhorrent (cf. infra, pp. 182, 190) but of the Paulicians. There were Paulicians in Macedonia probably by the ninth century (cf. supra, pp. 147, 153), and

that Bogomilism was prevalent in Macedonia in the twelfth and thirteenth centuries and thus confirm the conclusion that the cradle of the sect was situated in this province.

The *Life of Saint Hilarion of Moglena* by Euthymius, Patriarch of Trnovo, shows that Bogomilism was rampant in the diocese of Moglena, a town in the valley of the lower Vardar in southeastern Macedonia, in the middle of the twelfth century.[1] The early thirteenth-century Bulgarian *Synodicon of the Tsar Boril* curses the Bogomil 'dyed of Sredets'; this can be taken as a proof that an organized Bogomil community existed by that time in Sredets (present-day Sofia), not far from the northern borders of Macedonia.[2]

Finally, important evidence for the location of centres of Bogomilism in Macedonia is supplied by place-names, many of which retain to the present day roots derived from the several names under which the Bogomils were known in the Middle Ages. The most common of these is the root *babun*, which is synonymous with Bogomil. A Slavonic Nomocanon of 1262 contains a section 'concerning the Massalians, who are now called Bogomils, Babuns'.[3] The word *babun* is derived by the celebrated Croatian scholar G. Daničić from *baba*, meaning 'an old woman'. According to him, the Bogomils were called *Babuni* because their faith was frequently associated in the popular conception with superstition and even magic, and hence they were said to believe

---

only fifteen years before Bohemond's atrocities in Pelagonia, they fought in the Byzantine army near Kastoria (cf. infra, p. 190). It would thus seem more satisfactory to regard the Pelagonian heretics as Paulicians.

Blagoev's attempt (Беседата на Презвитер Козма, *G.S.U.* (1923), vol. XVIII, pp. 35–6) to show that they were Bulgarian Orthodox, considered to be heretics by the Crusaders, is quite unsatisfactory. To medieval writers of western Europe the members of the Eastern Church were not heretics, but schismatics.

[1] An analysis of the *Life of Saint Hilarion*, which is one of the principal sources for a study of Bogomilism, is given below. (See infra, pp. 223–6.)

[2] See infra, pp. 242–5, for an interpretation of this passage of the *Synodicon* and for an explanation of the term 'dyed'.

[3] ѡ масалиянѣхь иже соуть ныня глаголемии Богомили Бабоуни (Περὶ Μασσαλιανῶν τῶν νῦν Βογομίλων): *Krmčaja Ilovička* cap. 42, ed. by V. Jagić, *Starine* (1874), vol. VI, pp. 100–1 (*Jugoslavenska Akademija Znanosti i Umjetnosti*, Zagreb). The same title recurs in most south Slavonic Nomocanons of the fifteenth and sixteenth centuries: see R. Karolev, За Богомилството, loc. cit. p. 65, n. 1.

in 'old wives' tales'.[1] The Babuna Mountain in central Macedonia between Prilep and Veles, and the River Babuna which flows out of this mountain into the Vardar, probably derived their name from the Babuni or Bogomils, who are frequently mentioned in this region in the Middle Ages.[2] It is very probable that the towns of Veles and Prilep, respectively north and south of the Babuna Mountain, were also nuclei of Bogomilism.[3]

On the slopes of the Babuna Mountain, the locality of Bogomili is mentioned in the fourteenth century.[4] It was probably situated in the *Bogomilsko Polje*, a small valley on the slopes of the same mountain.[5]

[1] Dj. Daničić, *Rječnik hrvatskoga ili srpskoga jezika* (Zagreb, 1880), vol. I, p. 136: 'Ime će im biti od babe, i tako će biti nazvani za to što se uzimalo da su im u vjerovańu bapske gatńe.' Karolev (loc. cit.), quoting the opinion of F. Miklošić, asserts that the expression бабоуньска рѣчь ('babun talk') means heresy in a general sense; the root *babun* can be found in several east European languages, with the meanings of 'incantatio', 'superstitio', 'idolum' (ibid.). Daničić (Рјечник из књижевних старина српских, Belgrade, 1863, vol. I, p. 21) translates *babun* as 'carmen magicum'. The name *babun* is particularly common in Serbian fourteenth-century sources (*Rječnik*, ibid.).

[2] This etymology, however, cannot be considered as certain, as authoritative scholars do not go beyond asserting its mere probability. V. Ćorović in the Народна енциклопедија српско-хрватско-словеначка (ed. S. Stanojević, Zagreb), vol. I, p. 213, writes: 'Можда је и име бабуни за богумиле, врло често у српским изворима, дошло од географске ознаке.' Daničić (*Rječnik*, ibid.) is likewise uncertain whether the name of the mountain Babuna is derived from the name of the Babun heretics, or directly from *baba*: 'postańem od babe (stare žene) kao što je i sama *baba* ime planinama, ili može biti od *babuna*.' Golubinsky asserts, on the contrary, that the Bogomils were named Babuns after the mountain and the river Babuna (Краткий очерк истории православных церквей, p. 156); but his view fails to take into account the correct etymology of *babun*.

[3] According to Golubinsky (op. cit. pp. 156–7) Veles was one of the main centres of the Bogomil sect.

[4] This locality was part of the land owned by the monastery of Treskavats. See *Glasnik Društva Srbske Slovesnosti* (Belgrade, 1859), vol. XI, p. 134; (1861), vol. XIII, p. 371; стась оу богомилѣхь, у богомили в бабунѣ. Cf. Daničić, *Rječnik*, vol. I, pp. 136, 492.

[5] See Balaschev, Климентъ епископъ словѣнски, p. xxxii. According to Balaschev (op. cit. pp. xxxiv–xxxv) the centre of Bogomilism was in western Macedonia, in the region of Polog, between Kishevo, Tetovo and the Shar Planina. He bases himself on a letter of Demetrius Chomatianus, archbishop of Ochrida at the beginning of the thirteenth century, which states the region of Polog was occupied by 'the power of the Dragovichi' (ὅτε δηλονότι καὶ τῆς τοῦ Πολόγου χώρας ἡ Δρουγουβιτικὴ κατεχόρευσεν ἐξουσία: J. B. Pitra,

Another name which was given to the Bogomils in Macedonia was *Kudugeri*. Symeon, archbishop of Thessalonica (1410–29), entitled one of the chapters of his *Dialogus contra haereses*: Κατὰ τῶν δυσσεβῶν Βογομύλων, ἤτοι Κουδουγέρων.[1] According to him, the *Kudugeri* or Bogomils, who lived in the neighbourhood of Thessalonica, were dualists, rejected the dogmas and practices of the Church, but called themselves Christians and were the descendants of the ancient Iconoclasts.[2] The Kudugeri are also mentioned, as living in Bosnia, by Gennadius Scholarius, patriarch of Constantinople, in a letter 'to the monk Maximus Sophianus and to all the monks of Sinai', written between 1454 and 1456.[3] The root of *Kudugeri* has remained to the present day in the names of two Macedonian villages, one in the extreme north of this province (Kutugertsi, in the district of Kustendil), the other in the south (Kotugeri, in the district of Vodena).[4] This shows that the Bogomil sect must have spread throughout the whole of Macedonia. The origin of the term *Kudugeri* is unknown.[5]

Finally, the Bogomils were also called *torbeshi*, from the Bulgarian 'torba' (торба) meaning 'a bag'; this name originated from their alleged custom of carrying a bag on their shoulders, which con-

Analecta sacra et classica (Paris, 1891), vol. VII, col. 410). But Balaschev's identification of these Macedonian Dragovichi with the members of the heretical *Ecclesia Dugunthiae* is somewhat arbitrary. If the former were really heretics, Demetrius Chomatianus would scarcely have failed to state this explicitly, particularly as the region of Polog was part of his diocese. Until the contrary is proved, the expression ἡ Δρουγουβιτική ἐξουσία can only be given a purely ethnological significance. However, there is no reason to doubt that the region of Polog was a Bogomil centre. Its situation in the very heart of Macedonia, surrounded on three sides by high mountain ranges, rendered it difficult of access and hence potentially a favourable ground for heresy.

[1] *P.G.* vol. CLV, cols. 65–74, 89–97; cf. infra, p. 267.
[2] οἱ Βογόμυλοι, ἀνθρώπια δυσσεβῆ, οἱ καὶ Κουδούγεροι καλούμενοι. (Ibid. col. 65.)
[3] *Œuvres complètes de Gennade Scholarios* (publiées par L. Petit, X. Sideridès, M. Jugie; Paris, 1935), t. IV, p. 200.
[4] See I. Ivanov, Сѣверна Македония (Sofia, 1906), p. 320; idem, Богомилски книги и легенди, p. 36; K. Jireček, *Cesty po Bulharsku* (Prague, 1888), p. 419.
[5] Runciman's view that the Kudugeri were so called 'probably from the name of the village that was their centre' (op. cit. pp. 97, 184) seems to me improbable. The fact that the Kudugeri are attested both in Macedonia and in Bosnia suggests that this was not simply a 'local name'.

tained the book of the Gospels and the alms they received.[1] *Torbeshi* is the exact Slavonic equivalent of φουνδαγιαγήται, a name which served to designate the Bogomils in certain parts of Asia Minor.[2] The name *torbeshi* is still applied to-day in a purely ethnological sense to the *pomaks*, or Moslem Bulgarians of central Macedonia, in the districts of Debar, Skoplje, Kishevo and Shar Planina.[3]

It is significant that the roots *babun*, *bogomil*, *kudugeri* and *torbeshi* cannot be found in either central, northern or eastern Bulgaria. On the other hand, a number of place-names in these regions are derived from the roots 'Armenian' and 'Paulician'.[4] This is an added proof that the centre of Bogomilism was Macedonia, while the Paulicians lived mainly in Thrace and eastern Bulgaria.

All the preceding evidence regarding the cradle of Bogomilism can be summed up as follows: the original home of the Bogomil sect was undoubtedly Macedonia. General historical and geographical data as well as the combined evidence of Bulgarian, Byzantine and Latin sources show its prevalence from the tenth to the fourteenth centuries in the region bounded in the east by the Rhodope Mountains and in the west by the Lake of Ochrida and the Black Drina.

[1] See I. Ivanov, Българитѣ въ Македония, p. 55, n. 1.
[2] Cf. infra, p. 177.
[3] See K. Jireček, *Das Fürstentum Bulgarien* (Prague, 1891), pp. 102–8; I. Ivanov, Българитѣ въ Македония, ibid.
[4] See Богомилски книги и легенди, p. 12; K. Jireček, *Cesty po Bulharsku*, p. 659; Yu. Trifonov, Бесѣдата на Козма Пресвитера, loc. cit. pp. 55–6.

CHAPTER V

BYZANTINE BOGOMILISM

I. *The spread of Bogomilism from Macedonia to Byzantium:* Bulgaria as a Byzantine province in the eleventh century. Role of the Paulicians and of the Bogomils in Bulgarian national resistance. Spread of Bogomilism to Asia Minor. John Tzurillas and the Phundagiagitae. The Bogomils and the Massalians in Thrace towards 1050.

II. *Byzantine Bogomilism:* The Thracian Paulicians in the eleventh century. Their revolts against Byzantium. Their disputations with Alexius Comnenus and their conversion to Orthodoxy. Penetration of Bogomilism into Byzantium. Arrest, trial and execution of Basil the Bogomil. The first systematic account of Bogomilism: the *Panoplia Dogmatica*. Evolution of Bogomilism in Byzantium in the spheres of doctrine, ethics and ritual.

III. *Repercussions of Byzantine Bogomilism in Bulgaria:* A new wave of Bogomilism in the middle of the twelfth century. Its effect on Bulgaria. St Hilarion of Moglena. The *Secret Book* of the Bulgarian Bogomils. Spread of Bogomilism throughout the Balkans.

An event which had far-reaching repercussions on the history of the Bogomil sect was its penetration into the Byzantine Empire in the course of the eleventh century. The true character of Bogomilism cannot be understood without taking into account its gradual evolution from the time of its rise in the tenth century to its final disappearance in the fourteenth and fifteenth centuries. This evolution took place under several influences, among the most important of which was Byzantine Christianity. In Byzantium, by contact with the Orthodox theology of the Churchmen and with the religious philosophy of the cultured secular classes, Bogomilism assimilated a number of new features which were later transmitted to the Bulgarian Bogomils in the twelfth and thirteenth centuries. The penetration of the sect into Byzantium was primarily due to the fact that after the fall of Samuel's Empire in 1018 Macedonia became a province of the Byzantine Empire; hence Bogomilism, unrestricted by national frontiers, could freely spread from its original home over the entire south-eastern part of the Balkan peninsula.

Moreover, one of the most important problems raised by a study of Bogomilism in the eleventh century is the exact relation of this sect to the Paulicians. An attempt to solve it is all the more necessary as scholars in the past have been prone to confuse the roles played by both sects in Bulgarian history. The distinction between the Bogomils and the Paulicians can only be made clear by an account of the history of the latter sect in the eleventh century; this, in its turn, should help to bring out several important features of Bogomilism.

The incorporation of Macedonia, Thrace and the north-eastern provinces of Bulgaria into the Byzantine Empire initiated a process of violent Byzantinization of these regions which continued until 1186. During this period the Bulgarian national resistance to the Greeks was largely in the hands of the sectarians, especially the Paulicians and, at least to some extent, the Bogomils.

The Byzantinization of Bulgaria in the eleventh century was an acceleration of the process already initiated in the tenth century. The independent Bulgarian patriarchate of Ochrida, established by Samuel, was abolished by Basil II; it is noteworthy, however, that Basil, by three imperial chrysobulls (1020), granted a nominal autonomy to the Bulgarian Church:[1] the newly appointed archbishop of Ochrida, under whose authority were thirty diocesan bishops, was recognized by the emperor as the rightful successor of the Bulgarian patriarchs of the First Empire. In practice, however, this autonomy did not amount to much and came to an end under Basil's successors: after 1037 the Bulgarian archbishops as well as the great majority of bishops were Greeks and their nomination and activities were strictly controlled by the Oecumenical See; Greek became the official and liturgical language of the archdiocese and the structure of the Bulgarian Church a medium for Hellenization of an extreme kind.[2]

As in the days of the Tsar Peter, this ecclesiastical imperialism of Byzantium had an adverse effect on the religious life of Bulgaria. The Greek episcopate, which enjoyed wide privileges in the tenure of land and possession of *paroikoi*, had little contact with the Bulgarian parish priests and their flocks. The gulf which separated

---

[1] See *Acta et diplomata res Albaniae mediae aetatis illustrantia*, ed. L. de Thallóczy, K. Jireček and E. de Sufflay (Vindobonae, 1913), vol. I, nos. 58, 59, pp. 15–16.

[2] See K. Jireček, *Geschichte der Bulgaren*, pp. 201 et seq.; Tsukhlev, op. cit. pp. 841 et seq.; M. Spinka, *A History of Christianity in the Balkans*, pp. 91–2.

the higher clergy from the people and the mutual hatred between Bulgarians and Greeks can be well judged from the expressions of contemptuous disgust with which the Greek Archbishop Theophylact of Euboea, who occupied the see of Ochrida approximately from 1078 to 1118, refers to his Bulgarian flock.[1] This gulf was in itself favourable to the proselytism of the heretics, who levelled their fiercest attacks at the behaviour of the Orthodox clergy.

Alongside the ecclesiastical domain, every sphere of Bulgarian life was invaded by Byzantine institutions, particularly in Macedonia, where the Greek domination after 1018 was absolute.[2] Bulgaria, on the model of the Empire, was divided into Themes; the military and civil power in each Theme belonged to a *Strategos* or *Dux*;[3] after 1041 all the officials in the country were Greeks. The period between the death of Basil II (1025) and the rise of the Comneni (1081) was one of severe crisis for the Empire; the dynastic struggles within and, above all, the constant pressure on every frontier from Normans, Pechenegs and Seljuq Turks,[4] together with the military reverses, frequently endangered its very existence; moreover, the decay of peasant and military holdings produced a severe decline in the State revenues and in the military resources of the Empire. To restore in some measure the military and financial structure of the State it was necessary to resort to wholesale recruitment and taxation. A conquered country like Bulgaria could provide large contingents of mercenaries as well as money for the imperial chest. It is hence not surprising that the extremely burdensome taxation to which the entire population of the Byzantine Empire (with the exception of the privileged minority which enjoyed rights of immunity) was subjected in the

---

[1] Theophylact was a celebrated theologian and scholar and had formerly been a tutor to the son of the Emperor Michael VII (see K. Krumbächer, *Geschichte der byzantinischen Litteratur*, 2nd ed., München, 1897, pp. 133–5, 463–5). His correspondence (*P.G.* vol. CXXVI, cols. 307–558) is full of complaints of the rude manners of the Bulgarians. He complains of being 'condemned to associate with those monsters' (col. 308) and refers to Bulgaria as 'a filthy marsh' and to its inhabitants as 'the frogs which emerge from it' (col. 309). He goes as far as to say that 'the Bulgarian character is the nurse of all evil' (col. 444).

[2] See V. N. Zlatarski, *Geschichte der Bulgaren*, vol. I, pp. 90–1.

[3] See Jireček, op. cit. p. 202.

[4] See F. Chalandon, *Essai sur le règne d'Alexis Ier Comnène*, pp. 2 et seq.

eleventh century[1] was applied in Bulgaria with particular rigour. The system of taxation in the reign of Samuel had been payment in kind.[2] Basil II had maintained and confirmed this practice. But in the reign of Michael IV the Paphlagonian (1034–41), the Byzantine officials in Bulgaria, to increase the State revenue, decided to levy taxes in money instead of in kind; the dissatisfaction aroused by this measure led to an open revolt of the Bulgarians in 1041, which was successfully crushed.[3] The rural population of Bulgaria experienced the brunt of unbridled taxation; in all parts of the Empire, apart from the land and poll taxes, the liabilities of the peasants included innumerable regular and extraordinary dues, such as labour services, providing for the needs of the army, etc. Although direct evidence is lacking, it seems probable that taxation in Bulgaria was even more severe than in the other provinces of the Empire, since it was a newly conquered territory. The misery and dissatisfaction of the Bulgarian rural population was, furthermore, increased by the rapacity of the tax officials, whose high-handed extortions were a subject of continual complaint throughout the Empire in the eleventh century.[4] Theophylact of Ochrida frequently complains of these 'robbers' in his letters.[5] Their treatment of the common people can be inferred from the fact that they did not even spare the archbishop, a Greek with high connections at court.

Apart from the method of taxation, other Byzantine social and economic institutions became firmly rooted in Bulgaria in the eleventh century. Thus the considerable development of *latifundia* and the economic domination of the landowning nobility, characteristic features of the Byzantine Empire after the death of Basil II, extended to the province of Bulgaria, where all the administrative power was vested in the local Byzantine governors. In the eleventh century a class of powerful landowners arose in Bulgaria, which even survived the Turkish invasion.[6] The *pronoia* system, the germs

[1] See G. Ostrogorsky, 'Agrarian conditions in the Byzantine Empire in the Middle Ages', *C.E.H.* vol. I, pp. 211 et seq.
[2] Runciman, *A History of the First Bulgarian Empire*, p. 231: [in the reign of Samuel] 'every man to possess a yoke of oxen was obliged to pay yearly a measure of corn, a measure of millet, and a flagon of wine'.
[3] See Spinka, op. cit. p. 93. [4] See *C.E.H.* ibid.
[5] See, in particular, *P.G.* vol. CXXVI, cols. 405, 416.
[6] See A. Pogodin, История Болгарии, p. 63.

of which can be found in tenth-century Bulgaria,[1] became definitely established in the second half of the eleventh, the tenure of land being coupled with the duty of defending the northern frontiers against the constant attacks of Pechenegs and Cumans. At the same time the peasants became attached to the land in increasing numbers and assumed the feudal status of *paroikoi*.[2]

The economic misery due to taxation, the systematic repression and exploitation of the Slavonic element by the Greeks, and the uprooting of the Slavonic agrarian community by the introduction of Byzantine feudal institutions produced a state of acute discontent in Bulgaria. From 1040 onwards a series of revolts broke out, all of which were suppressed.[3] Although they were too sporadic and unorganized to achieve any real success, they testify to the readiness of the Bulgarians to rally round any centre of opposition to the rule of their foreign masters.

In these circumstances it was natural that the Bulgarian people, failing to find adequate support in their own secularized Church and Hellenized government, sought protection among the Paulician and Bogomil sectarians, who consistently opposed both the political and economic exploitation by the Greeks and the authority of the Byzantine Church. The Paulicians were traditional enemies of the Empire; those who had been settled in Thrace by John Tzimisces were very ephemeral allies and soon showed their open hostility to Byzantium.[4] The Bogomils, whose popular and democratic tendencies brought them into close touch with the people, appealed particularly to the Bulgarian peasants, who suffered from Byzantine oppression more than any other class. It was hence inevitable that many Bulgarians, driven into active opposition to the Greeks or simply seeking protection against their unscrupulous exploitation, looked to Bogomilism as to the only force, at once religious and Slavonic, capable of overcoming the evils of the world by its doctrines of brotherhood and equality of all men. This role of defender of the people, so successfully played

[1] Cf. supra, pp. 99–100.
[2] See Pogodin, op. cit. pp. 62–3; V. N. Zlatarski, *Geschichte der Bulgaren*, p. 91. Bulgarian *paroikoi* are mentioned in a chrysobull of Basil II (see Jireček, op. cit. p. 202) and by Theophylact of Ochrida (loc. cit. passim).
[3] See Jireček, op. cit. pp. 203 et seq.; Pogodin, op. cit. pp. 67 et seq.; Spinka, op. cit. pp. 93 et seq.
[4] Cf. infra, pp. 188 et seq.

by Bogomils, explains the considerable growth of their teaching in Bulgaria during the period of Byzantine domination (1018–1186).[1] It is unfortunately impossible to determine the precise character of the Bogomil opposition to the Greeks at that time, owing to the lack of any positive evidence. But one should beware of making the unjustifiable assumption that the Bogomils ever formed an organized nationalistic party in Bulgaria. As it has already been pointed out, the Bogomils were essentially a religious sect, and if and when they participated in secular events it was for temporary and contingent reasons.[2]

These religious, political and social factors explain the great development of Bogomilism in the eleventh century, its spread from Macedonia over the greater part of the Byzantine Empire and its penetration into Constantinople. All our information about the sect during that century is derived from Byzantine sources, which is largely due to the fact that Bogomilism, no longer confined within the territorial boundaries of Bulgaria, soon secured the attention of the ecclesiastical authorities and theologians in Constantinople. Nevertheless, many of these sources also testify to the prevalence of the sect in Bulgaria.

[1] See Jireček, op. cit. pp. 211–12.
[2] The nationalistic and anti-Byzantine features of Bogomilism are frequently emphasized by scholars, with particular reference to the eleventh century. See Gilferding, op. cit. p. 226; V. Levitsky, Богомильство—болгарская ересь, loc. cit. p. 391; R. Karolev, За Богомилството, loc. cit. p. 60; V. N. Zlatarski, *Geschichte der Bulgaren*, p. 92; Ivanov, op. cit. p. 31. This view of Bogomilism, legitimate within its own limits, should not, however, be exaggerated. The Bogomils were undoubtedly opposed to the growth of Byzantinism in Bulgaria, as this entailed the domination of the Byzantine Church and the rule of aristocracy and officialdom, which they condemned on principle. But there is no ground for accepting Spinka's statement that 'they formed the best organized anti-Byzantine element in the country' (op. cit. p. 94). Still less is it possible to maintain that Bogomilism was essentially nothing but a nationalistic revolt of Slavdom against Byzantium. This view is upheld in particular by N. Blagoev (Беседата на Презвитер Козма против богомилите, *G.S.U.* 1923, vol. xviii), who emphasizes the political significance of Bogomilism to the extent of altogether denying its existence as a heresy; this leads him to conclusions which are manifestly absurd. That Bogomilism was above all a system of religious and ethical teachings and that its political, anti-Byzantine aspect was secondary and almost accidental is shown by the fact that the Bogomils of Constantinople, in the eleventh and twelfth centuries, were not opposed to the Byzantine government.

The earliest Byzantine document concerning the eleventh-century Bogomils is a letter of Euthymius, monk of the monastery τῆς Περιβλέπτου in Constantinople, addressed to his compatriots of the diocese of Acmonia in the province of Phrygia, in Asia Minor.[1] The time in which Euthymius lived and the approximate date of his letter can be deduced from the following autobiographical episode: Euthymius tells us that in the joint reign of Basil II and Constantine IX (976–1025), when the future Emperor Romanus III Argyrus (1028–34) was judge in the Theme of Opsikion, he himself with his mother came to Acmonia for the purpose of participating in a lawsuit.[2] It was there that he came into contact with the heretics. The period when Romanus Argyrus was judge in Asia Minor is not known exactly, so the episode described by Euthymius must be placed some time between 976 and 1025. His letter, on the other hand, must have been composed after the death of Romanus (11 April 1034), since he refers to him as ὁ μακαρίτης κῦρις.[3] Ficker, accordingly, places the composition of Euthymius's letter around 1050 and his stay in Acmonia at the beginning of the eleventh century.[4]

During his stay in Acmonia, Euthymius witnessed the trial and condemnation of a heretic, John Tzurillas, accused among other things of unlawfully assuming the monastic garb, forcing his wife

[1] Edited by G. Ficker (*Die Phundagiagiten: Ein Beitrag zur Ketzergeschichte des byzantinischen Mittelalters*, Leipzig, 1908) under the title: Ἐπιστολὴ Εὐθυμίου μοναχοῦ τῆς περιβλέπτου μονῆς σταλεῖσα ἀπὸ Κωνσταντινουπόλεως· πρὸς τὴν αὐτοῦ πατρίδα στηλιτεύουσα τὰς αἱρέσεις τῶν ἀθεωτάτων καὶ ἀσεβῶν πλάνων τῶν Φουνδαγιαγιτῶν ἤτοι Βογομίλων. See the detailed review of Ficker's book by M. Jugie, 'Phoundagiagites et Bogomiles', *Échos d'Orient* (Paris, 1909), t. XII, pp. 257–62.

[2] See Ficker, op. cit. pp. 66–7. For the situation of Acmonia, see W. M. Ramsay, *The Cities and Bishoprics of Phrygia* (Oxford, 1897), pp. 621–30.

[3] Ibid. p. 67.

[4] F. Cumont ('La date et le lieu de la naissance d'Euthymios Zigabénos', *B.Z.* 1903, vol. XII, pp. 582–4) identifies Euthymius of Acmonia with Euthymius Zigabenus, who was entrusted by Alexius Comnenus with the task of composing a general refutation of all heresies, including that of the Bogomils (cf. infra, p. 205). Ficker, however (op. cit. pp. 182–91), has convincingly proved that Cumont's theory is untenable. The main argument against this identification is the chronological one: Euthymius Zigabenus wrote his *Panoplia Dogmatica* after the trial of the Bogomils in Constantinople which took place around 1110 (see Appendix III), while Euthymius of Acmonia could not have been born after 1000.

to enter a nunnery and living unchastely.[1] According to Euthymius he was the first teacher of a 'newly appeared' heresy (πρῶτος τῆς νεολέκτου ἀσεβοῦς...θρησκείας) which he had preached for three years 'in Thrace' (ἐν τοῖς τῶν Θρᾳκῶν μέρεσιν), in the region of Smyrna and in many other places, gaining numerous disciples. The centre of Tzurillas's activity seems to have been the village of Χίλιοι Καπνοί in the Theme of Opsikion, where of all the inhabitants scarcely ten remained Orthodox.[2]

The location of these centres of Tzurillas's proselytism raises several important problems. In the first place, does the expression ἐν τοῖς τῶν Θρᾳκῶν μέρεσιν refer to the Balkan province of Thrace? Ficker does not think so, and suggests that the correct reading is not Θρᾳκῶν, but Θρᾳκησίων, i.e. the Thracesian Theme, in the south-west corner of Asia Minor.[3] This interpretation would seem to be confirmed by the fact that the town of Smyrna, juxtaposed in Euthymius's text with the region τῶν Θρᾳκῶν, is, in fact, contiguous to the Thracesian Theme; moreover, the whole of this passage appears to concern only Asia Minor. However, circumstantial evidence can also be found in support of the opposite opinion. Ivanov considers it very probable that John Tzurillas was a Bulgarian Bogomil who came from the Balkans to the Theme of Opsikion with the aim of spreading his heresy among the population of Asia Minor.[4] This hypothesis seems quite acceptable, especially in view of the close connections which existed in the Middle Ages between the Balkan Slavs and the north-western Themes of Asia Minor, particularly Opsikion and Bithynia.[5] The

[1] Ficker, op. cit. pp. 66, 68.
[2] Ibid. pp. 67–8.   [3] Op. cit. p. 249, n. 2.
[4] In support of this opinion, Ivanov puts forward the hypothesis that the name Τзουρίλλας is of Bulgarian origin. He derives it from чоурила, meaning 'house' or 'household', corresponding to the fiscal unit of καπνικόν. In his opinion, the name of the village of Χίλιοι Καπνοί, the centre of Tzurillas's proselytism, is a translation of the Old Bulgarian Тысяща чоурила, from which John Tzurillas would have derived his name. (Богомилски книги и легенди, p. 38.)
[5] A Slavonic population existed in Bithynia already in the seventh century: at that time a Slavonic bishopric of Gordoserba, whose name is indicative of its Serbian origin, is attested in Bithynia, south-east of Nicaea and north of Dorylaeum. (See L. Niederle, *Slovanské Starožitnosti*, vol. II, pt 2, p. 399; F. Dvorník, *Les Slaves, Byzance et Rome*, pp. 102–3.)
From the seventh century onwards the Slavonic population in Bithynia and

population of those districts in Asia Minor where John Tzurillas spread his heresy must have comprised a large number of Slavs from Thrace and Macedonia, and it is hence not at all improbable that Tzurillas, after teaching in Thrace, crossed the Bosporus and continued his work among his compatriots in Opsikion. On purely historical grounds, however, Ivanov's theory remains merely probable and cannot be fully proved, but its validity is confirmed by an analysis of the teachings of John Tzurillas, which bear an unmistakable stamp of their Bulgarian origin.

Euthymius's letter contains a detailed exposition of the heretical doctrines taught by Tzurillas and his disciples. He derived his knowledge of them directly from the heretics. One of them had once been a travelling companion of his and had even attempted to convert him. Moreover, on returning from a pilgrimage to Jerusalem Euthymius had discovered that the heretics had penetrated into his own monastery in Constantinople and had even corrupted a disciple of his. The heretics—four in number—were

Opsikion was numerous, owing to transplantations carried out by Byzantine emperors and to peaceful immigration.

In 688 Justinian II transported some 80,000 Slavs to the Theme of Opsikion (see V. I. Lamansky, О славянах в Малой Азии, в Африке и в Испании, *Uchenye zapiski Vtorogo Otdel. Imperator. Akad. Nauk*, 1859, vol. v, pp. 2–3; A. A. Vasiliev, *Byzance et les Arabes*, vol. I, p. 24). In 762 Constantine V deported some Slavs from Thrace and Macedonia to Bithynia. (See Dvorník, op. cit. p. 18.) Bithynia and Opsikion became the centres of Slavonic colonization in Asia Minor in the seventh to ninth centuries. (See Niederle, ibid. pp. 458–68; В. A. Panchenko, Памятник славян в Вифинии VII века, *I.R.A.I.K.* 1902, vol. VIII, pp. 15–62; cf. L. I. Dorosiev, Българскитѣ колонии въ Мала Азия, *S.B.A.N.* 1922, vol. XXIV, pp. 32–192.)

Lamansky has shown, furthermore (loc. cit. pp. 6–17), that these colonists, in spite of the constant attempts to Hellenize them, retained their Slavonic characteristics at least up to the fifteenth century. Moreover, they were frequently reinforced by waves of Slavonic immigrants from the Balkans, the majority of whom settled in Bithynia. Apart from the presence of their compatriots in this region, the factors which favoured the Slavonic immigration to Bithynia and Opsikion were the fertility of the soil, the facility of communications, the flourishing commercial relations between the Balkans and Asia Minor, and the fact that these districts lay on the way of the pilgrims travelling from Bulgaria to the Holy Land. According to Lamansky, the Slavonic colonies were 'an attraction and an enticement to all their dissatisfied fellow-countrymen in Europe'. It is surely most likely that these 'dissatisfied' elements included at least some Bulgarian Bogomils who were only too willing to exchange persecution in their home for greater security and the chance of proselytizing among their compatriots across the Bosporus.

apprehended and questioned and revealed their doctrines to Euthymius.[1]

Euthymius asserts that this heresy has two different names: in the Theme of Opsikion the heretics are known as Phundagiagitae, while in the Cibyrrhaeot Theme, in 'the West' and in other places they are called Bogomils.[2] The Cibyrrhaeot Theme lay in the southern extremity of Asia Minor, by the gulf of Antalya; the attested presence of the Bogomils in this region at the beginning of the eleventh century, together with the mention by Euthymius of Opsikion and the district of Smyrna as centres of Bogomilism, show that the sect must have spread at that time over the entire western part of Asia Minor. The fact that it was known in the Cibyrrhaeot Theme under its Bulgarian name provides additional evidence that Bogomilism penetrated to Asia Minor from Bulgaria, probably in the second half of the tenth century, and also indirectly strengthens Ivanov's hypothesis regarding the Bulgarian origin of John Tzurillas. As for the term 'the West' (ἡ Δύσις), it was the traditional name applied by the Byzantines to the Balkan provinces of the Empire,[3] which included in the eleventh century Thrace and Macedonia. Thus Euthymius's letter can be regarded as the earliest Byzantine document directly referring to the Bulgarian Bogomils.

The name Phundagiagitae[4] is generally derived from φούνδα, itself a Greek form of the Latin 'funda', meaning a bag or scrip. The heretics are supposed to have acquired this name from their life of poverty, which compelled them to beg for their living.[5]

---

[1] These details are to be found in the Vatican MS. of Euthymius's *Liber invectivus contra haeresim exsecrabilium et impiorum haereticorum qui Phundagiatae dicuntur* (*P.G.* vol. CXXXI, cols. 48–57), falsely attributed to Euthymius Zigabenus.

[2] Οἱ τοῦ Ὀψικίου λαοὶ καλοῦσι τοὺς τὴν κακίστην ταύτην ἀσέβειαν μετερχομένους Φουνδαγιαγίτας, εἰς δὲ τὸν Κιβυρραιώτην, καὶ εἰς τὴν Δύσιν καὶ εἰς ἑτέρους τόπους καλοῦσιν αὐτοὺς Βογομίλους. (Ficker, op. cit. p. 62.)

[3] See G. Schlumberger, *Un Empereur Byzantin au Xe siècle, Nicéphore Phocas* (2nd ed.; Paris, 1923), p. 263.

[4] For the various spellings of this name see Ficker, op. cit. p. 192 and Puech, op. cit. p. 281. The most common are Φουνδαγιαγῆται and Φουνδαῖται.

[5] See P. Lambecius, *Commentaria de augustissima Bibliotheca Caesarea Vindobonensi* (2nd ed.; Vindobonae, 1778), vol. v, col. 85: 'Illi haeretici adpellabantur Phundaitae et Saccophori, quod ob austeram paupertatem, quam publice profitebantur, in saccos et crumenas stipem collegerint.' Cf. J. C. Wolf, *Historia Bogomilorum* (Vitembergae, 1712), p. 7; J. Engelhardt, *Kirchen-*

Although the historical sources are not very explicit on this point, the truth of this etymology cannot be denied.[1]

The doctrines and practices ascribed by Euthymius to John Tzurillas and his followers exhibit all the main features of Bogomilism. Not only is Euthymius's evidence in complete agreement on all essential points with that of Cosmas, but his account of the heresy of the Phundagiagitae reveals that fusion of Paulician and Massalian teachings which is characteristic of Bogomilism. Moreover, his information is particularly important, as it shows that by the middle of the eleventh century Bogomilism had probably evolved some novel features, particularly in the realms of doctrine and ritual.

Euthymius's evidence, for the purposes of the present study, can be examined from a threefold aspect: (1) some of the features he ascribes to the Phundagiagitae can already be found in an identical form in the *Sermon against the Heretics*; (2) others, although mentioned by Cosmas, are described by Euthymius in greater detail and thus illustrate and extend our knowledge of the Bogomils derived from tenth-century sources; (3) others, finally, are ascribed to the Bogomils for the first time, and represent either some borrowings from Paulicianism and Massalianism which are not attested by the tenth-century sources, or a further development of Bogomilism in the direction of greater complexity.

(1) Many external traits of the Phundagiagitae are attributed by Cosmas to the tenth-century Bogomils: according to Euthy-

---

geschichtliche Abhandlungen, pp. 205 et seq.; G. Rouillard, 'Une étymologie (?) de Michel Attaliate', *Rev. de philol., litt. et d'hist. anciennes* (Paris, 1942), 3ᵉ série, vol. XVI, p. 65; Runciman, op. cit. p. 184. Puech (op. cit. p. 281, n. 3) also suggests a possible connection between the Phundagiagitae and Φουνδᾶς, the name of a heretic described by Euthymius as a disciple of Mani (Ficker, op. cit. p. 42).

[1] Ficker, however, alleging that there is no evidence that the Phundagiagitae ever begged, denies their connection with 'funda' and claims that their name must be derived from some unknown non-Greek root (op. cit. pp. 193–4). But his statement can be refuted by remarking that this derivation exists in Bulgarian as well as in Greek: the name *torbeshi*, given in the Middle Ages to the Bogomils in certain parts of Macedonia, is derived from the Bulgarian 'torba', meaning a bag. (Cf. supra, p. 166.) Moreover, a life of poverty is ascribed to the Bogomils by Cosmas (cf. supra, pp. 136–7) and the practice of begging probably existed among them as a result of Massalian influence.

mius they call themselves Christians,[1] conform in their outward behaviour to all the rules of the Church,[2] thus incurring the usual accusation of hypocrisy,[3] and are particularly dangerous owing to the great difficulty of distinguishing them from the Orthodox[4] and to the absolute impossibility of reconverting those who had embraced their doctrines, 'even if the whole world were to instruct them'.[5] A number of their doctrines were already held by the tenth-century Bogomils. Thus the Phundagiagitae rejected the Old Testament,[6] the Order of Priesthood,[7] the cult of the saints,[8] all the prayers of the Church with the exception of the Lord's Prayer,[9] and denied the efficacy of the Cross,[10] the validity of Baptism and of the Eucharist.[11]

(2) Among the Bogomil doctrines which are set out more fully

[1] See Ficker, op. cit. pp. 4, 30, 31. In one MS. of the letter of Euthymius of Acmonia, published by Migne and falsely attributed to Euthymius Zigabenus, the heretics are also said to call themselves Χριστοπολῖται (P.G. vol. cxxxi, col. 48).

[2] Ficker, op. cit. pp. 25, 26, 28, 31.

[3] Ἡμεῖς δὲ εἰ ποιοῦμεν πάντα, ἀλλ' οὖν πίστει οὐ ποιοῦμεν, οὔτε βάπτισμα, οὔτε ἱερωσύνην, οὔτε μοναχικὴν οὔτε ἄλλο τι τῶν Χριστιανῶν· ἀλλ' ἐπιδεικτικῶς, μᾶλλον δὲ καὶ ἐμπαικτικῶς πάντα ποιοῦμεν πρὸς τὸ λανθάνειν. (Ibid. p. 25–6.)

[4] Ibid. pp. 35, 210.   [5] Ibid. p. 57.

[6] Οἶδα...ὅτι τὴν παλαιὰν γραφὴν οὐκ ἀναγινώσκετε. (Ibid. p. 40.) Ficker, however, disbelieves this: 'Aber von einer Verwerfung des Alten Testaments ist nicht die Rede; vielmehr meint der Verfasser nur, wenn sie das Alte Testament richtig läsen, würden sie ihre Anschauung von der Schöpfung aufgeben müssen.' (Ibid. p. 205.) This opinion seems substantially correct. The wholesale rejection of the Old Testament, characteristic of the Paulicians (cf. supra, p. 39), is nowhere ascribed to the Bogomils. Cosmas merely accuses them of spurning the Mosaic Law and the prophets (see supra, p. 127). The attitude of the Bogomils to the Old Testament was essentially eclectic: they resorted to it whenever a given passage could be twisted into accordance with their own views. This is particularly clear in the case of their interpretation of the Book of Genesis (see infra, pp. 207–9). The Phundagiagitae themselves based their teaching on the origin of man on a combination of the Biblical story with dualistic legends.

[7] This they held to be superfluous: καὶ τί ἐστι πρεσβύτερος; τοῦτο περιττόν ἐστιν (ibid. p. 76).

[8] For 'God alone is holy'. (Ibid.)

[9] Οὔτε τρισάγιον, οὔτε δόξα πατρὶ καὶ υἱῷ, οὔτε τὸ κύριε ἐλέησον, οὔτε ἄλλο τι ἐκδιδάσκονται ψάλλειν ἢ εὔχεσθαι, εἰ μὴ γυμνὸν καὶ μόνον τὸ πάτερ ἡμῶν. (Ibid. p. 33.) They described all other prayers as 'babblings' or 'vain repetitions' (βαττολογίαι). This word, derived from the Gospels (cf. Matt. vi. 7), was used in the same sense by the tenth-century Bogomils. (See supra, p. 134.)

[10] Ibid. p. 74.   [11] Ibid. pp. 28, 74.

by Euthymius than by Cosmas is the cosmological dualism. According to Euthymius, the Phundagiagitae taught that there are eight heavens. God has created seven of them over which He rules; the eighth, which corresponds to our visible world, is the creation of 'the prince of this world', who is the Devil. To him belong the sky, the earth, the sea and everything in them. The Devil also created Paradise[1] and made Adam, the first man. There are, however, two things in this visible world which are not the Devil's, but God's creation: the sun[2] and the soul of man, which the Devil stole from God when he was expelled from His sight.[3] Between this soul, polluted by the Devil, but still retaining the mark of its divine origin, and the body of man, formed by the Devil, there is an absolute duality which is illustrated by the Bogomil myth of the creation of man, described by Euthymius. According to the heretics, the Devil, having made the body of Adam, tried to animate it by means of the soul, which he had stolen from God. However, in spite of his repeated efforts to unite the two, the soul would not remain in Adam's body. For three hundred years the body lay lifeless, abandoned by the Devil. At the end of this period the Devil, having eaten of the flesh of all the unclean animals, returned to it and forced the soul to remain in the body by stopping up with his hand the anus, through which the soul had been wont to escape. He then disgorged his repast over the soul. In this manner the soul remained in the body and Adam came to life.[4] This myth, for all its crudeness, is an interesting example of a 'Bogomil legend'; it unites

[1] The idea that the Garden of Paradise was created by the Devil to bring about man's downfall is also expressed in the Bogomil *Liber Sancti Johannis*, known to us only through its Latin translation (cf. infra, pp. 226–8): 'Sententiator malorum ita cogitavit cum ingenio suo ut faceret paradisum, et introduxit homines, et praecepit adducere.' See J. Benoist, *Histoire des Albigeois et des Vaudois* (Paris, 1691), vol. I, p. 288; I. Ivanov, Богомилски книги и легенди, p. 78. Cf. the Bogomil belief that the vine, identified with the Tree of the Knowledge of Good and Evil, was planted by the Devil in Paradise (supra, pp. 128–9).
[2] Puech (op. cit. pp. 184–5) has pointed out the discrepancy in the Bogomil teachings on the sun attributed to them respectively by Euthymius and Cosmas (cf. supra, p. 122). We cannot be sure whether these contradictory views correspond to two different currents in Bogomilism, and whether any Bogomils really excluded the sun from the 'visible world' ruled by the Devil.
[3] Ficker, op. cit. pp. 33–4. [4] Ibid. pp. 35–7.

the tendency towards gross materialization with an attempt to satisfy popular curiosity in the interpretation of Biblical events.

Further details concerning the behaviour of the Bogomils towards the Orthodox Church are supplied by Euthymius. Cosmas tells us that in order to avoid detection they went to church and outwardly venerated the Crucifix and the images.[1] The followers of John Tzurillas, it seems, went even further in their simulation of Orthodoxy: according to Euthymius, they took part in the Church services,[2] had their children baptized, partook of the Sacrament, even built churches, painted icons and made crosses, all for the sake of ostentatious deceit.[3]

Euthymius paints an impressive picture of their systematic and zealous proselytism which, according to him, extended to the four corners of the Byzantine Empire and indeed over the whole of Christendom.[4] The heretics, he says, following the example of the Apostles, drew lots among themselves for the zones of activity allotted to each.[5] For the sake of spreading their teaching they overcame any difficulty or danger.[6] There is no doubt a certain measure of exaggeration in Euthymius's description of the widespread proselytism of the Bogomils. However, its success in tenth-century Bulgaria and the spread of the sect in the eleventh century throughout the south-eastern part of the Balkan peninsula and the west of Asia Minor show that by 1050 Bogomilism was a very serious menace to the Orthodox Church.

(3) The following doctrines and practices ascribed to the Bogomils by Euthymius are not expressly mentioned by previous sources and hence extend our knowledge of Bogomilism.

According to Euthymius, the heretics rejected the Christian dogmas of the Resurrection of the Dead, of the Second Coming and of the Last Judgement.[7] This was a logical consequence

---

[1] See supra, pp. 141–2. [2] Op. cit. p. 78.
[3] Ibid. pp. 26–8. According to Euthymius, the Phundagiagitae conformed to the rules of the Christian life οὐ πίστει, but ἐν ὑποκρίσει and διὰ τὸ λανθάνειν.
[4] Περιτρέχουσιν πᾶσαν τὴν τῶν Ῥωμαίων ἐπικράτειαν καὶ εἰς ὅσους ὁ ἥλιος ἐφορᾷ Χριστιανούς. (Ibid. p. 63.)
[5] Ibid. p. 64.
[6] ὑπομένοντες κόπους καὶ φόβους καὶ θλίψεις καὶ στενοχωρίας, πολλάκις καὶ κινδύνους. (Ibid. p. 64.) The same zeal for proselytism is ascribed to the Paulicians by Peter of Sicily. (See supra, pp. 30, 42.)
[7] Ibid. p. 38.

of the Bogomil view of matter, as the principle of evil and corruption.[1]

Both the *pop* Bogomil and John Tzurillas used the New Testament for exegetical purposes; however, while Cosmas tells us that the Bogomils relied above all on the Gospels and the Acts of the Apostles,[2] Euthymius mentions also the epistles of St Paul as an object of their particular veneration. The Phundagiagitae apparently claimed that the words of the Gospels and of St Paul 'breathed again' (ἀναπνέειν) owing to their own interpretation of them.[3] Their cult of St Paul suggests Paulician influence.

The letter of Euthymius is, moreover, the earliest document clearly showing the monastic orientation of Bogomilism. It cannot be doubted that the Bogomils borrowed this important feature from the Massalians.[4] John Tzurillas himself became a monk, dispatched his wife to a nunnery and taught his disciples to do the same.[5] His followers assumed the monastic habit and were noted for their 'insidious and humble bearing'.[6] They were also forbidden to shed blood.[7]

Finally, Euthymius provides some valuable information on their manner of holding prayer-meetings. His evidence reveals that by the middle of the eleventh century the Bogomils had a definite though rudimentary ritual. He describes these prayer-meetings as follows: 'the presiding member of the community rises and begins with the words: "let us adore the Father, the Son and the Holy Spirit" (προσκυνοῦμεν πατέρα καὶ υἱὸν καὶ ἅγιον πνεῦμα); the congregation replies: "it is meet and just" (ἄξιον καὶ δίκαιον); then they recite the Lord's Prayer, making prostrations (μετανοίας) in a prescribed manner and bobbing their heads

---

[1] It was the Resurrection of the Body that the heretics denied: this is clear from the fact that they based their rejection of this dogma on the words of St Paul: 'Flesh and blood cannot inherit the kingdom of God; neither doth corruption inherit incorruption.' (I Cor. xv. 50.) See Ficker, op. cit. pp. 7, 13.
[2] See supra, p. 127.
[3] Op. cit. p. 40.
[4] Cf. supra, p. 105.
[5] Op. cit. p. 66.
[6] Σχῆμα μοναχικὸν καὶ ὄνομα καὶ πρᾶξιν Χριστιανοῦ καὶ ἦθος ὕπουλον καὶ ταπεινόν. (Ibid. p. 30.) The humble and modest behaviour of the Bogomils is also attested by Cosmas. (See supra, p. 121.)
[7] Τὸ μὴ σφάʒειν (ibid. p. 59).

up and down like men possessed (ὡς οἱ δαιμονιζόμενοι);[1] and when they pray they do not look towards the East, but in whatever direction they happen to be facing.'[2]

We can thus conclude from the evidence of Euthymius of Acmonia that by the middle of the eleventh century the Bogomil sect had spread not only over the Balkan provinces of the Byzantine Empire, but also over the western part of Asia Minor. In the latter region its adherents were called Phundagiagitae.[3] The Bulgarian origin of the teaching of John Tzurillas, indirectly suggested by historical evidence, is confirmed by its analysis, which reveals the double influence of Paulicianism and Massalianism.[4] As it has been shown, the fusion of these two heresies into Bogomilism took place in Bulgaria.

Euthymius's evidence is corroborated by another contemporary Byzantine document, the *Dialogus de daemonum operatione* by Michael Psellus.[5] This work was composed towards the middle of the eleventh century and is thus approximately contemporaneous with the letter of Euthymius of Acmonia.[6] Its contents show that at that time news had reached Constantinople that the sect of the Euchitae was pursuing its unlawful activities within the borders

[1] This behaviour is strongly reminiscent of the 'sacred delirium' of the Massalians (cf. supra, p. 49).

[2] Ibid. p. 77. It is interesting to compare this Bogomil ritual with that used by the French Cathars in the thirteenth century. The comparison reveals several points of similarity, in particular the recitation of the Lord's Prayer and of the formula 'adoremus Patrem et Filium et Spiritum Sanctum'. See L. Clédat, *Le Nouveau Testament traduit au XIIIe siècle en langue provençale, suivi d'un Rituel Cathare* (Paris, 1887), pp. ix–xxvi.

[3] Ficker, however, thinks that Bogomilism first arose in Asia Minor and thence spread to Bulgaria (op. cit. pp. 271–3). But, in his ignorance of the Slavonic sources, he was unaware that at the time when John Tzurillas taught in Phrygia Bogomilism had been rife in Bulgaria for at least half a century.

[4] The clearest evidence of Paulician influence among the Phundagiagitae is their cult of St Paul, and Massalian influence appears in their monastic mode of life.

[5] The principal editions of this work are by J. Boissonade (Nuremberg, 1838) and Migne (*P.G.* vol. cxxii, cols. 820 et seq.). A French sixteenth-century translation by Pierre Moreau is given by E. Renauld, 'Une traduction française du Περὶ ἐνεργείας δαιμόνων de Michel Psellos', *Revue des Études Grecques* (1920), vol. xxxiii, pp. 56–95.

[6] See the chapter concerning Psellus in Hussey's *Church and Learning in the Byzantine Empire*, pp. 73–88.

of the Byzantine Empire. It is possible to infer that these activities were centred in Thrace.[1]

The identification of the doctrines and the practices of the eleventh-century Thracian Euchitae, as described in this dialogue by the personage called Θρᾷξ, is a complex problem, especially as Psellus seems to have had no direct contact with the heretics and probably wrote from hearsay or rumour. Thrace was still a great centre of Paulicianism in the eleventh century.[2] But the teachings and practices ascribed by Psellus to the Euchitae bear no direct resemblance to the tenets of this sect. They can be divided into three groups: (1) those which can legitimately be described as Bogomil; (2) those which cannot be so defined, as they contradict both previous and later evidence on Bogomilism, but which are suggestive of Massalianism; (3) those, finally, which appear alien to both these sects.[3]

(1) Psellus states that the basic doctrine of the Euchitae was borrowed from the Manichaean dualism, with the important difference, however, that to the two principles taught by Mani they added a third:

'This pernicious doctrine derives its premises from Mani the madman. From his teaching, as from a stinking fountain, the Euchitae extracted their plurality of principles. Now the accursed

---

[1] Although Psellus is extremely vague about the region inhabited by the heretics he describes, the following considerations seem to prove that it was Thrace: (1) one of the interlocutors of the dialogue exclaims, with reference to the Euchitae: δεινόν γε.. εἰ τοιοῦτον μύσος εἰς τὴν καθ' ἡμᾶς οἰκουμένην ἐπεχωρίασεν (*P.G.* vol. cxxii, col. 836). We know, on the other hand, from the testimony of Cedrenus that at the end of the eleventh century the Massalians, or Euchitae, were widespread in Thrace (cf. supra, p. 94). (2) The personage in the dialogue who describes the Euchitae relates his encounter with one of them in the Chersonese: μονάζοντι δέ τινι περὶ χερρόνησον τὴν ὅμορον Ἑλλάδος ξυγγέγονα (ibid. col. 840). The Chersonese was the name given by the Greeks to the strip of Thrace which runs along the Hellespont. (3) This personage is given in the dialogue the significant name of Θρᾷξ. It appears from his words that he had just returned to Byzantium after an absence of more than two years (ibid. col. 821), and that he was a provincial military commander. This explains the modernized title of 'Monsieur le Capitaine de Thrace' given him by Pierre Moreau. (See Renauld, loc. cit. pp. 60 et seq.)

[2] Cf. infra, pp. 188 et seq.

[3] It is all the more necessary to make this distinction as some scholars, in particular Levitsky (loc. cit. pp. 41 et seq.), Karolev (loc. cit. pp. 61 et seq.) and Puech (op. cit. p. 326), have rashly assumed that the Euchitae described by Psellus are simply the Thracian Bogomils of the eleventh century.

Mani laid down that there are two principles of being, and directly opposed one God to the other—the worker of evil to the creator of good, the prince of all evil which is on earth to the good Prince of heaven. These wretched Euchitae, however, have added yet a third principle. *Their principles consist of the Father and His two Sons, the elder and the younger; they assign to the Father only the supramundane things, to the younger Son the heavenly things, and to the elder the rule over this world.*' (Πατὴρ γὰρ αὐτοῖς υἱοί τε δύο, πρεσβύτερος καὶ νεώτερος, αἱ ἀρχαί· ὧν τῷ μὲν πατρὶ τὰ ὑπερκόσμια μόνα, τῷ δὲ νεωτέρῳ τῶν υἱῶν τὰ οὐράνια, θατέρῳ δὲ τῷ πρεσβυτέρῳ τῶν ἐγκοσμίων τὸ κράτος, ἀποτετάχασιν.)

On this basic principle, according to Psellus, all the Euchitae were agreed. But as regards its application they were divided into three groups, each holding its own opinion:

'Some of them worship both Sons: for they say that although at present they differ from each other, yet they are to be worshipped equally, since, proceeding from one Father, they will become reconciled to one another in the future. Others serve the younger Son as the ruler of the better and superior part, but without despising the elder and while being on their guard against him, as he is capable of working mischief. But the most impious of them separate themselves completely from heaven and embrace the *earthly Satanael* alone (αὐτὸν δὲ μόνον τὸν ἐπίγειον Σαταναὴλ ἐνστερνίζονται). They extol him with the finest-sounding names, *calling him the first-born of the Father and the creator of trees, animals and other compound bodies* (πρωτότοκον τὸν ἀλλότριον ἐκ πατρὸς καλοῦσι, φυτῶν τε καὶ ζώων καὶ τῶν λοιπῶν συνθέτων δημιουργόν), when in reality he is ruinous and destructive. Wishing to honour him still more,... they say that the heavenly [ruler] is jealous of him and... envies him his good arrangement of the earth and that, smouldering with envy, he sends down earthquakes, hailstorms and plagues. For this reason they curse him.'[1]

The belief in the supreme God, Lord of the supramundane spheres (τὰ ὑπερκόσμια), and in His two Sons, the one the prince of the heavens (τὰ οὐράνια), the other the ruler of the visible world, is typically Bogomil. It is already alluded to by John the Exarch at the beginning of the tenth century[2] and expressly mentioned by Cosmas[3] and, at the beginning of the twelfth century, by Euthymius Zigabenus.[4] Moreover, the name Satanael (Σαταναήλ),

---
[1] *De daemon. oper.*, P.G. vol. CXXII, cols. 824–5. [The italics are mine.]
[2] See supra, p. 95. [3] See supra, p. 122. [4] See infra, pp. 207 et seq.

as applied to the elder Son of God and the lord of this world, is Bogomil in origin and character.¹ Its prevalence among the Bogomils is attested by Zigabenus in the early twelfth century and by a number of Bogomil apocrypha and literary productions.² Finally the term πρωτότοκος, applied by the Euchitae to Satanael, was used in the same context by the Byzantine Bogomils at the beginning of the twelfth century.³

The distinction made by Psellus between those Euchitae who worship equally both Sons of God, those who worship the younger while honouring the elder for reasons of safety and those who worship Satanael alone, is confirmed by no other source. Hence it is not possible to say whether this distinction existed among the Bogomils. However, the teachings of the first two groups are quite compatible with Bogomil doctrine: the Bogomils, while regarding the Devil or Satanael as the origin of all suffering and evil, held nevertheless that it was necessary to propitiate him; for life on this earth would be precarious without lip-service to him who was in their opinion 'the prince of this world'. This belief clearly appears in the testimony of Zigabenus that the Byzantine Bogomils taught that the demons had unlimited power in this world and that men should consequently honour them in order to guard against their harmful action.⁴ Bogomil demonology, often connected, particularly in Bulgaria, with popular magic, was probably influenced by Massalianism as well as by paganism.⁵

(2) A characteristic of the Euchitae which is certainly not Bogomil but which may be indicative of the Massalians, are the orgiastic rites ascribed to them by Psellus. We are told of dreadful ceremonies performed at night, in which the ashes and blood of

¹ Boissonade's edition (p. 198) gives Σατανακί as a variant. The origin of the name Satanael is discussed by Ivanov (op. cit. p. 25, n. 1). It is derived from the Hebrew 'Satan', meaning 'adversary'; Satana-el is literally 'the adversary of God'. According to Ivanov, in certain pre-Bogomil and early Bogomil legends Satanael is identified with Samael, a name which occurs in the Talmud and in Jewish apocrypha. (Ibid. pp. 260–1; cf. Puech, op. cit. p. 189, n. 3.)
² In particular in the *Book of Enoch* (Ivanov, op. cit. pp. 172, 177), the *Apocalypse of Baruch* (ibid. p. 196), the *Elucidarium* (ibid. pp. 260–1), *The Sea of Tiberias* (ibid. pp. 290 et seq.) and the Greek legend Περὶ κτίσεως κόσμου καὶ νόημα οὐράνιον ἐπὶ τῆς γῆς (ibid. pp. 313–16).
³ Cf. infra, p. 207.   ⁴ Cf. infra, pp. 213–14.
⁵ See P. Kemp, *Healing Ritual*, pp. 167 et seq.

infants conceived in incest were consumed and wanton sexual promiscuity was practised.¹ Even if these lurid stories contain a measure of truth, they are totally incompatible with what we know of the morals of the Bogomils, whose rigid austerity was recognized by their greatest enemies, at least until the fourteenth century. They are, on the other hand, reminiscent of accusations levelled in the past against the Massalians.²

(3) Among the doctrines described by Psellus, which are alien to the Bogomils and probably also to the Massalians, is the exclusive worship of Satanael, attributed to the third group of the Euchitae. Although the Bogomils claimed to pay lip-service to the 'prince of this world', they were certainly in no sense 'Satanists'.³ Likewise the extremely complex and intricate demonological science, to the exposition of which the greater part of Psellus's dialogue is devoted, is foreign to the teachings both of the Bogomils and the Massalians.⁴

Unfortunately Psellus's evidence is not sufficiently clear to permit of a precise definition of the doctrines and practices of the Thracian Euchitae. The most that can be said is that of the heretics described by Psellus some were, in all probability, Bogomils, while the others no doubt belonged to the ancient sect of the Massalians, or Euchitae, which penetrated from Asia Minor into the Balkans between the eighth and ninth centuries.⁵

¹ *De daemon. oper.* (ibid. cols. 828–33).
² Cf. supra, p. 50.
³ This third trend among the Euchitae has been studied by M. Wellnhofer, *Die Thrakischen Euchiten und ihr Satanskult im Dialoge des Psellos:* Τιμόθεος ἢ περὶ τῶν δαιμόνων, *B.Z.* (1929–30), vol. xxx, pp. 477–84.
⁴ Against C. Zervos (*Un philosophe néoplatonicien du XIe siècle: Michel Psellos*, Paris, 1919, p. 202), who claims that this system of demonology was part of the teaching of the Euchitae, J. Bidez has shown ('Michel Psellus.' *Catalogue des manuscrits alchimiques grecs* (Bruxelles, 1928), vol. vi, pp. 100 et seq.) that Psellus based his exposition of this system on his knowledge of Chaldaean teaching, which he derived from the study of the Neoplatonists, mainly of Porphyry and Proclus. Cf. K. Svoboda, 'La Démonologie de Michel Psellos', *Spisy Filosofické Fakulty Masarykovy University v Brně* (Brno, 1927), no. 22.
⁵ The direct connection between the eleventh-century Thracian Euchitae and the ancient Massalians of Asia Minor is recognized by most scholars, following Gieseler and Döllinger. (See supra, p. 94, n. 4.) It is, however, denied by Schnitzer ('Die Euchiten im 11. Jahrhundert', *Studien der evangelischen Geistlichkeit Württembergs* (Stuttgart, 1839), vol. xi, pt 1, pp. 169–86), but his arguments are insufficiently conclusive.

The importance of the *Dialogus de daemonum operatione* from the point of view of the present study lies, first in the fact that it supplies some new information on eleventh-century Bogomilism, and secondly in that it shows that the Massalian and Bogomil sects were still at that time largely distinct from one another.[1]

It is now necessary to consider the relation between the Bogomils and the Paulicians. This problem acquires particular significance in the eleventh century, since it is in this period that the historical sources permit of the clearest differentiation between the two sects. Moreover, it is precisely with reference to the eleventh century that the greatest number of confusions have been made by scholars concerning this relation, with the result that several false conclusions have been drawn about the character and history of Bogomilism. In order to dispel some of these confusions, we must study the relation of the Paulician sect to the Byzantine Church and State in the eleventh century. This will not only illustrate more clearly the differences, already indicated, between the Paulicians and the Bogomils,[2] but will also shed new light on the general character of the latter sect.

During the first three-quarters of the eleventh century, with the exception of the general studies of Euthymius of Acmonia and Psellus, the Byzantine historical sources make no mention of the Paulicians or the Bogomils. Rather than suggesting, as Zlatarski does,[3] any toleration of the Orthodox Church towards their teaching, this silence is doubtless due to the pressure of more urgent political and military problems. All the energies of the imperial government were directed towards the protection of the frontiers against foreign invaders, and so long as the heretics were not actively aggressive towards the Empire they were left in peace. Towards the end of the century, however, the Thracian heretics became a considerable menace to the Byzantine government. The Paulicians, settled round Philippopolis by John Tzimisces, owing to successful proselytism and the arrival of fresh contingents of heretics had greatly increased in number. Moreover, they were showing distinctly hostile intentions towards the remaining

[1] Cf. supra, p. 145.
[2] Cf. supra, pp. 143–4.
[3] *Geschichte der Bulgaren*, vol. I, p. 92.

Orthodox population of that region. Anna Comnena tells us that

'the Manichaeans, being naturally free and unruly, soon... reverted to their original nature. For, as *all the inhabitants of Philippopolis were Manichaeans except a few*, they tyrannized over the Christians there and plundered their goods, caring little or naught for the envoys sent by the emperor. They increased in numbers until *all the inhabitants around Philippopolis were heretics*. Then another brackish stream of Armenians joined them and yet another from the most polluted source of James. And thus, metaphorically speaking, it was a meeting-place of all evils.'[1]

Anna also mentions the Bogomils among the heretics living in Philippopolis: 'For the Armenians[2] took possession of the city and the so-called Bogomils, and even those most godless Paulicians, an offshoot of the Manichaean sect.'[3]

The most troublesome of these heretics were undoubtedly the Paulicians, on account of their numbers and military strength. Their agelong hostility to the Byzantine Empire drove them into an alliance with its enemies, Pechenegs and Cumans, nomadic tribes of Turkish origin which from their encampments on the Danube wrought periodic devastations in Thrace and Macedonia.[4] This shows the complete failure of the imperial policy of transporting Eastern heretics to Thrace. The imperial government, which liked to oppose its subject races one against the other, had sought to gain allies against the Pechenegs by settling the Paulicians in Thrace; instead of this, however, it merely increased the number of the enemies of Byzantium on the northern frontier. The restlessness of the Paulicians drove them into several revolts against the Empire.

In 1078 a certain Lecus, a Greek Paulician from Philippopolis who was married to a Pecheneg woman, incited the population round Sredets (present-day Sofia) and Niš to revolt against the Byzantine rule. He was joined by another group of insurgents,

[1] *The Alexiad*, C.S.H.B. lib. xiv, cap. 8, vol. ii, pp. 299–300; tr. by E. Dawes, p. 385. [The italics are mine.]
[2] These 'Armenian' heretics were undoubtedly Monophysites. Their presence in Thrace and Macedonia is frequently attested in the Middle Ages.
[3] Ibid. p. 384.
[4] See V. Vasilievsky, Византия и печенеги (1048–94), *Zh.M.N.P.* (1872), vol. clxiv, pp. 116–65, 243–332; cf. F. Chalandon, *Essai sur le règne d'Alexis Ier Comnène*, pp. 2–5, 103 et seq.

led by Dobromir, an inhabitant of Mesembria, and together they secured the aid of the Pechenegs and Cumans. Their army, some 80,000 strong, sacked Niš and Sredets; the bishop of the latter city, who exhorted his flock to remain faithful to Orthodoxy, was killed by Lecus. The rebels were routed by the Byzantine general and future emperor, Alexius Comnenus. Lecus and Dobromir were captured, but were released in 1080 and, for some unknown reason, given rich presents and high positions.[1]

It is generally considered that Dobromir was a Bogomil, simply, it would seem, on account of his Slavonic name. In reality, however, this view is unjustifiable: neither of the two Byzantine chroniclers, Michael Attaliates and Joannes Scylitzes, from whom our knowledge of this event is derived, mentions the Bogomil origin of Dobromir.[2] Moreover, it is most improbable that a Bogomil could ever have commanded a group of insurgent forces. There is no evidence to suggest that the Bogomils ever indulged in warfare, which was a favourite occupation of the Paulicians. There is no doubt that their austere and ascetic mode of life and their ideal of evangelical poverty forbade them to shed human blood.[3]

The Paulicians of Philippopolis are mentioned again in 1081, fighting in the army of Alexius Comnenus against the Norman troops commanded by Robert Guiscard and his son Bohemond, who had invaded the Balkan peninsula and were aiming at Constantinople. The Paulicians, however, soon deserted, returned to Philippopolis, and, in spite of repeated injunctions from the emperor, stubbornly refused to go back to the army.[4]

[1] See K. Jireček, *Geschichte der Bulgaren*, pp. 208–9; Vasilievsky, loc. cit. p. 153; Spinka, op. cit. p. 94.

[2] Michael Attaliates, *Historia*, C.S.H.B. p. 302; Joannes Scylitzes, *Historia*, C.S.H.B. p. 741.   [3] Cf. supra, p. 182.

[4] Anna Comnena, *Alexiad*, C.S.H.B. lib. v, cap. 3, vol. I, p. 232. The Paulician forces were commanded by Xantas and Culeon (cf. Chalandon, op. cit. pp. 76 et seq.). It is interesting to note that the motley collection of troops which composed Alexius's army included a number of English soldiers who had emigrated to Byzantium after the conquest of England by William the Conqueror. (See E. A. Freeman, *The History of the Norman Conquest of England*, Oxford, 1871, vol. IV, p. 628.) Thus Englishmen fought for a time side by side with the Thracian Paulicians.

William of Apulia gives an interesting piece of information concerning those Paulicians who fought under the banners of Alexius: he says that they made the sign of the cross with one finger ('et fronti digito signum crucis imprimit uno'): *Gesta Roberti Wiscardi*: M.G.H. Ss. vol. IX, p. 248.

Here again it has been falsely asserted, without any evidence, that the Bogomils fought together with the Paulicians in the Byzantine army.[1]

The unruly and troublesome character of the Paulicians of Philippopolis is illustrated by the following episode: the Emperor Alexius, after his victory over the Normans at Kastoria (1083), resolved to punish the defection of the Paulicians from his army. However, their strength in their own home was such that Alexius was afraid to risk a punitive expedition against Philippopolis. Instead he summoned a number of representative Paulicians to Constantinople, had them arrested and imprisoned and their property confiscated. Those of them who consented to be baptized were later released and allowed to return home, the others were banished.[2]

Alexius's action in summoning the Paulicians to Constantinople and then sending some of them back to Philippopolis, after an abjuration doubtless largely prompted by fear, was ill-advised: he had shown himself too weak to repress the Paulicians in their own home; moreover, he provided the heretics of Philippopolis with leaders who had now acquired a halo of martyrdom for the suffering they had incurred for the sake of their faith. It can be supposed that this action infused fresh courage and strength not only into the Paulicians, but also into the other heretics of Philippopolis, including the Bogomils, whose presence in that town in the second half of the eleventh century is attested by Anna Comnena.[3]

The consequences of Alexius's punishment of the Paulicians were not slow in becoming manifest. In the same year (1084)[4] a mutiny against the emperor broke out under the leadership of a certain Traulus, who was a personal servant of Alexius and a baptized Paulician. According to Anna Comnena he rebelled out of anger against the emperor who had caused his four sisters to be driven from their homes by reason of their Paulician faith.[5] Traulus rallied his former co-religionists and from the fortified

[1] See Spinka, op. cit. p. 95. This error was made by such an authority on Bogomilism as Ivanov (op. cit. p. 31).
[2] For further details see Anna Comnena, *Alexiad*, lib. VI, cap. 2, pp. 272 et seq.
[3] Cf. supra, p. 189.
[4] The chronology of these events is studied by Chalandon (op. cit. pp. 105–6).
[5] *Alexiad*, lib. VI, cap. 4, pp. 279–80.

castle of Belyatovo[1] carried out raids into the neighbourhood (1085-6). Following the example of Lecus, Traulus allied himself with the Pechenegs by marrying a daughter of one of their chiefs, and provoked an invasion of Thrace by a force of 80,000 Pechenegs and Cumans. For two years the emperor was unable to expel these formidable invaders, who succeeded in defeating a Byzantine army sent against them.[2]

It is not improbable that Traulus received some support from the other heretical sects in Thrace, which were doubtless glad of a useful ally against the hated Byzantine domination. Although evidence is lacking on this point, it is legitimate to suppose that the Bogomils of Philippopolis supported the Paulicians, at least passively, against their common enemy. But again it is false to conclude that they participated in any armed insurrection.[3]

The frequent contacts between Paulicians and Pechenegs in the eleventh century, which included marital relations, suggest the possibility of the spread of dualistic doctrines among the latter. Attaliates paints a picture of the motley collection of peoples of different races who lived at that time on the Danube and had adopted the nomadic life of the Pechenegs.[4] The conditions in this region were undoubtedly favourable to the spread of heresy: all those who harboured a grudge against Byzantium or were prompted by ambition and the desire for adventure would seek refuge among the Pechenegs on the Danube. Moreover, the ecclesiastical administration, centred in Ochrida, could exercise little control over this distant borderland. In these circumstances it is not unlikely that Paulician missionaries, always eager for any opportunity to proselytize, found their way to the shores of the

[1] The exact geographical position of Belyatovo is not known. Jireček (op. cit. p. 209) and Chalandon (op. cit. p. 107, n. 1) place this fortress in the Balkan mountains, to the north of Philippopolis.

[2] See Chalandon, op. cit. pp. 107 et seq.; Spinka, op. cit. p. 95.

[3] Jireček unjustifiably ascribes the rebellion of 1084-6 to the Bogomils: 'gleichzeitig erhoben sich die Bogomilen und besetzten...das Bergschloss Beljatovo...und brandschatzten von da aus ganz Thrakien'. (Ibid.) Ivanov (op. cit. p. 31) makes the same mistake. The use of warfare is totally incompatible with our knowledge of Bogomil ethics.

[4] Παράκεινται...τῇ ὄχθῃ τούτου πολλαὶ καὶ μεγάλαι πόλεις, ἐκ πάσης γλώσσης συνηγμένον ἔχουσαι πλῆθος, καὶ ὁπλιτικὸν οὐ μικρὸν ἀποτρέφουσαι. (*Historia*, p. 204.)

Danube and spread their doctrines among the 'Scythians', as the Byzantines called the inhabitants of this region.[1] The possibility of contact between the Bogomils and the Pechenegs is more problematic, since direct evidence is lacking, yet it cannot altogether be excluded; even if Bogomilism did not penetrate at that time to the Danubian settlements, the Pechenegs may well have encountered its teachers during their frequent raids into Macedonia and Thrace. It may be noted, in this connection, that after the battle of 29 April 1091, in which the greater part of the Pecheneg hordes was slaughtered by the Byzantines, Alexius Comnenus settled their remnants to the east of the Vardar, in the region of Moglena,[2] where the Bogomils were particularly numerous in the twelfth century.[3]

The contacts of Alexius with the heretics of Philippopolis were not solely of a military nature. When preparing in 1114 for a campaign against the Cumans, Alexius established his headquarters at Philippopolis and began a systematic attempt at converting the Paulicians to the Orthodox faith. The method to which he now resorted was theological disputation. The traditional role of the Byzantine Basileus as the supreme upholder and protector of Orthodoxy was always assumed by Alexius with great earnestness. On every possible occasion he set himself up as the champion of the true faith.[4] For this reason his daughter Anna

[1] Vasilievsky (op. cit. pp. 150 et seq.) has put forward the hypothesis that between the Pechenegs and the Paulicians of Philippopolis there existed similarities in faith and customs due to their common Manichaean inheritance. According to him, Manichaean beliefs spread among the Pechenegs through the Cumans, who lived in the tenth century in the neighbourhood of Khorasan and Turkestan, countries occupied at that time by the Turkish Manichaeans.

In support of this hypothesis, Ivanov (op. cit. pp. 19–20) adduces the evidence of an Arabic source, according to which the original religion of the Pechenegs was 'Zarathustrian dualism'; later, however, some of them became Moslems, fought against the remaining dualistic tribes and compelled them to accept Islam.

However, this theory cannot be considered as proved, as the evidence adduced in support of it is derived from vague or insufficiently reliable sources. Chalandon describes Vasilievsky's hypothesis as 'plus ingénieuse que vraie' (op. cit. p. 104, n. 1).

[2] See Jireček, op. cit. p. 209; Chalandon, op. cit. pp. 132–4.
[3] Cf. infra, p. 223.
[4] See Chalandon, op. cit. pp. 309 et seq.; L. Oeconomos, *La Vie religieuse dans l'Empire Byzantin au temps des Comnènes et des Anges* (Paris, 1918), pp. 48–9.

claimed for him the title of the thirteenth apostle, or at least of the fourteenth, if priority must be given to Constantine the Great.[1]

Alexius's disputations with the Paulicians, as they appear from Anna's vivid account, are most interesting for a student of Bogomilism: their behaviour during these theological jousts reveals certain similarities with as well as differences from that generally attributed to the Bogomils. Features common to both sects were a profound acquaintance with the Scriptures, an astonishing dialectical skill and a never-failing ability to interpret the Holy Writ in accordance with their own doctrines. Attended by Nicephorus Bryennius (his son-in-law and the husband of Anna Comnena), by Eustratius, metropolitan of Nicaea and a celebrated theologian,[2] and by the bishop of Philippopolis, Alexius held lengthy disputations with the Paulicians: 'from the morning till afternoon or even evening, and sometimes till the second or third watch of the night he would send for them and teach them the Orthodox faith and refute their distorted heresies.'[3] The emperor succeeded in converting a number of them to Orthodoxy. The more adamant ones, however, and especially the three Paulician leaders, Culeon,[4] Cusinus and Pholus, stubbornly withstood the emperor's arguments. Anna asserts that 'they were... exceedingly able in pulling the Scriptures to pieces and in interpreting them perversely'. She vividly describes their heated arguments with Alexius: 'The three stood there sharpening each other's wits, as if they were boar's teeth, intent on rending the emperor's arguments. And if any objection escaped Cusinus, Culeon would take it up; and if Culeon was at a loss, Pholus in his turn would rise in opposition; or they would, one after the other, rouse themselves against the emperor's premises and refutations, just like very large waves following up other large waves.' Cusinus and Pholus persisted in their faith till the end; Alexius finally wearied of them and had them im-

---

[1] *Alexiad*, C.S.H.B. lib. XIV, cap. 8, vol. II, pp. 300–1.

[2] An account of this important personage in the history of the Byzantine Church at the beginning of the twelfth century can be found in Th. Uspensky, Богословское и философское движение в Византии XI и XII веков, *Zh.M.N.P.* (September, 1891), vol. CCLXXVII, pp. 145–7.

[3] *Alexiad*, ibid. p. 301; Dawes's tr. p. 386.

[4] Culeon had commanded a Paulician detachment in Alexius's army in 1081. Cf. supra, p. 190, n. 4.

prisoned in Constantinople, where they were 'allowed... to die in company with their sins alone' as Anna euphemistically puts it; Culeon, however, was eventually won over by the emperor's arguments, whose potency had doubtless been greatly increased by the use of force. Together with him a considerable number of Paulicians were converted: 'every day he brought to God, may be a hundred, may be even more than a hundred; so that the sum total of those he had captured before and those whom he won now by the words of his mouth would amount to thousands and tens of thousands of souls.' Great material benefits were bestowed on the converts: the more eminent received 'great gifts' and high military positions; for the smaller fry Alexius built a new city, Alexiopolis, more commonly known as Neocastrum, to which he transferred the converted Paulicians of Philippopolis, granted them land and by special chrysobulls secured them in their possessions for all time.[1]

The emperor's treatment of the Paulicians, which combined theological controversy with the occasional display of force, is characteristic of the attitude of the Byzantine Church towards heretics under the Comnenian dynasty and also illustrates Alexius's behaviour towards the Bogomils.[2] These sectarians do not figure in Anna's account of her father's disputations in Philippopolis in 1114; and yet she herself asserts that the Bogomils existed in Philippopolis at that time. It can only be concluded that in 1114 the Bogomils of Philippopolis escaped the notice of Alexius; this fact can be explained by remarking that they were probably far less numerous in the city than the Paulicians, that they lacked the warlike instincts of the latter and, in contrast to their open and fearless proselytism, worked more by concealment and subtle infiltration. This is confirmed by Anna's opinion that the Bogomil heresy 'probably existed even before my father's time, but in secret; for the sect of the Bogomils is very clever in aping virtue'.[3]

The evidence of the Byzantine sources relating to the late eleventh and early twelfth centuries thus shows that the identification so often made in this period between the Bogomils and the Paulicians is not legitimate. The roles played by both sects in Bulgarian and Byzantine history were in many respects very

[1] *Alexiad*, C.S.H.B. lib. XIV, caps. 8–9, vol. II, pp. 301–6; Dawes's tr. pp. 386-9.
[2] See infra, pp. 203 et seq. [3] Ibid. p. 351.

different. Thus it cannot be maintained that the Bogomils either instigated or actively participated in the numerous revolts against Byzantium, in which the Paulicians played so active a role. Moreover, their methods of proselytism were entirely different: the Paulicians publicly professed their teaching and fearlessly maintained their ground against the emperor and the highest ecclesiastical and civil authorities of the Empire. The Bogomils, on the contrary, held no public disputations, claimed to be Orthodox Christians and only revealed their secret teaching under dire necessity. The difference in the very nature of both sects shows that of the two the Bogomils were undoubtedly the more dangerous for the Church: the Paulicians formed turbulent military colonies, mainly of foreign origin, no doubt troublesome subjects of Byzantium, but easy to locate and combat on their own ground. The Bogomils, on the contrary, were often almost indistinguishable from the Orthodox Christians, as they were generally Bulgarians or Greeks and outwardly obeyed all the rules of the Church. This largely explains the fact that, whereas a large number of Paulicians renounced their doctrines owing to the efforts of Alexius Comnenus, the Bogomils, who were generally reputed to be incapable of conversion, resisted the strongest persecution.

Although the spread of Bogomilism in Thrace and Asia Minor in the eleventh century is attested by Psellus, Anna Comnena and Euthymius of Acmonia, we possess no unimpeachable and contemporary evidence of the prevalence of the sect in Macedonia.[1] Yet there can be little doubt that in the eleventh century the

[1] Two letters of Theophylact, archbishop of Ochrida, to Adrian, brother of Alexius Comnenus, and to Nicephorus Bryennius, the emperor's son-in-law, contain perhaps an allusion to the Bogomils (*Epistolae a J. Meursio editae*, P.G. vol. CXXVI, cols. 441–52, 453–60). He bitterly complains in them of a certain Lazarus who, harbouring some grudge against him, incited the inhabitants of Ochrida to oppose their archbishop. He also went round other districts of Macedonia and sought 'with great assiduity' (λίαν ἐπιμελῶς) to rally all those who bore any resentment against Theophylact, particularly *any one who had been condemned for heresy* or imprisoned for other transgressions. He succeeded in discovering a large number of such malcontents (εὑρὼν...πολλοὺς τοιούτους). Lazarus then left Ochrida and went to Pelagonia, where he continued his seditious activities. (Ibid. cols. 444–5.) It will be remembered that Pelagonia, or the region of Bitolj, was at that time a centre of Bogomilism.

Bogomil sect was solidly rooted in its original home. Conditions in Macedonia at that time were most favourable to its development: not only did the widespread hatred of the Greeks increase the prestige of the Bogomils as defenders of the people, but the political confusion and economic decline due to the constant invasions of Macedonia by Normans, Latin Crusaders, Pechenegs and Cumans produced in the people a restlessness and dissatisfaction which so frequently resulted in a recrudescence of Bogomilism. The Archbishop Theophylact of Ochrida, whose correspondence paints an eloquent picture of the constant troubles brewing in Macedonia in the late eleventh century, compared his diocese to David's vineyard, laid open to the plunder of all the passers-by.[1]

But the most far-reaching event in the history of Bogomilism in the eleventh century is its penetration into Constantinople. The study of this penetration is important for a complete understanding of Bulgarian Bogomilism for several reasons: first, the growth of Bogomilism in Byzantium affords a good example of the methods of propagation and the success enjoyed by the sect in the eleventh and twelfth centuries. Secondly, the information on the Byzantine Bogomils supplied by the twelfth-century Byzantine writers deepens, extends and illustrates the evidence on the Bulgarian Bogomils provided by the tenth- and eleventh-century sources. This method of investigation is justified by the direct contact which existed in the eleventh and twelfth centuries between the Bogomils of Constantinople and their Bulgarian co-religionists. Thirdly, in view of this contact, a study of Byzantine Bogomilism explains a number of features in the future development of the sect in Bulgaria during the twelfth and thirteenth centuries.

Our knowledge of Byzantine Bogomilism in the late eleventh and early twelfth centuries is almost exclusively derived from the *Alexiad* of Anna Comnena[2] and the *Panoplia Dogmatica* of Euthymius Zigabenus.[3]

The *Alexiad*, completed in 1148, gives the fullest account of the discovery and prosecution of the Bogomils in Constantinople.

---

[1] *Epistolae a J. Lamio editae*, ibid. col. 529. Cf. Ps. lxxx. 12–16.
[2] Lib. xv, caps. 8–10, pp. 350–64.
[3] Tit. 27, *P.G.* vol. cxxx, cols. 1289–1332. A slightly different version of Zigabenus's account of the Bogomils was published by Ficker (op. cit. pp. 89–111) under the title: *Euthymii Zigabeni de haeresi Bogomilorum narratio*.

Unfortunately, this event has remained undated in the manuscript, owing no doubt to a lacuna in Anna's memory. It can, however, be placed approximately in 1110.[1] Towards that date, in Anna's words,

'A very great cloud of heretics arose, and the nature of their heresy was new and hitherto quite unknown to the Church. *For two very evil and worthless doctrines, which had been known in former times, now coalesced; the impiety, as it might be called, of the Manichaeans, which we also call the Paulician heresy, and the shamelessness of the Massalians. This was the doctrine of the Bogomils compounded of those of the Massalians and the Manichaeans.*'[2]

Of these three features of Bogomilism described by Anna—its magnitude, its novelty and its composition—the second is incorrect, as the sect had in fact already been attracting the attention of the Bulgarian and Byzantine authorities for some 150 years.[3] The other two, however, are fully confirmed by our past knowledge of the Bogomil sect: the reference to 'a very great cloud of heretics' is scarcely an exaggeration, when viewed in the light of the considerable success which Bogomilism had already gained in Bulgaria and was then gaining, judging by Anna's account, in Byzantium; finally, it is interesting to note that Anna's analysis of Bogomilism is identical with that of the Patriarch Theophylact who, in the middle of the tenth century, defined it as 'Manichaeism mixed with Paulicianism'.[4]

[1] The date of the trial of the Bogomils in Byzantium is discussed in Appendix III.
[2] *Alexiad*, lib. xv, cap. 8, pp. 350–1. [The italics are mine.]
[3] Anna, however, qualifies her statement by the remark that 'probably it existed even before my father's time, but in secret'.
[4] As has been shown (see supra, pp. 114–15) the term 'Manichaean' is used by Theophylact in the sense of Massalianism; for Anna, on the contrary, it is synonymous with Paulicianism.

Anna says that Bogomilism combined the 'impiety' (δυσσέβεια) of the Paulicians with the 'shamelessness' (βδελυρία) of the Massalians. Her use of these terms is significant: δυσσέβεια (or ἀσέβεια) had for the Byzantines a definitely doctrinal connotation, designating a teaching contrary to that of the Church (see J. M. Hussey, *Church and Learning in the Byzantine Empire*, pp. 84, 89, 120, 153); while βδελυρία implied above all moral depravity. This distinction illustrates the fact that the Bogomils derived most of their doctrines from the Paulicians and much of their ethical teaching and social behaviour from the Massalians; it justifies to some extent the statement of G. Buckler that the Bogomils 'may be said to have been Paulicians in dogma and Massalians in morals' (*Anna Comnena*, Oxford, 1929, p. 339).

The vivid picture drawn by Anna of the outward appearance of the Bogomils is strikingly similar to Cosmas's description of the Bulgarian heretics:[1]

'The sect of the Bogomils is very clever in aping virtue. And you would not find any long-haired worldling belonging to the Bogomils, *for their wickedness was hidden under the cloak and the cowl.* A Bogomil looks gloomy and is covered up to the nose and walks with a stoop and mutters, but within he is an uncontrollable wolf. And this most pernicious race, which was like a snake hiding in a hole, my father lured and brought out to the light by chanting mysterious spells.'[2]

The accusation of hypocrisy and pharisaic humility is levelled against the Bogomils by nearly all their opponents. This behaviour was probably due not so much to their attempts to deceive the Orthodox into believing that they were good Christians, as to a genuine, though exaggerated, preoccupation with asceticism and moral purity, based on a hatred of the material world. Both Anna and Zigabenus explicitly state that the Bogomils dressed as monks and led the monastic life.[3]

According to Anna, Alexius became aware of the existence of the heretics in the capital of his Empire from the fact that 'by this time the fame of the Bogomils had spread everywhere'. A certain Diblatius, member of the sect, was arrested and questioned; he revealed under torture the names of the leading Bogomils and of the supreme head of the sect, Basil.. 'And Satanael's arch-satrap, Basil, was brought to light, in a monk's habit, with a withered countenance, clean-shaven and tall of stature.' Anna also describes him as follows: 'Basil, a monk, was very wily in handling the impiety of the Bogomils; he had twelve disciples whom he called "apostles", and also dragged about with him some female disciples, wretched women of loose habits and thoroughly bad, and disseminated his wickedness everywhere.'[4] Euthymius Zigabenus asserts that he was a doctor (ἰατρός);[5]

---

[1] Cf. supra, p. 121.   [2] *Alexiad*, ibid. p. 351. Dawes's tr. p. 412.
[3] Zigabenus writes: 'they dress after the fashion of monks, wear the habit as a bait...and thus avoid suspicion and by their unctuous speech inject their venom into the ears of those who listen to them.' (*Pan. Dog.*, *P.G.* vol. cxxx, cap. 24, col. 1320.)
[4] *Alexiad*, ibid. pp. 351–2.   [5] *Pan. Dog.* ibid. col. 1289.

this statement is repeated by Zonaras[1] and Glycas[2] and by the thirteenth-century *Synodicon of the Tsar Boril*.[3] Zigabenus also states that Basil had studied the Bogomil doctrines for fifteen years and then had taught them for fifty-two years.[4] Hence, if the trial of Basil in Constantinople is placed *c.* 1110, it can be inferred that he became a teacher in the Bogomil sect not later than 1070. It is not known whether between 1070 and 1110 he preached in Byzantium or in other parts of the Empire; but judging by the widespread nature of the sect at the time of his arrest, it seems most probable that Basil had taught in Constantinople for at least several years previously.[5]

The name of 'apostles', given to the immediate disciples of Basil, may have existed already among the Bulgarian Bogomils of the tenth century.[6] Their symbolic number of twelve is reminiscent not only of the apostles of Christ, but of the twelve disciples of

---

[1] *Epitome historiarum*, l. xviii, c. 23, *C.S.H.B.* vol. iii, p. 743.

[2] *Annales*, pars iv, *C.S.H.B.* p. 621.

[3] Cf. infra, p. 240. G. Kiprianovich (Жизнь и учение богомилов по Паноплии Евфимия Зигабена и другим источникам, *P.O.* July 1875, vol. ii, p. 380, n. 2) thinks, however, that Basil's title of ἰατρός is fictitious and is simply due to contemporary popular belief which endowed him with magical powers.

[4] *Narratio*, p. 111; this figure is confirmed by Zonaras (loc. cit.). The *Panoplia Dogmatica* gives the figure as 'more than forty years' (ibid. col. 1332).

[5] N. Kalogeras ('Ἀλέξιος Α' ὁ Κομνηνός, Εὐθύμιος ὁ Ζιγαβηνὸς καὶ οἱ αἱρετικοὶ Βογόμιλοι, *Athenaion*, Athens, 1880, vol. ix, p. 259) supposes that Basil was of Bulgarian origin. Tsukhlev (История на българската църква, vol. i, pp. 1032–4) and Klincharov (op. cit. p. 74) identified him with an unnamed monk, mentioned in a letter of Theophylact of Ochrida, who renounced his vows, lived unchastely and conspired with the archbishop's enemies (*P.G.* vol. cxxvi, cols. 513–16). This theory has prompted Spinka to state of Basil, the leader of the Byzantine Bogomils: 'he was a Bulgar born in Macedonia. Having become a monk in some monastery of the Ohrid diocese, he learned Greek there and became proficient in the rudiments of the healing art, so that he was later spoken of as a physician. Later he left the monastery for some unknown reason and returned to lay life, but soon passed over to the Bogomil community as one of the "perfect" (!).' (Op. cit. p. 97.)

This hypothesis, attractive as it may be, rests on absolutely no conclusive evidence. The identification of Basil with the unfrocked monk mentioned by Theophylact is entirely arbitrary, since we do not even know this monk's name and he is nowhere accused of heresy. Moreover, neither Zigabenus, nor Anna Comnena, nor the other Byzantine chroniclers, who are the only source of our knowledge of Basil, say a word concerning his origin.

[6] Cf. supra, p. 133.

Mani. But Anna's assertion that Basil was surrounded by women of loose morals is difficult to believe. In no other source were the Bogomils accused of sexual immorality, at least until the fourteenth century. On the contrary, their rigid asceticism was recognized by all their opponents. Nor is there any ground for maintaining that at that time 'their principles may have been so lofty as to impose a strain on human nature which few men could have been expected to bear'.[1] The equality between the sexes which existed in the Bogomil sect[2] and the presence of women among Basil's followers probably gave rise to Anna's unjustified suspicions.[3]

A most interesting and novel feature of Bogomilism appears in Anna's remark that the heresy had numerous adherents in the aristocratic families in Byzantium (καὶ εἰς οἰκίας μεγίστας).[4] This suggests a distinct evolution in the character of the sect. In tenth-century Bulgaria, as far as can be judged from Cosmas's

---

[1] Buckler, op. cit. p. 344.   [2] Cf. supra, p. 135.
[3] Anna's accusations of immoral conduct levelled against the followers of Basil may also be due to a confusion between the Bogomils and the Massalians; the latter were commonly accused of the foulest practices (cf. supra, pp. 50, 186–7). But the identification of both sects occurs only later (cf. infra, pp. 251 et seq.), and Zigabenus still recognizes the distinction between them.

Mrs Buckler (op. cit. pp. 339–44) expresses astonishment at the 'unexplained hatred' of Anna Comnena and other contemporary Byzantine writers towards the Bogomils and at the violent expressions which the former uses to describe them. She thinks that they may be due to 'the prevalence among the Bogomils of unholy and awful rites', or to the fact that 'under the protection of the monastic habit these heretics wormed their way into families for evil purposes'. In reality, these hypotheses are irrelevant and can be substantiated by no evidence. Their principal weakness lies in the fact that they appear to underestimate the paramount importance which the preservation of the purity of the Orthodox doctrine had in the eyes of the Byzantines. Theological heresy for them was more reprehensible than immorality, although the two were frequently connected in their minds. Anna's hatred of the Bogomils is primarily aimed at their false teaching, and the accusation of immorality is of secondary importance. Moreover, the abhorrence for the Bogomils which she shows, contrasted with the comparative mildness with which she refers to the Paulicians, can be explained by the fact that the former were the more dangerous to the Church (cf. supra, p. 196) and that many of the latter had been converted to Orthodoxy by the time the *Alexiad* was written. That is why Basil was for Anna 'Satanael's arch-satrap', 'abominable', 'accursed' (*Alexiad*, lib. xv, cap. 8, p. 352; cap. 10, pp. 362, 364), and for Zigabenus— 'pernicious and pestilent, full of corruption and the instrument of all evil' (*Pan. Dog., P.G.* vol. cxxx, col. 1289).

[4] *Alexiad*, lib. xv, cap. 9, p. 358.

evidence, Bogomilism was predominantly an ethical teaching with a dualistic foundation and a popular movement with a particular appeal to the peasant masses. Its doctrinal and speculative aspects remained comparatively undeveloped, which caused Cosmas to deride the contradictions and inconsistencies in the teachings of the Bogomils.[1] In Byzantium, on the contrary, Bogomilism came into contact with the upper classes, always eager for theological speculation, and with various philosophical theories of an unorthodox nature.[2] Under these influences, its theology and cosmology were reshaped into a coherent system, and Bogomilism assumed the character of a philosophical sect.

It seems, moreover, that the success that Bogomilism enjoyed at that time in the educated society of Byzantium was also due to its increased contact with certain mystic trends of Massalian origin. Anna tells us of a certain Blachernites, a priest who 'had consorted with the Enthusiasts [i.e. the Massalians] and became infected with their mischievous doctrines, led many astray, undermined great houses in the capital, and promulgated his impious doctrines'. Unlike most of the Paulicians, Blachernites proved to be impervious to the exhortations of Alexius and was finally condemned to 'a perpetual anathema'.[3] The fact that the Bogomils and the Massalians were working hand-in-glove among the upper classes in Constantinople would naturally have facilitated a more intimate contact between their teachings.

But Anna does not concern herself with the doctrines of the Bogomils, not wishing 'to defile her tongue', and refers the curious to the *Panoplia Dogmatica* of Zigabenus, where they are fully set out.[4] She only mentions briefly some of the tenets professed by Basil: rejection of Orthodox theology, criticism of the ecclesiastical hierarchy, scorn of churches as the abode of demons and

---

[1] Cf. supra, p. 126.

[2] In particular, there appears to have been some contact between Bogomilism and the teaching of the disciples of John Italus, especially of the monk Nilus. An examination of these connections would require a special study of the philosophical movements in eleventh- and twelfth-century Byzantium, which lies outside the scope of the present work. See Th. Uspensky, Богословское и философское движение, loc. cit. passim; Chalandon, op. cit. pp. 309 et seq.; Hussey, op. cit. pp. 89 et seq.

[3] *Alexiad*, lib. x, cap. 1, vol. II, p. 4; Dawes's tr. p. 236.

[4] *Alexiad*, lib. xv, cap. 9, p. 357.

denial of the Real Presence in the Eucharist,¹ all of which are familiar from the earlier sources on the Bogomils.

The major part of chapters 8, 9 and 10 of the fifteenth book of the *Alexiad* describes the trial of the Bogomils in Byzantium and the execution of Basil. The leader of the Bogomils was summoned to the Palace, where his confession of faith was extorted from him by the following trick: the Emperor Alexius and his brother Isaac the Sebastocrator feigned an interest in Basil's teaching and professed a desire to become his disciples. Basil was lured into giving a full exposition of his teaching, after which Alexius dramatically threw back the curtain separating the room from the adjoining one and revealed the presence of a secretary who had taken down the heresiarch's confession.² Formally charged with heresy and threatened with torture and death, Basil, however, refused to abjure and 'remained the same, an inflexible and very brave Bogomil'. 'He clung to the demon with closed teeth and embraced his Satanael.' The emperor had him imprisoned and pleaded with him again and again, but Basil refused to renounce his doctrines. Alexius, meanwhile, had ordered a general round-up of all the Bogomils. According to Anna, he 'had summoned Basil's disciples and fellow-mystics from all over the world (ἁπαν-ταχοῦ γῆς),³ especially the so-called twelve disciples, made trial of their opinions and found that they were openly Basil's followers'. As some of them had recanted out of fear, Alexius resorted to another ruse to separate the sheep from the goats: he ordered all those suspected of Bogomilism to be burnt alive, but allowed them to choose between a pyre with a cross and a pyre without one. Those who chose the first pyre were released as having proved their orthodoxy; the rest were imprisoned and again subjected to daily exhortations by Alexius; those who persisted in their heresy were imprisoned for life, but, Anna adds unctuously, 'were amply supplied with food and clothing'.⁴

¹ Ibid. lib. xv, cap. 8, p. 354.
² The same trick had been used by Flavian, patriarch of Antioch, at the end of the fourth century to obtain the confession of the leader of the Syrian Massalians. See Georgius Cedrenus, *Hist. Compend.*, C.S.H.B. vol. I, pp. 514–16.
³ 'The whole world' clearly means in the present context the Byzantine Empire and, no doubt, Bulgaria in particular.
⁴ *Alexiad*, lib. xv, cap. 8–9, pp. 352–60.

By the unanimous decision of the Holy Synod, the Patriarch Nicholas and the chief monks, Basil was condemned to the stake. Anna's account of his execution is uncommonly realistic:

'The emperor...after conversing with him several times and recognizing that the man was mischievous and would not abandon his heresy...finally had an immense pyre built in the Hippodrome. A very large trench was dug and a quantity of wood, all tall trees piled up together, made the structure look like a mountain. When the pile was lighted, a great crowd slowly collected on the floor and steps of the circus in eager expectation of what was to happen. On the opposite side a cross was fixed and the impious man was given a choice, for if he dreaded the fire and changed his mind, and walked to the cross, then he should be delivered from burning. A number of heretics were there watching their leader Basil. He showed himself contemptuous of all punishment and threats, and while he was still at a distance from the fire he began to laugh and talk marvels, saying that angels would snatch him from the middle of the fire, and he proceeded to chant these words of David's: 'It shall not come nigh thee; only with thine eyes shalt thou behold.' But when the crowd stood aside and allowed him to have a free view of that terrifying sight, the burning pyre (for even at a good distance he could feel the fire, and saw the flames rising high and as it were thundering and shooting out sparks of fire which rose to the top of the stone obelisk which stands in the centre of the Hippodrome), then the bold fellow seemed to flinch from the fire and be disturbed. For as if wholly desperate, he constantly turned away his eyes and clapped his hands and beat his thigh. And yet in spite of being thus affected by the mere sight he was adamant. For the fire did not soften his iron will, nor did the messages sent by the emperor subdue him. For either great madness had seized him under the present stress of misfortunes and he had lost his mind and had no power to decide about what was advantageous; or, as seems more likely, the devil that possessed his soul had steeped it in the deepest darkness. So there stood that abominable Basil, unmoved by any threat or fear, and gaped now at the fire and now at the bystanders. And all thought him quite mad, for he did not rush to the pyre nor did he draw back, but stood fixed and immovable on the spot he had first taken up. Now many tales were going round and his marvellous talk was bandied about on every tongue, so the executioners were afraid that the demons protecting Basil might perhaps, by God's permission, work some wonderful new miracle, and the wretch be seen snatched unharmed from the middle of the mighty fire and transported to some very frequented place. In that case the second state would be worse than the first, so they decided to make an experiment. For, while he was talking marvels and boasting that he would be seen unharmed in the middle of the fire, they took his cloak and said, 'Now let us see whether the fire will touch your garments', and they threw it right into the middle of the pyre. But Basil was so uplifted by the demon that was deluding him that he said, 'Look at my cloak floating up to the sky!' Then they, 'recognizing the web from the edge', took him and pushed him, clothes, shoes and all, into the middle of the pyre. And the flames, as if deeply enraged against him, ate the impious man up, without any odour arising or even a fresh appearance of smoke, only one thin smoky line could be seen in the midst of the flames.... Then the people looking on clamoured loudly and demanded that all the rest who belonged to Basil's pernicious sect should be thrown into the fire as well, but the emperor did not allow it but ordered them to be confined in the porches and verandahs of the largest palace. After this the concourse was dismissed.'[1]

Alexius's treatment of the Bogomils was comparatively mild. Basil was the only one to be punished by death; those of his followers who refused to be converted were, it is true, sentenced to perpetual imprisonment, but their treatment in prison was not harsh. It is to Alexius's everlasting credit that in his dealings with heretics he used the weapon of persuasion in preference to any other.

For an exposition of the doctrines of the Byzantine Bogomils we must turn to the *Panoplia Dogmatica* of Euthymius Zigabenus. The circumstances in which it was composed are described by Anna Comnena:

'There was a monk called Zigabenus, known to my mistress, my maternal grandmother, and to all the members of the priestly roll, who had pursued his grammatical studies very far, was not unversed in rhetoric, and was the best authority on ecclesiastical dogma; the emperor sent for him and commissioned him to expound all the heresies, each separately, and to append to each the holy Fathers' refutations of it; and amongst them too the heresy of the Bogomils, exactly as that impious Basil had interpreted it. The emperor named this book the Dogmatic Panoply, and that name the books have retained even to the present day.'[2]

[1] *Alexiad*, lib. xv, cap. 10, pp. 361–4; Dawes's tr. pp. 417–18.
[2] Ibid. cap. 9, p. 357; Dawes's tr. p. 415; cf. *Pan. Dog.*, *P.G.* vol. cxxx, col. 1292.

It is clear from this that Zigabenus had first-hand information on the Bogomil sect, which makes his *Panoplia* on the whole a very reliable document.[1] Its primary importance lies in the fact that it is by far the fullest and most systematic account of the Bogomil doctrines that we possess. It also brings out a number of new features of Bogomilism and thus contributes to our knowledge of both the past and the subsequent development of the sect in Bulgaria.

Zigabenus's definition of Bogomilism, without being as explicit as that of the Patriarch Theophylact or of Anna Comnena, agrees with them as to the double derivation of this heresy from Massalianism and Paulicianism. In his words, 'the Bogomil heresy is not much older than our generation; it is part of the heresy of the Massalians (μέρος οὖσα τῆς τῶν Μασσαλιανῶν) and agrees with its doctrines on most points; however, it added to it some other

---

[1] Before the discovery and publication of the Slavonic sources relating to the Bogomils, the trustworthiness of Zigabenus's evidence was frequently denied. G. Arnold viewed his account of Bogomilism with great suspicion (*Kirchen- und Ketzer-Historien*, Schaffhausen, 1740, Th. I, Bd XII, cap. 3, § 2, vol. I, p. 374). J. L. Oeder devoted a treatise to an attempt to prove that the Bogomils were maliciously slandered by the representatives of the Church, who were jealous of their influence over the people (*Dissertatio...prodromum historiae Bogomilorum criticae exhibens*, Gottingae, 1743). In his opinion, the Byzantine Church at that time was entirely decadent and the very fact that Zigabenus was a monk invalidates his evidence: 'nullam...paene auctoritatem Graeculi huius agnosco, dum *monachum* cogito, i.e. hominem superstitiosum, orthodoxitam, crudelem, credulum' (p. 8).

But already in 1712 J. C. Wolf devoted an extensive and erudite treatise to the vindication of Zigabenus (*Historia Bogomilorum*). His main argument is that Zigabenus's evidence is confirmed by that of later Byzantine chroniclers, particularly of Nicetas Choniates and Constantine Harmenopulus.

The best modern work on the subject is that of Kiprianovich (Жизнь и учение богомилов по Паноплии Евфимия Зигабена, *P.O.* (1875), vol. II, pp. 378–407, 533–72). He points out (pp. 386–8) that Zigabenus, according to Anna Comnena, described the Bogomil heresy 'exactly as that impious Basil had interpreted it' (καθὼς ὁ ἀσεβὴς...Βασίλειος ὑφηγήσατο) (*Alexiad*, lib. xv, cap. 9, p. 357); this probably means that Zigabenus consulted a copy of Basil's profession of faith, recorded by a scribe during the heresiarch's conversation with Alexius. It is also very likely that he possessed at least part of a Bogomil written commentary on the Gospels (cf. infra, p. 217, n. 5).

To-day the problem of the reliability of Zigabenus has lost most of its actuality owing to the publication of Bulgarian sources, especially of the *Sermon against the Heretics* and the *Synodicon of the Tsar Boril*, which on all main points confirm the evidence of the *Panoplia Dogmatica*.

doctrines and this increased the evil'.[1] Most of these 'other doctrines' were Paulician in origin, which is recognized by Zigabenus, particularly as regards the Bogomil doctrine of the Creation.[2] Nevertheless, he does not identify or confuse the three heresies, but devotes a separate chapter of the *Panoplia* to the refutation of each.[3]

Zigabenus gives great prominence to the basic doctrine of the Bogomils—the dualistic cosmology. According to him, the Byzantine Bogomils taught that the Devil, or Satan, is the first-born son of God the Father and the elder brother of the 'Son and Logos'. His original name was Satanael.[4] Like the Bulgarian Bogomils of the tenth century, they held him to be the 'unjust steward' of the parable in St Luke.[5] Second to the Father in dignity, Satanael was clad in the 'same form and garments' as He, and sat on a throne at His right hand. Stricken with pride, he decided to rebel against his Father and persuaded the 'ministering powers' (τῶν λειτουργικῶν δυνάμεων) to shake off their yoke and to follow him.[6] The similarity of this doctrine with the Christian teaching on the fall of Satan is obvious.[7] The difference, however, lies in the Bogomil belief that Satanael was not an angel, but the elder Son of God and the creator of the visible world.

The Bogomil teaching on Satanael is described in the *Panoplia* in much greater detail than in any other source. Satanael, together with those 'ministering powers' who had followed him in his rebellion, was cast out of heaven.[8] However, he retained

[1] *Pan. Dog.*, P.G. vol. cxxx, col. 1289.   [2] Ibid. cols. 1300–1.
[3] Κατά...Παυλικιάνων, tit: 24, cols. 1189–1244; Κατά Μασαλιανῶν, tit. 26, cols. 1273–89; Κατά Βογομίλων, tit. 27, cols. 1289–1332.
[4] For the origin of this name, see supra, p. 186, n. 1.
[5] Εἰς πίστιν...τῆς ληρωδίας ταύτης παράγουσι τὴν ἐν τῷ κατὰ Λουκᾶν Εὐαγγελίῳ παραβολὴν τοῦ οἰκονόμου τῆς ἀδικίας, τὰ τῶν ὀφειλόντων χρέη μειώσαντος. Τοῦτον γὰρ τὸν Σαταναὴλ εἶναι, καὶ περὶ τούτου γεγράφθαι τὴν τοιαύτην παραβολήν. (Ibid. col. 1296.) Cf. supra, p. 123.
[6] Ibid. cols. 1293–6.
[7] It should be noted that the Bogomil view of Satanael, as described by Zigabenus, combines the notions of the fallen angel and the unjust steward, current among the Bulgarian Bogomils of the tenth century.
[8] According to Zigabenus's *Narratio*, after the fall of Satanael his place in heaven and his seniority (τὰ πρωτοτόκια) passed over to his younger brother by birth, the 'Son and Logos'. (See Ficker, op. cit. p. 95.)

his creative power, the attribute of his divine origin, represented by 'el', the last syllable of his name. Assisted by his fallen companions, Satanael created the visible world, with its firmament, its earth and its products. This, according to the Bogomils, was the creation of the world, described in the first chapter of the Book of Genesis and falsely attributed by the Christians to God Himself. The motive of Satanael's creation of the world was to imitate his Father, and the world he created is in fact an imitation of the celestial world over which God reigns. Satanael next created Adam's body out of earth and water. But when he set the body upright, the water flowed out of the big toe of the right foot and assumed the shape of a serpent. Satanael then tried to animate the body by breathing into it, but his breath went out by the same channel as the water and entered into the body of the serpent, which thus became a minister of the Devil. Seeing his failure to give life to Adam's body, Satanael begged his Father to send down His Spirit on Adam and promised that man, a mixture of good and evil, should belong to both of them.[1] To this God agreed, and Adam came to life, a compound of a divine soul and a body created by Satanael. Eve was then created and animated in the same manner. Satanael seduced Eve, who bore him a son Cain, and a daughter Calomena. Only after Eve's intercourse with Satanael did she bear Adam a son, Abel.[2]

From this crudely anthropomorphic myth an important Bogomil view can be deduced. Satanael or the Devil is considered as the imitator of the Father, the ape of God. This conception, inherent in Christianity,[3] appears clearly in Bogomil cosmology. Satanael's motives in rebelling against his Father are identical with those attributed to Lucifer in the Book of the Prophet Isaiah (xiv. 13, 14): 'I will exalt my throne above the stars of God.... I will be like the most High.'[4] In the *Liber Sancti Johannis*, Satan is expressly

---

[1] This dualism between the body and the soul of man, symbolized by the inability of Satanael to animate Adam's body without the help of his Father, is expressed in a very similar form in the myth of the creation of man, ascribed to the Bogomils by Euthymius of Acmonia. (See supra, p. 180.)

[2] *Pan. Dog.*, *P.G.* vol. cxxx, cap. 7, cols. 1296–7.

[3] In the Judaeo-Christian view, Satan, or 'the adversary', is the one who reverses the proper relations and who uses the methods of God for his own evil purposes. Hence he appears as the ape of God.

[4] Ibid. cap. 6, col. 1296.

called 'imitator Patris',[1] and the same idea can be found in several Bogomil books and legends.[2]

The Bogomil story of the creation is clearly based on the Book of Genesis, but the Biblical story is distorted by individualistic interpretation and enlarged by an admixture of unorthodox mythology. This free use of the Scriptures, typical of the Bogomils, is further illustrated by Zigabenus's account of their interpretation of the Old Testament. According to the Bogomils, the fallen angels, seeing that they were betrayed by Satanael, took the daughters of men as their wives and from this intercourse arose the race of giants[3] who were to fight for mankind against Satanael. This incensed Satanael, who in his anger sent down the Flood upon mankind and thus destroyed almost all living flesh. Noah alone was saved, apparently for a purely accidental reason: he had no daughter, and therefore was ignorant of the struggle of mankind against Satanael, to whom he remained faithful. Satanael in reward permitted him to build the Ark and to save himself and its occupants.[4] The Bogomils considered the greater part of the Old Testament to be the revelation of Satanael.[5] Moses, according to them, was led astray by him and in his turn deceived the Jewish people through the power given him by the Demiurge. The Law given to Moses on Mount Sinai came from Satanael.[6] Owing to

[1] See J. Benoist, *Histoire des Albigeois et des Vaudois*, Paris, 1691, vol. I, p. 284; Ivanov, op. cit. p. 74.

[2] Particularly in *The Sea of Tiberias* (see Ivanov, op. cit. p. 304); a trace of this view can also be found in the *Elucidarium*, where God is said to have created the Devil from His own shadow, reflected in the waters. (Ibid. pp. 260, 270.)

[3] Cf. Gen. vi. 2, 4.

[4] *Pan. Dog.* cap. 9, col. 1305.

[5] The fundamental antithesis between the God of the Old Testament, inconsistent and wrathful, identified with the creator of the world, and the God of the Gospel, loving and merciful, was borrowed by the Bogomils from the Paulicians (cf. *Pan. Dog.* col. 1305). This antithesis is eminently characteristic of dualism in general and existed in Gnosticism and especially in the teaching of Marcion. See A. Harnack, *Lehrbuch der Dogmengeschichte* (4th ed.; Tübingen, 1909), vol. I, pp. 243–309 and *Marcion: das Evangelium vom Fremden Gott* (2nd ed.; Leipzig, 1924), pp. 93–143.

[6] *Pan. Dog.* cap. 10, cols. 1305–8. The Bogomils based their rejection of the Mosaic Law on the Pauline antithesis between Law and Grace, which they isolated from the body of St Paul's theology. Zigabenus says that to support their view they adduced the words in Rom. vii. 7, 9. The same use of St Paul's antithesis was made by Marcion.

the power held by Satanael over the human race, only very few of the Old Testament Fathers escaped perdition and ascended to the angels. Those who were saved were the ancestors of Jesus Christ, enumerated in the genealogies in Matt. i. 1–16 and Luke iii. 23–38. At last the Father took pity on the human soul, imprisoned in the body, and, in the year 5500 from the creation of the world, He brought forth from His heart the Word, who is 'the Son and God'.

The Logos, or Son of God, had, according to the Bogomils, three names: (1) He was the Archangel Michael, for the prophet Isaiah said of Him: καλεῖται τὸ ὄνομα αὐτοῦ, Μεγάλης βουλῆς ἄγγελος.[1] (2) He was Jesus, because He cured every sickness and every disease. (3) He was Christ, because He was 'anointed with the flesh' (Χριστὸν..., ὡς χρισθέντα τῇ σαρκί).[2] Christ descended from heaven (ἄνωθεν), passed through the right ear of the Virgin and assumed a body, but a non-material, 'seeming' one (ἐν φαντασίᾳ). Clothed in His non-material body, Christ performed His mission on earth as described in the Gospels, was crucified, died and rose from the dead. During His descent into hell, He cast aside His mask and bound His enemy Satanael in Tartarus with heavy chains. Then Satanael was deprived of the last syllable of his name together with his divine attributes and became Satan.[3] His mission accomplished, Christ returned to His Father, to sit on the throne formerly occupied by Satanael, and was resolved into the Father from whom He had proceeded.[4]

The evidence supplied by Zigabenus on Bogomil Christology is particularly valuable, as the other sources give only very brief indications on this point. First, it should be remarked that the Bogomils limited Christ's separate existence from the Father to the

---

[1] Cf. Isa. ix. 6. (Septuagint version.)   [2] *Pan. Dog.* col. 1301.

[3] The theme of Satanael's loss of the syllable *el* of his name and of his defeat by the Archangel Michael is fairly common in Bogomil literature. After Satanael's defeat, Michael is considered to have become an Archangel and to have received the full prerogatives attached to the final syllable *el* of his own name. Cf. the Bulgarian Bogomil legend of *The Sea of Tiberias*: Ютна ся от Сотонаила и дасть Господь илъ Михаилу, и о томъ нарече ся Михаилъ архаггелъ, а Сотонаилъ сотона (Ivanov, op. cit. p. 291); cf. also the Greek legend Περὶ κτίσεως κόσμου καὶ νόημα οὐράνιον ἐπὶ τῆς γῆς, partly of Bogomil inspiration. (Ibid. p. 316.)

[4] *Pan. Dog.* cap. 8, cols. 1301–4.

period beginning in the year 5500 after the creation of the world, and ending with His resurrection, ascension to heaven and resolution back into the Father. Thus they completely denied the fundamental Christian dogma of the Logos, eternally subsisting in the Blessed Trinity. Closely connected with this heretical doctrine was their Docetism, whereby they claimed that the body of Christ was of a non-material nature and thus rejected the dogma of the Incarnation. Docetic Christology is one of the doctrines most frequently attributed to the Bogomils[1] and is a logical consequence of their dualistic view of matter. The Logos was for them not the Second Person of the Blessed Trinity, the Eternal Word incarnate, but merely the spoken word of God, manifested in the oral teaching of Christ.[2] Hence the Bogomils taught that Christ was not really born of the Blessed Virgin, but entered through her ear, just as the spoken word enters the ear of the one who hears it.[3] It followed that Christ's redemption of mankind consisted not in His death on the Cross and His resurrection, but solely in His teaching, aimed at the liberation of man's soul from his body.

The significance of Christ in Bogomil theology is, furthermore, determined by His position within the Blessed Trinity. Zigabenus's account of the Bogomil conception of the Trinity is confused and on some points contradictory.[4] It would seem that the Bogomils recognized two distinct Trinities, or rather two separate aspects of the Trinity: on the one hand the Father, Satanael and Christ, on the other the Father, the Son and the Holy Spirit. The first is a familiar Bogomil notion and is most clearly described in the *Dialogus de daemonum operatione* of Michael Psellus.[5] The second is mentioned by no other source except Zigabenus. According to

[1] Cf. supra, pp. 113, 126, and infra, p. 238.
[2] This view was also held by the Paulicians (see supra, p. 40). Zigabenus accuses the Bogomils of failing to understand the difference between ' ἐνυποστάτου καὶ ζῶντος Λόγου' and ' ἁπλῶς λόγου προφορικοῦ'. (Ibid. cap. 22, col. 1317.)
[3] The belief that the Logos entered into the Blessed Virgin through her ear is not peculiar to the Bogomils. The Orthodox used it from the third century as a metaphor to illustrate the will of God spoken to Our Lady by the angel, and it can be found both in Eastern and Western liturgical and patristic texts. See the references in C. Schmidt, *Histoire et doctrine de la secte des Cathares ou Albigeois*, vol. II, pp. 41-2.
[4] Cf. Kiprianovich, loc. cit. pp. 399-402. [5] Cf. supra, pp. 184-5.

him, the Bogomils taught that the Son and the Holy Spirit are not distinct hypostases, but different names of the Father; they are emanations of the Father, two rays proceeding from the two lobes of His brain. This emanation is an event in time, which took place between the years 5500 and 5533 from the creation of the world; before and after these dates, the Son and the Holy Spirit had no separate existence outside the Father. Zigabenus attributes this Bogomil doctrine to the influence of Sabellianism.[1] The Father, according to the Bogomils, was incorporeal in essence, but capable of assuming human form (ἀσώματον μέν, ἀνθρωπόμορφον δέ).[2]

The attitude of the Byzantine Bogomils towards the canon of the Scripture should be mentioned, as it is somewhat different from that attributed to the Bulgarian sectarians. The latter, according to the evidence of Cosmas, rejected the Mosaic Law and the Old Testament Prophets.[3] The Byzantine Bogomils on the other hand, while spurning, in common with their Bulgarian co-religionists, the Mosaic books, accepted the sixteen books of the Prophets. Their canon consisted of seven books corresponding to the seven pillars of the House of Wisdom (Prov. ix. 1): (1) The Psalter, (2) the sixteen Books of the Prophets, (3-6) the four Gospels, (7) the Acts of the Apostles, all the Epistles and the Apocalypse.[4] There seems to have been a real difference on this point between the Bulgarian and Byzantine Bogomils, probably due to the greater proclivity of the latter towards an allegorical and rationalistic interpretation of the Scriptures in accordance with their own doctrines.[5]

[1] Sabellianism, or Modalist Monarchianism, a heresy current in the East and the West from the end of the second century to the end of the third and generally associated with the name of Sabellius, denied the distinction of the Three Persons within the Blessed Trinity. See H. Hagemann, *Die Römische Kirche und ihr Einfluss auf Disciplin und Dogma in den ersten drei Jahrhunderten* (Freiburg, 1864), pp. 129 et seq.: A. Harnack, 'Monarchianismus', *R.E.* vol. XIII, pp. 303-36. [2] *Pan. Dog.* caps. 2-5, cols. 1292-3.
[3] See supra, p. 127. [4] Ibid. cap. 1, col. 1292.
[5] Kiprianovich asserts that the highly individualistic and allegorical manner in which the Bogomils interpreted the Old Testament caused them to be accused of rejecting it in its entirety (loc. cit. pp. 390-5). But this is incorrect: the Bogomils were not generally accused of rejecting the whole of the Old Testament canon, but only certain parts of it, especially the Mosaic books. (Cf. supra, p. 179, n. 6). The greater eclecticism of the Byzantine Bogomils regarding the Books of the Prophets is probably due to Christian influence. Cf. Puech, op. cit. pp. 168-72.

One of the most interesting features of Bogomilism, described by Zigabenus in greater detail than by the other sources, is its demonology. It has already been pointed out that the important part played by the demons in the Bogomil doctrines and popular belief is partly due to Massalian influence.[1] According to Zigabenus, 'the Bogomils say that the demons fly from them alone like an arrow from a bow; they inhabit all other men [i.e. non-Bogomils] and instruct them in vice, lead them to wickedness and after their death dwell in their corpses, remain in their tombs and await their resurrection in order to be punished together with them and not to desert them in their torments. The belief that each man is inhabited by a demon they hold from the Massalian heresy'.[2] Satanael's demons live in waters, fountains, seas and subterranean places.[3] The association of the agents of the Demiurge with the aquatic element is a common feature of Bogomil belief and can perhaps be explained by the fact that water is the image of universal passivity, the symbol of the plastic principle from which the world was created, and hence essentially the realm of Satanael and his servants.[4] This was probably one of the reasons why the Bogomils rejected Baptism by water.[5] But although they claimed that the demons fled from their approach, the Bogomils feared them, 'for they have great and invincible power to harm, which not even Christ and the Holy Spirit can withstand, because the Father still spares them and has not deprived them of their power, but has granted them sovereignty over the whole world until its consummation'. Zigabenus asserts that a precept

---

[1] See supra, p. 186.   [2] *Pan. Dog.* cap. 13, col. 1309.
[3] Ibid. cap. 7, col. 1300.
[4] The significance of water as the plastic and passive principle in creation, distinct from and, in a sense, opposed to the Spirit, which is the active principle, can be seen in Gen. i. 2: 'And the Spirit of God moved upon the face of the waters.' The connection between Satanael and the element of water is a common Bogomil theme. Zigabenus says that, according to Bogomil belief, Satanael, after having rebelled against his Father, was cast out of heaven into a universe of water (*Pan. Dog.* col. 1296). The same notion can be found in the *Liber Sancti Johannis*: 'et transcendens, invenit universam faciem terrae coopertam aquis.' See Benoist, op. cit. p. 284; Ivanov, op. cit. p. 74. According to the Bogomil legend of *The Sea of Tiberias*, the world was created by God from elements taken by Satanael from the sea; see Ivanov, ibid. pp. 290, 304–5.
[5] *Pan. Dog.* cap. 16, col. 1312.

taught by Basil was: 'honour the demons, not in order to receive help from them, but lest they should harm you.'[1] This respect paid to the demons apparently went as far as actual worship (θεραπεύειν διὰ προσκυνήσεως).[2]

When Basil was asked why his followers honoured the relics of the Christian saints whom they rejected, he apparently answered that it was on account of their being inhabited by demons.[3]

The attitude of the Bogomils towards the tradition and sacraments of the Orthodox Church, as described by Zigabenus, entirely confirms the evidence of previous sources. The Byzantine Bogomils, as well as their Bulgarian co-religionists, rejected Baptism,[4] the Eucharistic Sacrifice,[5] churches,[6] the Cross,[7] saints,[8] images[9] and the Orthodox ecclesiastical hierarchy.[10]

Their ethical rules show the same dualistic condemnation of matter which we find among the tenth-century Bulgarian Bogomils, and even appear more rigorous: the Byzantine Bogomils not only rejected marriage and the eating of meat, but were also forbidden to eat cheese, eggs and 'other things of this kind'.[11] Zigabenus does not tell us whether these rules were equally obligatory for all members of the sect or whether, on the contrary, extreme asceticism was only required of the minority of elect. The latter, however, appears probable, since, as it will be shown, the distinction

[1] *Pan. Dog.* cap. 20, col. 1316.
[2] Ibid. cap. 38, col. 1325. They claimed to base this teaching on the words of Our Lord: 'agree with thine adversary quickly' (Matt. v. 25).
[3] Ibid. cap. 12, col. 1309. [4] Ibid. cap. 16, col. 1312.
[5] Ibid. cap. 17, col. 1313: 'They call the Holy Communion of the Lord's Body and Blood a sacrifice of the demons who dwell in the temples.'
[6] They claimed to base their rejection of churches on Matt. vi. 6: ibid. cap. 42, col. 1328. The churches were dwelling-places of the demons, who were supposed to draw lots for them according to their rank and power. Satan himself, according to the Bogomils, had first dwelt in the Temple of Jerusalem and, after its destruction, had made his abode in St Sophia of Constantinople. (Ibid. cap. 18, col. 1313.)
[7] Ibid. caps. 14–15, cols. 1309–12.
[8] Ibid. cap. 11, col. 1308. They recognized, however, as saints the ancestors of Our Lord, mentioned in Matt. i. 1–16 and Luke iii. 23–38, the sixteen prophets of the Old Testament and the martyred Iconoclasts.
[9] Their hatred of images led them to 'call the Iconoclasts alone orthodox and faithful, and especially Copronymus'. (Ibid.) Cf. supra, p. 131, n. 2.
[10] Ibid. cap. 28, col. 1321; cap. 49, col. 1329.
[11] Ibid. caps. 37, 39, col. 1325.

between the 'perfect' or 'chosen' on the one hand and the 'believers' or 'hearers' on the other, prevalent among the early Manichaeans and the Cathars of western Europe, existed also among the Byzantine Bogomils.

The Bogomil elect were considered to be the bearers of the Holy Spirit and were consequently called θεοτόκοι, a name which the Christian Church reserves exclusively for the Mother of God. The Bogomils, who denied the reality of the Incarnation and replaced the Christian conception of the Logos by that of the spoken word of God, consequently called θεοτόκοι all those who 'give birth to the word' by their teaching. The θεοτόκοι claimed a mystic vision of the Trinity. The Father, they said, appeared to them as an old man with a flowing beard, the Son as a young man and the Holy Spirit as a beardless youth.[1] These θεοτόκοι were supposed to discard at death their earthly bodies, which dissolved into dust never to rise again; they put on the immortal garment of Christ, assumed the same body and form as He and entered the Kingdom of the Father.[2]

Zigabenus gives a brief description of the ceremony of initiation into the Bogomil sect, which is particularly valuable, being the only account we possess of initiatory rites among the Balkan sectarians. This ceremony they called the Baptism of Christ through the Spirit, and carefully distinguished it from Orthodox Baptism, which they rejected as being of St John and by water.

'Therefore they rebaptize those who come to them. First they appoint him [i.e. the catechumen] a period for confession (εἰς ἐξομολόγησιν), purification (ἀγνείαν) and intensive prayer (σύντονον προσευχήν). Then they lay the Gospel of St John on his head, invoke their (παρ' αὐτοῖς) Holy Spirit and sing the Lord's Prayer. After this Baptism they again set him a time for a more rigorous training, a more continent life, purer prayer. Then they seek for proof that he has observed all these things and performed them zealously. If both the men and the women testify in his favour, they lead him to their celebrated consecration (ἄγουσιν αὐτὸν ἐπὶ τὴν θρυλλουμένην τελείωσιν). They make the wretch face

---

[1] Ibid. cap. 23, col. 1320.
[2] Ibid. cap. 22, cols. 1317–20. The Bogomil teaching on the θεοτόκοι provides a good example of the esoteric character of Bogomilism, which forms a constant, though not always very apparent, background to its doctrines. See Puech, op. cit. pp. 161–3.

the East and again lay the Gospel on his...head. The men and women of the congregation place their foul hands on him and sing their unholy rite (τὴν ἀνόσιον ἐπᾴδουσι τελετήν): this is a hymn of thanksgiving for his having preserved the impiety transmitted to him.'[1]

It can be inferred from this account that full initiation into the Bogomil sect consisted of two distinct ceremonies, called βάπτισμα and τελείωσις, separated by a period of severe asceticism and inner preparation. However, from the description of Zigabenus they appear almost identical in nature and we are not told in what the difference between them consisted. The precise nature of this distinction can be understood by remarking that the French and Italian Cathars, whose ritual very probably developed under the direct influence of Bogomilism,[2] also possessed a double initiation: the first reception of the catechumen into the sect (the *abstinentia*) and the accession of the 'believer' to the rank of the 'elect' or 'perfect' (the *consolamentum*).[3] Zigabenus clearly refers to the same two ceremonies among the Byzantine Bogomils; their very names are suggestive of their nature: βάπτισμα, like the Christian Baptism, is the primary act, by which the neophyte becomes a member of the community of the faithful, or 'believers'; τελείωσις, a somewhat more solemn ceremony, implying by its meaning the idea of perfection, is the raising of the 'believer' to the rank of the 'perfect', or, in the language of the Byzantine Bogomils, of the θεοτόκοι.

Those who had received initiation, whether the βάπτισμα or the τελείωσις, were henceforth members of the true Church, outside which there was no salvation. The Byzantine Bogomils called their community (συναγωγή) Bethlehem, the birthplace of Christ; the Orthodox Church, on the contrary, was Herod, who

[1] *Pan. Dog.* cap. 16, col. 1312.

[2] A study of the relation between the ritual of the Bogomils and that of the Cathars lies outside the scope of this book. Most of the works on the Cathars suffer from an insufficient acquaintance with Bogomilism on the part of their authors. Many of their decisions could be modified or revised in the light of the evidence adduced by Ivanov, showing the considerable influence exerted by Bogomilism on Catharism, particularly in the sphere of ritual (see in particular op. cit. pp. 113 et seq.).

[3] See J. Guiraud, 'Le Consolamentum Cathare', *R.Q.H.* (1904), vol. LXXV, pp. 74–112.

had sought to slay the Word born among them.¹ The Bogomil community imposed a severe discipline on its members; according to Zigabenus, they were obliged to pray seven times a day and five times a night; in this their rule was stricter than that of the tenth-century Bulgarian Bogomils, who prayed four times a day and four times a night.² Following the practice of all other Bogomils, the Byzantine sectarians recognized only the Lord's Prayer; every time they prayed they recited it with genuflexions, some ten times, others fifteen, others more or less frequently.³ They fasted on the second, fourth and sixth day of every week, until the ninth hour.⁴ But here again Zigabenus does not tell us whether these rules applied only to the θεοτόκοι or to every member of the sect.

The last part of the *Panoplia Dogmatica* contains a systematic account of the Bogomil interpretation of certain Scriptural texts, all taken from the Gospel of St Matthew.⁵ Its great interest lies in the fact that it shows how extremely developed was the practice among the Byzantine Bogomils of interpreting Scriptural texts in accordance with their heretical doctrines by means of allegory. This practice was already widespread among the Bulgarian Bogomils of the tenth century.⁶ But it is very doubtful whether the use of allegory had then reached the degree of intricacy and ingenuity which it acquired among the Bogomils in Byzantium. It seems that this art developed among the more educated and theologically minded heretics of Constantinople. However, it cannot be doubted that after the temporary suppression of Bogomilism in Byzantium at the beginning of the twelfth century, when Bulgaria once more became the stronghold of this sect in the Balkans, the Scriptural commentaries of Basil's disciples became current among the Bulgarian Bogomils, who thus profited by the theological and exegetical labours of their Byzantine co-religionists.

¹ *Pan. Dog.* cap. 28, col. 1321. ² Cf. supra, p. 135.
³ Ibid. cap. 19, cols. 1313-16. ⁴ Ibid. cap. 25, col. 1320.
⁵ It is extremely probable, as Kiprianovich suggests (loc. cit. pp. 386-8), that Zigabenus, when composing this section of the *Panoplia*, used a written Bogomil commentary on the Gospel of St Matthew. This can be inferred both from the numerous quotations he cites from this Gospel and from the systematic order in which they are arranged.
⁶ Cf. supra, p. 127.

Zigabenus cites twenty-five examples of allegorical or unorthodox interpretations of Scriptural texts.[1] Some of them have already been mentioned. The following is a particularly characteristic example of Bogomil exegesis.

The text is taken from Matt. iii. 4: 'And the same John had his raiment of camel's hair, and a leathern girdle about his loins; and his meat was locusts and wild honey.' The Bogomils interpreted the passage as follows. The camel's hair are the commandments of the Mosaic Law: like the camel this law is impure, as it permits the eating of meat, marriage, oaths,[2] sacrifices, murders and the like; the girdle of leather is the Holy Gospel, since it was originally written down on sheep-skin; the locusts are again the commands of the Mosaic Law, which are incapable of distinguishing what is right and good; the wild honey is the Gospel which seems like honey to those who receive it (cf. Ps. cxix. 103) and wild to those who do not receive it. Thus the Forerunner is shown to be the intermediary between the Old Law and the New and to belong to both.[3]

It can be concluded from the preceding analysis that the *Panoplia Dogmatica*, while confirming in the main the evidence on the Bogomils supplied by tenth- and eleventh-century sources, gives at the same time a deeper and wider account of Bogomilism. This cannot be explained solely by the inadequacy of earlier sources: Cosmas in particular was admirably qualified to paint as complete a picture of Bogomilism as was possible in his time. Many features of the sect described by Zigabenus can be rightly understood only by recognizing the gradual development of Bogomilism from the time of its appearance in the tenth century to the beginning of the twelfth century. This development can be regarded as a gradual unification of the doctrinal and ethical aspects of Bogomilism, the first mainly Paulician in origin, the second a combination of Massalian and Christian elements. A comparison between the *Sermon against the Heretics* and the *Panoplia Dogmatica*

[1] *Pan. Dog.* caps. 28–52, cols. 1321–32.
[2] The prohibition of swearing seems to have been an essential part of Bogomil ethics.
[3] Ibid. cap. 30, col. 1324. This attitude to St John the Baptist is more tolerant than that ascribed to the Bulgarian Bogomils by Cosmas (cf. supra, p. 129). This is probably another example of the evolution of Bogomilism under the influence of Christianity.

shows that the ethical teaching of the Bogomils did not change very much between the tenth and twelfth centuries, except for a certain increase in austerity and asceticism, possibly due to the necessity of resisting persecution. Although it is dangerous to argue too much from Cosmas's silence, it seems that the Bogomils developed by the twelfth century a more complex ritual than they had hitherto possessed. But it is principally in the realm of doctrine that the evolution is most noticeable. In Byzantium, under the influence of the religious and philosophical speculation of the educated circles, Bogomilism acquired a complex theology and cosmology which it cannot have possessed at the time of Cosmas.

It can truly be said that at the time of the death of Alexius Comnenus (1118), Bogomilism, although suppressed in Byzantium, was at the beginning of its greatest development in the Balkans. Although Basil had been executed and his followers were in prison or had recanted, their labours were not lost to the Bogomil sect. In the course of the twelfth century the stronghold of the sect moved back to Bulgaria, where Bogomilism, enriched and fortified by its evolution in Byzantium, reached the summit of its development.

The efforts of Alexius Comnenus in rooting out the Bogomil sect in Constantinople were only partly successful. Certainly after Basil's execution we hear no more of any widespread or organized outbreak of Bogomilism in Byzantium. But the heresy was still rampant in the imperial provinces, particularly in Asia Minor, and, in spite of the drastic measures taken against it in the twelfth century, was still capable of sporadic outbursts; one of these led to a new outbreak of Bogomilism between the years 1140 and 1147.

In 1140, in the reign of John II Comnenus and in the Patriarchate of Leo Stypes, a Synod in Constantinople condemned as heretical the writings of a certain Constantine Chrysomalus. His doctrines were especially popular in monastic circles, particularly in the monastery of St Nicholas in Constantinople. The records of the Synod describe them as 'more absurd than the teachings of the Enthusiasts [i.e. the Massalians] and the Bogomils'. Constantine was accused of teaching the following heresies: Baptism, the confession of sins, the study of the Gospels, the singing of psalms, attending church services are of no avail until the Christian has received, through initiation, the gift of perceiving intellectually (νοερῶς) the Holy Spirit; those who have reached that stage are no longer subject to the law, have escaped the power of Satan and are incapable of sin; in every Christian there dwell two souls, the one sinless, the other sinful: this last doctrine, according to the records of the Synod, is held 'in express terms' by the Massalians and the Bogomils. Constantine's rejection of Christian Baptism, which he replaced by his own initiatory rite, is branded by the Synod as 'an undoubted sign of the Bogomil heresy'.[1] It must be admitted, however, that the latter teaching is far from being exclusively typical of Bogomilism; and the doctrine of the two souls is not attributed to the Bogomils by any other source: it seems to be, in fact, characteristically Massalian. It seems rather doubtful whether any direct connection existed between the teaching of Constantine Chrysomalus and that of the Bogomils. Constantine's heresy is much more suggestive of Massalianism, which at that time was still distinct from Bogomilism.

It is tempting, however, to seek for Bogomil influences in Constantine's alleged repudiation of all authority (ἀρχῆς ἀπάσης) and in his declaration that all who show honour and veneration to any ruler (ὁποιῳδήποτε ἄρχοντι) worship Satan. It may be that this social anarchism of Constantine is related to the practice of civil disobedience ascribed by Cosmas to the Bulgarian Bogomils. But in the absence of any close doctrinal similarities it seems inadvisable to push the comparison too far.[2]

On 20 August 1143, in the first year of the reign of Manuel Comnenus, in the Patriarchate of Michael II Kurkuas, a Synod in Constantinople deposed and excommunicated two bishops of the diocese of Tyana in south-eastern Asia Minor—Clement of Sosandra and Leontius of Balbissa, convicted of Bogomilism. The

[1] See L. Allatius, *De Ecclesiae occidentalis atque orientalis perpetua consensione* (Coloniae Agrippinae, 1648), cols. 644–53; G. Rhalles and M. Potles, Σύνταγμα τῶν θείων...κανόνων, vol. v, pp. 77–80; cf. F. Chalandon, *Jean II Comnène et Manuel Ier Comnène* (Paris, 1912), p. 23.

[2] Puech, while admitting the possible influence of Massalianism on the doctrines of Constantine Chrysomalus, regards his teaching of civil disobedience as Bogomil in character (op. cit. pp. 137, 275). The hypothesis is tempting, but the evidence seems rather inconclusive.

records of this Synod show that Bogomilism was rife in Cappadocia at that time.¹ Some of the doctrines professed by these two 'pseudo-bishops' are unmistakably Bogomil: for instance, they were convicted of teaching that the demons live in the bodies of the dead and work the miracles ascribed by Christians to the power of the Cross, and of spurning the Cross and the images. Other features of their teaching, however, are specifically Massalian, such as the belief that monks alone can be saved, and that after an abstention of three years from sexual intercourse, meat, milk and wine, a man can indulge in all of them without sin; the latter tenet is even in direct contradiction to the lifelong asceticism practised by the Bogomil 'elect'.

On 1 October 1143, a monk, Niphon, appeared before a Synod in Constantinople, charged with preaching Bogomilism in Cappadocia. Pending a final judgement, he was placed in solitary confinement in the monastery τῆς Περιβλέπτου.² On 22 February 1144, solemn judgement was passed on Niphon by the Synod. He was condemned as a Bogomil, excommunicated and, after his beard had been shaved off, was cast into prison. The only heretical doctrine ascribed to Niphon by the sources is his exclamation before the entire Synod: 'anathema to the God of the Jews'.³

Niphon's condemnation was destined to have grave repercussions in Constantinople. Cosmas II, who ascended the patriarchal throne in 1146, a man of charity and integrity, showed himself very lenient to Niphon, allowed him complete freedom and even treated him as a close friend. Cosmas's enemies seized this pretext to secure, on 26 February 1147, his condemnation and deposition

¹ See Allatius, ibid. cols. 671–8; J. D. Mansi, *Sacrorum conciliorum...collectio*, vol. XXI, cols. 583–90; D. Farlatus, *Illyricum sacrum* (Venetiis, 1817), vol. VII, p. 354; *Acta et diplomata res Albaniae...illustrantia*, vol. I, no. 85, p. 29; Chalandon, op. cit. p. 635. For the names of the sees of the two heretical bishops, see A. Papadopoulos-Kerameus, Βογομιλικά, *V.V.* (1895), vol. II, pp. 720–3.

² See Allatius, op. cit. cols. 678–81; A. Banduri, *Imperium Orientale* (Venetiis, 1729), p. 635.

³ See Joannes Cinnamus, *Historiae*, lib. II, *C.S.H.B.* pp. 63–6; Nicetas Choniates, *De Manuele Comneno*, lib. II, *C.S.H.B.* p. 107; Nicholas of Methone, *Orationes*, ed. A. K. Demetrakopoulos, Ἐκκλ. Βιβλιοθήκη (Leipzig, 1866), p. 267; Eustathius of Thessalonica, *Manuelis Comneni laudatio funebris*, cap. 36, *P.G.* vol. CXXXV, col. 1000; Allatius, ibid. cols. 681–3.

as a Bogomil.¹ The motives and circumstances of Cosmas's condemnation are not at all clear. It seems more likely that the patriarch was a victim of a political intrigue than that he really favoured the Bogomil heresy.²

Several deductions regarding twelfth-century Bogomilism can be drawn from these facts. In the first place, the sect continued to extend its influence in Asia Minor and particularly in Cappadocia. The heretical bishops Clement and Leontius, as well as Niphon, taught in this province. This prevalence of Bogomilism in Asia Minor in the twelfth century confirms the evidence of Euthymius of Acmonia with regard to the eleventh.³ Moreover, in Constantinople itself, after the sensational trial of Basil and his followers at the beginning of the century, Bogomilism inspired such a loathing and fear that any one suspected of the least sympathy towards it stood in immediate danger of punishment, not excluding the patriarch himself. Finally, a number of doctrines which passed at that time as Bogomil are not ascribed to the Bogomils in earlier sources and are really typical of the Massalians. This identification of Bogomil and Massalian teachings is significant; as it will be shown, the distinction between Bogomilism and Massalianism, still recognized theoretically by Zigabenus and Anna Comnena, tends to disappear in practice by the middle of the twelfth century; in the thirteenth and fourteenth centuries the two names are used synonymously in Byzantine and Bulgarian sources.⁴

The repercussions of the wave of Bogomilism which swept over the Byzantine Empire between 1140 and 1147 were immediately felt in Bulgaria. At that very time, the Bogomil sect raised its head in Macedonia. The connection between the sectarian movements in Constantinople and in Macedonia is illustrated by the fact that the repression of Bogomilism in the latter province was instigated by the Byzantine emperor himself. The activity of this

¹ See Cinnamus, op. cit. pp. 63–6; Nicetas Choniates, op. cit. p. 107; Allatius, op. cit. cols. 683–9.   ² See Chalandon, op. cit. pp. 635–9.
³ Cf. supra, pp. 177 et seq. Asia Minor remained a centre of Bogomilism in subsequent years. At the beginning of the thirteenth century, Germanos II, patriarch of Nicaea (1222–40), wrote against the Bogomils in Nicaea his polemical work: *In exaltationem venerandae crucis et contra Bogomilos*. (*P.G.* vol. CXL, cols. 621–44.)
⁴ Cf. infra, pp. 241, 254.

sect in south-eastern Macedonia in the middle of the twelfth century is attested by the Bulgarian *Life of Saint Hilarion of Moglena* written by Euthymius, patriarch of Trnovo, which is one of the important sources for a study of Bogomilism.[1] St Hilarion was a contemporary of Manuel Comnenus (1143–80) and was consecrated bishop of Moglena in the fourth decade of the twelfth century.[2]

A short time after Hilarion was elevated to the episcopate, the *Life* tells us, the heretics made their appearance:

'While the saint was zealously preaching and teaching, he discovered a considerable number of Manichaeans, Armenians and also Bogomils, who were plotting against him and were trying to pierce the Orthodox in the dark, corrupting and attacking the Orthodox flock like beasts of prey. Seeing that they were daily increasing in number, he suffered great sorrow and prayed earnestly from his heart to Almighty God that their...mouths might be closed. He often preached to his people, teaching them and strengthening them in the Orthodox faith. Hearing these sermons often, the heretics were enraged in their hearts, gnashed their teeth like wild beasts and caused him vexations. They were fond of disputing and wrangling with him, but Hilarion, the good shepherd of the...sheep of Christ, having made the Lord his habitation, tore their intrigues and whisperings to pieces like a spider's web, and at this the faithful rejoiced.'[3]

The heretics who confronted St Hilarion belonged, as the text clearly shows, to three distinct groups: the 'Manichaeans' were obviously Paulicians, both terms being used synonymously by the Byzantines.[4] The 'Armenian' heresy was a common name for Monophysitism.[5] As for Bogomilism, though mentioned as the

---

[1] Житіе и жизнь прѣподобнаго ѡтца нашего Иларіѡна, епископа мегленскаго, въ немже и како прѣнесенъ бысть въ прѣславный градъ Тръновъ, съписано Еѵѳиміемь патріархомъ Трьновьскыимь. The document has been edited by Dj. Daničić in *Starine* (Zagreb, 1869), vol. I, pp. 65–85 and by E. Kałużniacki, *Werke des Patriarchen von Bulgarien Euthymius: Leben Hilarions, Bischofs von Moglen* (Vienna, 1901), pp. 27–58. Kałużniacki mentions no less than twelve MSS. of the work (op. cit. pp. c–cxx). The author of the *Life of Saint Hilarion* occupied the patriarchal throne of Bulgaria from *c.* 1376 to *c.* 1402. His life and activity have been studied by V. S. Kiselkov, Патриархъ Евтимий (животъ и обществена дейность), *B.I.B.* (1929), vol. III, pp. 142–77.
[2] See Spinka, op. cit. p. 99.   [3] Kałużniacki, op. cit. p. 33.
[4] Cf. supra, pp. 31–2, 189.   [5] Cf. supra, p. 189.

object of St Hilarion's attacks at the end of the document, it does not occupy the central position frequently ascribed to it. The failure to recognize this threefold distinction has led some scholars to the erroneous conclusion that the *Life of Saint Hilarion* is an exclusively anti-Bogomil work.[1] Others, like Rački, neglect the obvious reference to the Paulicians and identify the 'Manichaeans' with the Bogomils.[2] Although Rački asserts that the Bogomils were often called Manichaeans, there is not a single example of this identification in any extant Bulgarian or Byzantine medieval source. Furthermore, most features of the 'Manichaeans' as described in the *Life of Saint Hilarion* are characteristic of the Paulicians and not of the Bogomils. We read, in particular, that 'the champions of the filthy Manichaean heresy, like wolves in sheep's clothing, approached him meekly and tempted him as the Pharisees tempted Our Lord. They wanted to catch him in his words.... They asked him: "When we teach that the good God made the heavens and that the earth and all that is on it was created by another, the evil creator, why do you not submit, but contradict the truth?"'[3] This insolent kind of proselytism is more consonant with the methods employed by the Paulicians, while the Bogomils, who were noted for their cautious and insidious behaviour,[4] would scarcely have used such outspoken language to a bishop of the Orthodox Church. Moreover, the doctrines ascribed to the 'Manichaeans' by St Hilarion's biographer are presented in a Paulician form.[5] But the most compelling argument is the fact that the exposition and the refutation of their doctrines are borrowed, often verbatim, from the *Adversus Paulicianos* of Euthymius Zigabenus.[6]

According to his biographer, the saintly bishop succeeded in

[1] Gilferding, in particular, more or less identifying all three heresies as 'Armenian', has drawn the unjustifiable conclusion that 'in the twelfth century the Bogomils formed the majority of the population in the eparchy of Moglena' (op. cit. vol. I, p. 227).
[2] Op. cit. *Rad*, vol. VII, p. 119.   [3] Kałużniacki, op. cit. p. 34.
[4] Cf. supra, p. 121.
[5] These are the cosmological dualism, the belief that Our Lady was the Higher Jerusalem, the rejection of the Old Testament and the Mosaic Law, the Docetic Christology, the hatred of the Cross, the extensive use of the Scriptures for exegetical purposes, the claim to follow the Apostolic tradition. (Ibid. pp. 34–42.)
[6] *Pan. Dog.* tit. 24, *P.G.* vol. CXXX, cols. 1200 et seq.

converting the Paulicians and the Monophysites to Orthodoxy.[1] Only then do the Bogomils appear. We are told that St Hilarion took action against them by the express command of the Emperor Manuel Comnenus:

'The pious emperor...wrote to him an edict, ordering him to purge the whole of the Bogomil heresy from his flock, to receive warmly those [Bogomils] who submit to the pious dogmas and to number them among the chosen flock, and to drive far away from the Orthodox fold those who do not submit and who remain in their impious and foul heresy.

On hearing this, they [i.e. the Bogomils] also entered the Universal Church and were found worthy of receiving Holy Baptism, spurning their heresy to the end.'[2]

Manuel's intervention against the Bogomils of Moglena is not at all improbable, as by the middle of the twelfth century Bogomilism was particularly widespread in the Byzantine Empire and had become for the authorities in Constantinople the heresy *par excellence*.[3]

It may be doubted whether all these conversions were sincere, since the Bogomils were notorious for their outward simulation of Orthodoxy whenever they were faced with persecution or any other necessity. In this case at least they were more refractory to conversion than the other heretics: some of them remained obdurate to the end and were condemned to banishment or imprisonment.[4]

The *Life of Saint Hilarion of Moglena* shows that in the middle of the twelfth century Bogomilism was a powerful force in southeastern Macedonia, that in this region the Church had also to fight against Paulicians and Monophysites, and that the Bogomil

---

[1] This conversion was not achieved without difficulty, particularly in the case of the Monophysites, who tried to kill St Hilarion and succeeded in stoning him almost to death (Kałužniacki, op. cit. pp. 42–3). The lengthy refutation of the Monophysite doctrines (ibid. pp. 43–51) is borrowed from the *Adversus Armenios* of Euthymius Zigabenus (*Pan. Dog.* tit. 23, *P.G.* vol. cxxx, cols. 1173–89).

[2] Kałužniacki, op. cit. pp. 52–3.

[3] Cf. supra, pp. 219–22.

[4] Нечьстивыя же и сквръннымя богѡмилскыя ереси поклѡнникы, елики благочьстіа приемшж сѣмя божій архіерей видѣ, въся съчета къ православныих стадоу, елики же непокорнѣ належжщж оусмотри, различными изгнаніи и заточенми ѿт благочьстиваго потрѣби стада. (Ibid. p. 54.)

heresy was probably the most dangerous of the three: for the Bogomils were apparently the only sectarians whom St Hilarion did not entirely succeed in converting,[1] and, moreover, they attracted the special attention of the Byzantine central authorities who were at that very time grappling with Bogomilism in Constantinople and in other parts of the Empire.

The development of Bogomilism in Bulgaria in the eleventh and twelfth centuries is indirectly confirmed by another document of considerable importance, which also supplies us with some further information on the doctrines professed at that time by the Bulgarian Bogomils. This is the so-called *Liber Sancti Johannis*, or the *Faux Evangile*, extant only in two, somewhat different, Latin translations.[2] Ivanov, who also edited them with a critical commentary,[3] has shown that they are both translations of an unknown Slavonic original, composed by a Bulgarian Bogomil, probably in the Bulgarian language, not earlier than the eleventh or the twelfth century. In the middle of the twelfth century, the *Liber Sancti Johannis* was brought to the Patarenes of Lombardy from Bulgaria by a certain Nazarius, who occupied a high position in the Bogomil sect.[4]

The *Liber Sancti Johannis*—referred to by Ivanov as the *Secret Book*—is essentially a doctrinal work, and its Bulgarian origin makes it a source of unique importance for our knowledge of the teaching of the Bulgarian Bogomils in the twelfth century. Moreover, the striking resemblances of some of the Bogomil doctrines expounded in the *Secret Book* to those set out in the *Panoplia Dogmatica* point to an early influence of Byzantine Bogomilism on the teachings of the Bulgarian sect.

[1] Tradition soon associated St Hilarion exclusively with the fight against Bogomilism. In a Bulgarian document of the thirteenth or fourteenth century, quoted by Ivanov (op. cit. p. 35), he is called 'the warrior against the Bogomils' (воинъ на богъмили).

[2] The two MSS. are generally known as the 'Carcassonne MS.' and the 'Vienna codex'. The first was edited by J. Benoist (*Histoire des Albigeois et des Vaudois*, Paris, 1691, vol. I, pp. 283–96), the second by I. von Döllinger (*Beiträge zur Sektengeschichte des Mittelalters*, München, 1890, vol. II, pp. 85–92). The best modern edition of both versions is by R. Reitzenstein, *Die Vorgeschichte der christlichen Taufe* (Leipzig und Berlin, 1929), pp. 297–311, where the work is entitled: *Interrogatio Iohannis et apostoli et evangelistae in cena secreta regni celorum de ordinatione mundi istius et de principe et de Adam*.

[3] Op. cit. pp. 60–87. [4] Cf. infra, pp. 242–3.

The *Secret Book* is a dialogue between Jesus Christ and His favourite disciple, John the Evangelist. At the Last Supper St John leans on the breast of his Master and questions Him on the origin of the world, the spiritual life, and the end of all things. The detailed answers which Christ gives His disciple are dualistic and typically Bogomil. Satan, He says, was once surrounded with such glory that he ruled the celestial powers, while He, Jesus, sat at the side of His Father.[1] Journeying through his domains, which extended from heaven down to the abyss, Satan, intoxicated by his own glory, was smitten with pride. Driven by envy, he decided to 'place his seat above the clouds of the heavens' and to become 'like the most High'.[2] He gained the adherence of one-third of all the angels of God, by promising to reduce the taxes which they owed their Lord, in the very terms used by the 'unjust steward' in St Luke's parable.[3] The rebellious angels were cast out of heaven by God the Father, Satan was deprived of his stewardship over the heavenly powers, and his countenance, no longer luminous with glory, became as red-hot iron and like the face of man. Cast down into the firmament, where he could find no rest, Satan implored the Father to grant him a respite, and God, out of compassion, conceded him a delay of seven days. In this breathing space Satan created the visible world—the sun, the moon, the stars and the earth with its animals and plants. Then he made man in his own image with a body of clay, and commanded the angel of the third heaven to enter the body of Adam and the angel of the second heaven that of Eve. Satan next created Paradise, in which he planted a reed, which was the 'tree of the knowledge of good and evil' referred to in the Book of Genesis. In this reed

[1] The term 'Satan' (Sathanas), which occurs throughout the *Secret Book*, is significant. It does not seem that the Bulgarian Bogomils of the tenth century used this term: the name 'devil', frequently mentioned by Cosmas, appears to have been the established one at that time. On the other hand, the epithet 'Satan' was used by the Byzantine Bogomils in the eleventh and twelfth centuries. It is not improbable that the widespread use of the name and notion of Satan, conceived as Satanael deprived of the last syllable of his name as a result of his defeat by Christ, first appeared among the Bogomils in Byzantium, whence it passed to the Bulgarian Bogomils, among whom it replaced the older name of 'devil'.

[2] Cf. supra, p. 208.

[3] The identification of Satan with the 'unjust steward' is typically Bogomil. Cf. supra, pp. 123, 207.

Satan hid himself, emerged in the form of the serpent and seduced Eve. Then he taught his angel who lived in Adam's body to commit sexual intercourse with her. A product of this intercourse, man, as a compound of a mortal body created by Satan and a soul originating from a fallen angel, was intended to perpetuate the rule of the Devil in the world until its consummation.

The interpretation of the Old Testament story in the *Secret Book* is on the usual Bogomil lines. To consolidate his power over mankind and to ensure that it worshipped him alone, Satan from time to time sent his servants into the world: of these the principal ones were Enoch[1] and Moses.

To save perishing mankind from the domination of Satan, God decided to send Christ, His Son, into the world. An angel of God was first sent to earth to receive Him: this was Mary.[2] Through her ear Christ entered and came out.[3] To counter God's action, Satan sent into the world his own emissary, the prophet Elijah, who is John the Baptist; John's baptism by water is opposed to the Baptism by the Spirit taught by Christ and His true disciples, who lead an unmarried life.

The last part of the *Secret Book* contains an account of the Bogomil teaching on the end of the world and the Last Judgement, which is very reminiscent of the Christian eschatology. Its dualistic basis, however, is revealed in the final separation of the spirit of man from the prison of his material body, which replaces the Christian dogma of the Resurrection of the flesh. After the complete destruction of the universe by fire, Satan and his cohorts will be bound with indissoluble fetters and cast into a lake of fire. 'And then the just will shine like the sun in the Kingdom of their Father.... And God will wipe away every tear from their eyes and the Son will reign with His Holy Father and His Kingdom shall have no end unto the ages of ages.'[4]

---

[1] Enoch, the father of Methuselah (Gen. v. 18–24; Heb. xi. 5) seems to have figured fairly prominently in Bogomil mythology. The Slavonic version of the apocryphal *Book of the Secrets of Enoch* (trans. W. R. Morfill, Oxford, 1896) influenced, as Ivanov has shown (op. cit. p. 72), the composition of the *Secret Book*.

[2] Cf. 'Nazarius...dixit, quod B. Virgo fuit Angelus' (infra, p. 242).

[3] Cf. supra, p. 211.

[4] Reitzenstein, op. cit. pp. 308–9. Apart from the brief references by Euthymius of Acmonia and the *Synodicon of the Tsar Boril* to the Bogomil rejection of the Resurrection of the Body (cf. supra, pp. 181–2, and infra, p. 241) and the

By the second half of the twelfth century Bogomilism became once again, as in the tenth century, a predominantly Slavonic movement. In Byzantium the sect had been temporarily suppressed by the vigorous ecclesiastical policy of Manuel Comnenus; but in the Slavonic Themes of the Empire, despite strong local action like that of St Hilarion, its influence remained unabated. Towards 1170–80 Theodore Balsamon alludes to whole regions inhabited by Bogomils (χωρία βωγομιλικά), including entire fortresses (ἀκέραια κάστρα).[1] It is probable that these fortresses were mainly situated in Macedonia, where Bogomilism is known to have flourished in the twelfth century.[2]

Moreover, in this period the Bogomil sect began to spread over the other south Slavonic countries and, from its home in Macedonia, penetrated westwards into Serbia, Bosnia and Hum, where it soon attracted the attention of the ecclesiastical and secular authorities.[3]

description by Euthymius Zigabenus of the entry of the θεοτόκοι into the Kingdom of God (cf. supra, p. 215), this is the only known account of Bogomil eschatology. Puech is probably right in thinking that eschatology always remained a comparatively undeveloped branch of the Bogomil teaching (op. cit. pp. 211–13).

[1] Theod. Balsamon, *Photii Patriarchae Constantinopolitani Nomocanon*, tit. x, cap. 8, in G. Voellus and H. Justellus, *Bibliotheca Juris Canonici Veteris* (Lutetiae, 1661), vol. II, p. 1042. Balsamon wrote his commentary on the *Nomocanon* in the last thirty years of the twelfth century; but his reference to the Bogomils must be dated before 1186, when the Byzantine domination of Bulgaria came to an end.

[2] Cf. supra, pp. 156 et seq.   [3] See Appendix IV.

CHAPTER VI

BOGOMILISM IN THE SECOND BULGARIAN EMPIRE

I. *Bulgarian Bogomilism in the thirteenth century:* The Paulicians in the first twenty years of Bulgarian independence. Their alliances with the Germans and the Bulgarians. The first anti-Bogomil council in Bulgaria (1211). The *Synodicon of the Tsar Boril*. Doctrines, customs and organization of the thirteenth-century Bogomils. John Asen II's toleration of the heretics.

II. *Decline of Bogomilism in the fourteenth century:* Hesychasm and the monastic revival. Bogomilism on Mount Athos. Heresy in Bulgaria. St Theodosius of Trnovo and two anti-Bogomil councils. Strong Massalian influence on Bogomilism. Disintegration of the Bogomil sect. The Turkish conquest and the disappearance of Bogomilism.

During the period of the Second Bulgarian Empire (1186–1393), the historical evolution of Bogomilism appears in two successive phases: the thirteenth century witnessed a great efflorescence of this sect in Bulgaria, the fourteenth century its decline and rapid disappearance. The causes of this twofold development must be sought in the basic features and in the history of Bogomilism.

It has already been pointed out that the influence exerted by the Bulgarian Bogomils on Byzantium in the tenth and eleventh centuries was reversed in the twelfth century, when the Byzantine sectarians transmitted in their turn many of their teachings and practices to their more primitive Bulgarian co-religionists. It is safe to assert that the new features displayed by Bulgarian Bogomilism in the thirteenth century, particularly in the spheres of doctrine, ritual and organization, are very largely due to the influence of the Byzantine heretics.

The recrudescence of Bogomilism at the beginning of the thirteenth century was largely due to internal conditions in Bulgaria during the twenty years following the revolt of 1186, which led to the shaking off of Byzantine domination and the establishment of the Second Bulgarian Empire in Trnovo. The times were reminiscent of the eleventh century: amid the general confusion arising out of the national revolt the dominant feeling in the people was the bitter hatred of the Greeks, in the fomenting of which the Paulicians played once again an active part. As in the eleventh century, the restlessness of the Bulgarian people was caused by the high-handed extortions of the Byzantine tax-collectors.[1] The mutual hatred of Bulgarians and Greeks was further inflamed in the reign of the Tsar Kaloyan (1196-1207). This bitter enemy of the Greeks, in deliberate contrast to Basil II Bulgaroctonus, liked to style himself 'Rhomaioctonus', whereas the Greeks called him 'Skylojoannes' ('John the Whelp'). The political instability of Bulgaria in those years was increased by the frequent invasions of the Cumans who, as in the eleventh century, allied themselves with the Bulgarians against the Greeks. In 1186 they supported the revolt of the Asen brothers and overran the whole of Thrace up to Adrianople. In 1205 they gave Kaloyan their armed assistance against the Franks.[2] Kaloyan himself was married to a Cuman, thus continuing the tradition of blood alliances with the Danubian nomads which existed among the Paulicians of the eleventh century.[3] Kaloyan's relations with the Cumans were a source of irritation to the Latin Crusaders in Constantinople: in 1205 they begged Pope Innocent III to declare a Crusade against the Bulgarian ruler, on the grounds that he had allied himself 'cum Turcis et ceteris Crucis Christi inimicis'.[4] These 'other enemies of the Cross of Christ' no doubt included the Paulicians and possibly also the Bogomils. However, there is no ground for maintaining that the latter played any active part in the revolt of 1186 or in Kaloyan's wars against the Greeks and the Latins.[5] As in the eleventh century, they probably contented themselves with a more or less passive opposition to the Greeks and supported the Paulicians against their common enemy by every means except the resort to warfare.

[1] Among the immediate causes of the Bulgarian revolt of 1186 was the taxation introduced by the Byzantine authorities in order to provide money for the nuptial celebrations of the Emperor Isaac II Angelus. See K. Jireček, *Geschichte der Bulgaren*, p. 225.
[2] See Jireček, ibid. pp. 226, 239.
[3] Ibid. p. 230.
[4] See A. Theiner, *Vetera monumenta Slavorum meridionalium*... (Romae, 1863), vol. I, p. 41.
[5] Jireček's opinion that the Bogomils played an 'important part' in the Asen revolt (op. cit. pp. 24-6) cannot be substantiated and is based on the same confusion between the Bogomils and the Paulicians as he makes with regard to the eleventh century (cf. supra, p. 192, n. 3.)

The Paulicians of Philippopolis, last mentioned in 1114–15,[1] appear again at the time of the Third Crusade. Nicetas Choniates, who was then governor of the Theme of Philippopolis, tells us that the Crusaders of Frederick Barbarossa, on entering Philippopolis in 1189, discovered that all the inhabitants had fled save for a few miserable beggars and the local 'Armenians', who welcomed the German invaders as friends.[2] It cannot be doubted that at least some of these Armenians were Paulicians, who were ever ready to welcome any enemy of their hated Byzantine oppressors.[3] As they had been transferred seventy-four years previously from Philippopolis to Neocastrum by Alexius Comnenus, it is to be presumed either that a number of them later returned to Philippopolis, or that some Paulicians of this city were only temporarily converted to Orthodoxy and returned to the faith of their fathers at the first opportunity.[4]

A similar role was again played by the Paulicians of Philippopolis in 1205. As a result of the Fourth Crusade, which led in 1204 to the establishment of the Latin Empire of Constantinople, the province of Thrace became subject to the French Baron Renier of Trit, newly appointed duke of Philippopolis. His power, however, weighed heavily on the local population, who appealed to Kaloyan to save them from the Latins. Kaloyan promptly invaded Thrace, was hailed as liberator by the Greeks themselves, and completely defeated the armies of the Emperor Baldwin I at Adrianople (14 April 1205). According to Villehardouin, the Paulicians of Philippopolis ('une grant partie des genz, qui estoient popelican') offered to surrender it to Kaloyan. Renier of Trit, who was then in Philippopolis, learnt of their scheme, had the Paulician quarter burnt to the ground and withdrew from the

[1] See supra, pp. 193–5.
[2] Nicetas Choniates, *Historia*, C.S.H.B. p. 527.
[3] From Nicetas's account they seem to be Monophysites: this appears from their name of 'Armenians' (generally applied by the Byzantines to the Monophysites), their use of azymes and their rejection of images. Monophysite colonies had existed in Thrace from the middle of the eighth century (cf. supra, p. 60) and Anna Comnena mentions them in Philippopolis. But these 'Armenians' must have included some Paulicians, for sixteen years later the Paulicians of Philippopolis again opened the gates of their city to the enemies of the Greeks.
[4] See D. E. Takela, Нѣкогашнитѣ павликяни и сегашнитѣ католици въ Пловдивско, *S.N.U.* (1894), vol. XI, pp. 107–8.

city. The Paulicians duly delivered the town to Kaloyan, who ordered the wholesale massacre or imprisonment of the Latin garrison and the destruction of the city.[1] The Bogomils are not expressly mentioned in Philippopolis at that time; but their presence in this city is attested a century previously by Anna Comnena and there is no reason to suppose that the Bogomil community of Philippopolis had disappeared in the course of the twelfth century. It is thus likely that the Bogomils as well as the Paulicians supported the cause of Kaloyan against the Franks, whose political and economic exploitation of Thrace was no less hateful to them than that of the Greeks.[2] It was easy for Villehardouin to confuse the two sects in view of the close similarity between their doctrines.

Kaloyan's popularity with the Bulgarian heretics bears some resemblance to the probable relations between Samuel and the Bogomils.[3] Both these Bulgarian rulers were rebels against the Byzantine Church and State and consequently fostered a policy of religious nationalism.[4] This policy could only succeed if the various religious and political parties rallied round the government and cemented their union by their common hostility to Byzantium. Thus both Samuel and Kaloyan must have sought the support of the Bulgarian Bogomils, who, as far as can be seen, responded to their call. Their motive in supporting the government was essentially opportunist, for at other times, when no

[1] *La Conquête de Constantinople*, éditée et traduite par E. Faral, t. II (1203–7), (Paris, 1939), p. 210. As a result of the contacts of the Crusaders with the Balkan Paulicians, the terms 'Poplicani', 'Populicani', 'Publicani', etc., were often used from the middle of the twelfth century to designate the French and English Cathars. See Schmidt, op. cit. vol. II, pp. 280–1; S. Runciman, *The Medieval Manichee*, pp. 121–3.

[2] Pogodin's statement, however, which is repeated by Klincharov (op. cit. p. 125), that 'Philippopolis was delivered to Kaloyan by the local Bogomils' (История Болгарии, p. 82) is incorrect, and is again due to the inability to distinguish between the Bogomils and the Paulicians.

[3] Cf. supra, pp. 149–51.

[4] Kaloyan's rebellion against the Byzantine Church led him to seek recognition of his authority from Rome. Lengthy negotiations between Kaloyan and Pope Innocent III resulted in the granting of the *pallium* to the Bulgarian Primate (1203) and the coronation of the Bulgarian ruler by the Papal Legate in Trnovo (1204). However, this temporary ecclesiastical union with Rome, based on purely political motives, was of no great consequence for Bulgaria. See Spinka, op. cit. pp. 102–6.

advantage could be gained by such a policy or when Bulgaria was governed according to Byzantine principles, they did not hesitate to teach civil disobedience.[1]

All these factors which favoured the spread of heresy during the first twenty years of the Second Bulgarian Empire—national hatred of the Greeks, the tension produced by continual wars, the religious toleration of the Bulgarian authorities—explain the considerable growth of Bogomilism in the reign of Kaloyan's successor Boril (1207-18). Boril usurped the throne and forced Kaloyan's rightful successor, John Asen II, to seek refuge in Russia. The whole of his reign reflects his unlawful accession: his position on the throne was never secure, and dissatisfied and separatist elements were continually working for his overthrow; one of his relatives wrested Macedonia from his realm; Boril was ultimately dethroned and blinded by the adherents of John Asen II.

The Bogomils were among the principal opponents of Boril. Their fresh appearance in Bulgarian history at the beginning of the thirteenth century coincides with the time when dualistic heresy was spread over all southern Europe, from the Black Sea to the Pyrenees.[2] Boril's reign is noted for the earliest known and most important legislation promulgated by the Bulgarian Church against the Bogomil heretics. These laws, issued in 1211, are contemporaneous with the measures taken by the Papacy to suppress the Albigensian heresy in southern France.[3] It seems natural to establish a connection between both these events, especially in view of the well-known intentions of Innocent III of extirpating all heresy in Europe. Some scholars have thought that Boril began the persecution of the Bogomils in his realm under direct pressure from the Pope.[4] Their hypothesis receives some confirmation from the fact that in 1206 a cardinal was sent from Rome to Bulgaria on an unknown mission.[5] However, as it will be shown, Boril's measures against the Bogomils were carried out according

[1] Cf. supra, p. 137.   [2] Cf. supra, p. 157.

[3] See A. Luchaire, *Innocent III et la Croisade des Albigeois* (Paris, 1905), pp. 115 et seq.

[4] See M. S. Drinov, Исторически прѣгледъ на българската църква, p. 83; M. G. Popruzhenko, Синодик царя Бориса, *I.R.A.I.K.* (1900), vol. v, Supplement, pp. 67-8.

[5] See *Chronicon Alberici monachi*, *M.G.H., Ss.* vol. XXIII, p. 886, ad an. 1206.

to the customs of the Eastern Church and differed essentially in character from the inquisitorial methods of the Papacy, which shows that the anti-Bogomil legislation in Bulgaria was not directly inspired by the Latin clergy.

On 11 February 1211, a council, convened in Trnovo, condemned the doctrines of the Bogomils and other kindred sects, anathematized their teachers and inflicted punishments on their adherents. The records of this council are preserved in a Bulgarian fourteenth-century manuscript, the *Synodicon of the Tsar Boril*.[1] The introduction describes the trial of the Bogomils as follows:

'After the accession to the throne... of the most pious Tsar Boril, there sprang up like evil thorns the thrice-accursed and hateful Bogomil heresy, whose leaders had been the most foul *pop* Bogomil and his disciples.... Having learnt this, the most pious Tsar Boril was inflamed with divine zeal and sent men to gather the heretics from the whole of his realm like tares are gathered into sheaves. Then he convened a council. There assembled all the bishops [of the land], priests, monks, and also all the boyars and a very great number of other folk. When the tsar saw that they were all gathered together, he came out, clad in bright purple robes, and took his seat in one of the great churches [in Trnovo]. Presiding over the assembled tribunal, he commanded that the sowers of impiety be brought before the council. He did not charge them at once, but entrapped them with great cunning: he told them to cast away all fear and to profess boldly their blasphemous teaching; and they, hoping to entice the tsar and those around him, exposed their ill-famed heresy in detail. They supported their arguments with many quotations from the Holy Scriptures, but the tsar and those with him questioned them with wisdom until their ill-famed sophistries were laid bare. Then the heretics were seized

---

[1] The MS. of the *Synodicon* was edited three times by M. G. Popruzhenko: (1) Синодик царя Бориса или Борила, *I.R.A.I.K.* (Odessa, 1896), vol. II, Supplement, pp. 1–82; (2) Синодик царя Бориса (Odessa, 1897); (3) Синодик царя Борила (Sofia, 1928) (Български Старини, vol. VIII). This edition, based on the earlier copies of Palauzov and Drinov, is the most critical and complete and also contains an account of the history of the MS. (pp. xix–xxvii).

See the analysis of the *Synodicon* by Popruzhenko in *I.R.A.I.K.* (1900), vol. v, Supplement, pp. 1–175. The principal defects of this otherwise useful work are its failure to bring out clearly the distinction between the Bogomils and other sects anathematized in the *Synodicon* and its unjustifiable conclusions regarding Bogomil ethics. Some important aspects of the *Synodicon* are studied by T. Florinsky (К вопросу о богомилах, *S.L.*, St Petersburg, 1883, pp. 33–40).

with confusion and were dumb as fish. The pious tsar, seeing that they were completely put to shame...was filled with joy and ordered that the heretical teachers and those whom they had seduced be detained. When they saw this, some of the heretics returned to the Universal Church; those, however, who did not submit to the Orthodox council were sent to prison or otherwise punished.'[1]

The evidence of the *Synodicon* precludes the supposition that the Latin clergy played any important part in the Council of Trnovo. We are told that the *Synodicon* was originally composed in Greek and later translated into Bulgarian by order of Boril.[2] The Orthodox character of the Council is particularly emphasized: the preface to the *Synodicon* contains three references to this 'Orthodox Council' (православныи съборъ) and states that it was the first of its kind to be convened in Bulgaria.[3] Moreover, its convocation and procedure were carried out according to the tradition of the Eastern Church.[4] The tsar himself convened it, conducted the interrogation of the heretics and decreed that the articles of the *Synodicon* should have the force of law. In his assumption of the threefold function of instigator, prosecutor and executive power, Boril followed the examples of Alexius Comnenus in his treatment of the Paulicians and the Bogomils[5] and of the Serbian Grand Župan Stephen Nemanja, who at the end of the twelfth century dealt most successfully with the Bogomils in his lands.[6] Finally, the punishment inflicted on the obdurate Bogomils was very different from the methods then employed by the Latin inquisitors. In the Orthodox Church imprisonment was generally the severest

[1] *Synodicon*, Popruzhenko's edition, 1928, pp. 77–80. The subsequent references to the *Synodicon* are taken from this edition.

[2] Повелѣ благочьстивыи царь Бориль прѣписати съборникь ѿ гръчьскаго на блъгарскыи свои языкь. (*Synodicon*, p. 80.)

[3] Прѣжде бо царства его никтоже инь сътвори православныи съи съборъ. (Ibid. pp. 80–2.)

[4] Cf. F. Dvorník, 'The Authority of the State in the Oecumenical Councils', *The Christian East* (London, 1933), vol. XIV, no. 3, pp. 95–108.

[5] The ruse ascribed to Boril in the *Synodicon*, which led to the confession of the apparently unsuspecting Bogomils, had already been resorted to by Alexius Comnenus and Flavian of Antioch (cf. supra, p. 203, n. 2). It is tempting to explain the somewhat improbable facility with which the Bogomils are said to have been tricked by Boril as a conscious reminiscence of those celebrated precedents by the later compilers of the *Synodicon*.

[6] See Appendix IV.

punishment for heresy, the death penalty being resorted to only in very exceptional circumstances.[1]

The *Synodicon* is the only source which gives us a direct knowledge of Bogomilism in the thirteenth century. It is, moreover, one of the most reliable of guides for a study of the sect, since, as an official document of the Orthodox Church, it is based on a careful and objective study of the doctrines and past history of the Bogomils. The material it contains is of a complex nature. That part of it which concerns Bogomilism can be classed into three sections:

(1) The condemnation, in the form of anathemas, of specific doctrines which are certainly Bogomil. A number of these appear in the same form as in the *Sermon against the Heretics*; others, on the contrary, while being confirmed by the evidence of the tenth- and eleventh-century sources, are nevertheless presented in a more complex and developed form, often traceable to the influence of Byzantine Bogomilism.

(2) Other heretical doctrines not originally Bogomil, but belonging to sects which were in constant contact with the Bogomils. Several of them had probably been assimilated by Bogomilism by the thirteenth century, owing to the evolution and adaptability of the sect.

---

[1] Some scholars, including Jireček, have taken the unflattering reference to Boril in the biography of Stephen Nemanja by his son, Stephen the First-Crowned, to imply a violent persecution of the Bogomils following on the Council of Trnovo: 'his soul found a sweet pleasure in shedding the blood of his race; he murdered countless other men, as though he wanted to destroy both land and sea.' (Šafařík, *Památky Dřevního Písemnictví Jihoslovanův*, Prague, 1873, p. 22.) But this view cannot be corroborated. Boril had many political enemies who were constantly seeking to destroy him, and he probably had to resort to several drastic political repressions. But there is no evidence of the 'blutige Bogomilenverfolgung' ascribed to him by Jireček (op. cit. p. 246, n. 27). Equally unjustifiable is D. Mishew's statement that after the Council of Trnovo 'a sort of inquisition was established' (*The Bulgarians in the Past*, Lausanne, 1919, p. 82), which Klincharov even describes as 'bloody' (op. cit. pp. 141–2).

The opposite error is committed by Blagoev, who thinks that Bogomilism was an imaginary label attached by Boril to his political enemies. (Беседата на Презвитер Козма против богомилите, *G.S.U.*, 1923, vol. xviii, pp. 43–4.) Blagoev's view of Bogomilism is vitiated by his complete incomprehension of the nature of heresy and by his preconceived desire to seek for purely political motives behind every religious movement in the Balkans in the Middle Ages.

(3) A number of features pertaining to the ritual, organization and general behaviour of the Bogomil communities, several of which again reveal the influence of Byzantine Bogomilism.

It should be noted that the *Synodicon* is the earliest Slavonic monument which refers to the heretics as Bogomils (богомилы) as distinct from the 'adherents of the Bogomil heresy'.[1]

The *Synodicon* gives the most explicit account we possess of the origins of the Bogomil sect in Bulgaria:

'Because our guileful foe spread the Manichaean heresy all over the land of Bulgaria, and mixed it with the Massalian heresy (манихеискѫѧ ересь развѣа смѣсивь сіѫ съ масаліанскоѫ) let the leaders (начѧлникы) of this heresy be cursed. May the *pop* Bogomil—who in the reign of Peter, tsar of Bulgaria, adopted this Manichaean heresy (въспріемшаго манихеискѫѫ сіѫ ересь) and spread it over the land of Bulgaria, and who added to them [i.e. these heresies] the belief that Christ Our Lord was seemingly (въ привидѣни) born of the Holy Mother of God and ever Virgin Mary, was seemingly crucified and ascended in His...body which He left in the air (на въздоусѣ остави)..., —and all his past and present disciples who are also called apostles (ученици и апостоли нареченіи), be cursed (anathema).'[2]

This resolution of Bogomilism into its two main component parts, 'Manichaeism' (which obviously means Paulicianism)[3] and Massalianism, is identical with the definitions of the Patriarch Theophylact and Anna Comnena.[4] The *Synodicon* is the only source which unequivocally shows that this compound of Paulicianism and Massalianism existed in fact in Bulgaria before the days of the *pop* Bogomil, for the heresiarch only 'adopted' these heresies. The docetic Christology attributed by this source to the *pop* Bogomil is indeed eminently characteristic of the teaching of his followers. But the statement in the *Synodicon* that the *pop* Bogomil was the first to adopt it cannot be true: docetism, being a Paulician doctrine, was no doubt known in Bulgaria long before the time of the heresiarch.

The doctrines ascribed to the Bogomils in the *Synodicon* which appear in an identical form in the *Sermon against the Heretics* are

---

[1] *Synodicon*, p. 44. Cosmas simply calls them heretics (cf. supra, p. 119) and St Hilarion's biographer—богѡмилскыѧ ереси поклѡнникы. (Cf. supra, p. 225, n. 4.)
[2] Ibid. p. 42.
[3] Cf. supra, p. 189.
[4] Cf. supra, pp. 112, 198.

the rejection of the Mosaic Law, of the Old Testament Prophets, as well as of churches, traditional prayers (with the exception of the Lord's Prayer), Holy Orders, the liturgy, the sacrament of the Eucharist, the Cross and the icons.[1] Others are more interesting, as showing the influence of Byzantine Bogomilism. An article of the *Synodicon* curses those who call *Satan* (сатану) the creator of all visible things and say that he is the steward (икшнωма) of rain and hail and of everything that comes from the earth.[2] Although this doctrine can be found in almost the same form in the *Sermon against the Heretics*,[3] Cosmas merely refers to him as the Devil, while the epithet Satan was used by the Byzantine Bogomils.[4] The influence of Byzantine Bogomilism is even more apparent in the article anathematizing 'those who say that Satan created Adam and Eve'.[5] This specific doctrine of the origin of man is ascribed to the Bogomils by no other Bulgarian source, but was held, on the other hand, by the Byzantine sectarians in the early twelfth century.[6] Likewise the *Synodicon* curses 'those who revile John the Baptist and say that he together with his baptism is from Satan; and who for this reason eschew Baptism by water and baptize without water, reciting only the Lord's Prayer'.[7] The opposition between Baptism by water and Baptism through the Spirit, as well as the so-called 'βάπτισμα', or ceremony of initiation into the sect, which included the recitation of the Lord's Prayer, are ascribed to the Byzantine Bogomils by Zigabenus.[8] They are not, however, mentioned by Cosmas, who speaks only in a very general manner of the Bogomil rejection of Baptism. Hence it is likely that the Bulgarian Bogomils, who in the tenth century simply rejected Baptism by water as a consequence of their denial that matter can be a vehicle for Grace, developed a more complex view of Baptism during the twelfth century under the influence of their Byzantine co-religionists; moreover, as the latter evolved a form of ritual probably unknown to the Bulgarian Bogomils in the tenth century, it can be supposed that the heretical 'baptism' mentioned in the *Synodicon* is of Byzantine origin.

[1] *Synodicon*, pp. 44–8.
[2] Ibid. p. 44.
[3] Cf. supra, p. 123.
[4] Cf. supra, p. 210.
[5] *Synodicon*, p. 44.
[6] Cf. supra, pp. 208–9.
[7] *Synodicon*, p. 46.
[8] Cf. supra, pp. 215–16.

The historical continuity between Byzantine Bogomilism and the teachings condemned at the Council of Trnovo is recognized by the *Synodicon*: one of its articles anathematizes the Byzantine heresiarch, Basil 'the physician'.[1]

Moreover, the *Synodicon* mentions by name some of the leaders of the Bulgarian Bogomils: the *pop* Bogomil and his successors in the tenth and eleventh centuries—Michael (his immediate disciple), Theodore, Dobry, Stephen, Basil and Peter.[2] An earlier article curses 'Alexander the smith, Avdin and Photin, Aphrigiy and Moses the Bogomil...Peter of Cappadocia, dyed of Sredets, Luke and Mandeley of Radobol'.[3] Of the heretics enumerated in the latter article, only Moses and Peter of Cappadocia[4] can be with certainty described as Bogomils. But as the whole of this paragraph in the *Synodicon* seems to refer to Bogomilism, it is probable that the others also belonged to the sect.

Among the doctrines of non-Bogomil sects condemned in the *Synodicon*, the most important are those of the Massalians. The doctrine of 'those who say that a woman becomes pregnant in her womb through the co-operation of Satan who remains there constantly even until the birth of the child and who cannot be driven away by Holy Baptism, but only through prayer and fasting'[5] is characteristically Massalian;[6] yet it is placed in the *Synodicon* among the Bogomil doctrines. Moreover, the direct participation of Satan in the act of procreation is already implied in the belief that Cain was born of the intercourse between Satanael and Eve and in the demonology ascribed to the Byzantine Bogomils by Zigabenus.[7] There can be no doubt that this Massalian doctrine condemned in the *Synodicon* was actually held by the Bulgarian Bogomils in the thirteenth century. The strong influence exerted by Massalianism on Bogomilism between the tenth and

---

[1] Василіа врача иже въ Кѡнстантінѣ градѣ въсѣавшаго сіѫ тръѡкааннѫѧ богомилскѫѧ ересь при Алеѯи православнѣмь цари Комнинѣ, анаѳема. (*Synodicon*, p. 48.)

[2] Cf. supra, p. 145.

[3] Алеѯандра ковача, Авдина же и Фѡтина, Афригіа же и Мѡѵсеа богомила...Петра кападокіискаго, дѣдца срѣдечьскаго, Лоукѫ же и Манделеа радоболскаго, анаѳема. (Ibid. p. 68.)

[4] The significance of Peter's title of *dyed* is discussed below (pp. 242–5).

[5] Ibid. pp. 44–6.   [6] Cf. supra, pp. 49–50.

[7] Cf. supra, pp. 208, 213–14.

twelfth centuries[1] led to both sects being identified by the Orthodox in the thirteenth century: the Patriarch Germanos II (1220–40) wrote: Μασσαλιανῶν ἤτοι τῶν Βογομίλων.[2] A Slavonic Nomocanon of 1262 condemns 'the Massalians who are now called Bogomils'.[3] By the fourteenth century all difference between them seems to have disappeared.[4]

Another teaching condemned in the *Synodicon* and not explicitly attributed to the Bogomils is the denial of the Resurrection of the Body.[5] In contemporary Christian circles this doctrine generally passed as Judaism, which was rampant in Macedonia owing to the proximity of Thessalonica, a great Jewish centre in the Middle Ages.[6] It is probable that Bogomilism and Judaism had points of contact in the thirteenth century, particularly as both were to become the object of a common persecution in the following century. The denial of the Resurrection of the Body is consonant with the Bogomil condemnation of matter and is indeed attributed to these heretics by Euthymius of Acmonia.[7]

Finally, the evidence of the *Synodicon* brings out several interesting features in the life of the Bogomil communities in the thirteenth century, which we do not find in earlier sources. Among these is the practice attributed to the Bogomils of reciting the Lord's Prayer 'wherever they happen to be' (на приключившим ся мѣстѣ).[8] The inference we can draw from this is that if a Bogomil happened to be travelling at a time appointed for prayer, he stopped to pray by the wayside. That the Bogomils had set hours in the day for prayer is attested by Cosmas and Zigabenus.[9] For those Bogomils who stayed at home during these hours, indoor communal prayer was prescribed. These prayer-meetings, according to the *Synodicon*, were held by night and were accompanied by a definite form of ritual: one of the clauses directly referring to the Bogomils curses 'their customs, *nocturnal meetings, mysteries*

[1] Cf. supra, pp. 114–15, 128, 136–7, 222.
[2] *Epistula ad Constantinopolitanos contra Bogomilos* (Ficker, *Die Phundagiagiten*, p. 116).
[3] Cf. supra, p. 164, n. 3.   [4] Cf. infra, p. 254.   [5] *Synodicon*, p. 70.
[6] See B. Melioransky, К истории противоцерковных движений в Македонии в XIV веке: Στέφανος: *Sbornik statey v chest' F. F. Sokolova* (St Petersburg, 1895), pp. 62–72. Cf. supra, p. 83, n. 1.
[7] Cf. supra, p. 182, n. 1.   [8] *Synodicon*, p. 46.
[9] Cf. supra, pp. 135, 217.

and harmful teachings'.¹ No information is given on the precise character of these meetings; but, judging from the scanty evidence of the Byzantine sources, they probably consisted of invocations to the Trinity and recitations of the Lord's Prayer with appointed prostrations.² In no other source are the Bogomils accused of meeting by night;³ in the thirteenth century, when the authorities were wide awake to the danger of the sect, surreptitious meetings were no doubt particularly necessary to escape detection, and the Bogomils were always past-masters in the art of concealment.

A particularly interesting passage of the *Synodicon* is the brief anathema against 'the dyed of Sredets'.⁴ Sredets was the medieval name for the town of Sofia. The significance of the title 'dyed' can best be understood by reference to the following Latin twelfth- and thirteenth-century sources:

A note in the Carcassonne manuscript of the *Liber Sancti Johannis*, one of the principal books of the Italian and French Cathars, says: 'hoc est secretum haereticorum de Concôrezio *portatum de Bulgaria Nazario suo Episcopo*, plenum erroribus.'⁵ This book has been conclusively shown by Ivanov to be of Bogomil origin and to have been translated into Latin from a Slavonic original, now lost.⁶ This Nazarius, clearly an important personage among the Bulgarian Bogomils, is also mentioned by Reinerius Sacchoni, who knew him personally: 'Nazarius vero quondam *eorum episcopus et antiquissimus* coram me et multis aliis dixit, quod B. Virgo fuit Angelus et quod Christus non assumsit animam humanam, sed angelicam, sive corpus coeleste. Et dixit quod habuit hunc errorem *ab episcopo et filio majore Ecclesiae Bulgariae* jam fere elapsis annis LX.'⁷ Reinerius wrote *c.* 1250⁸ and his

¹ *Synodicon*, p. 42. ² Cf. supra, pp. 182–3.
³ Psellus's assertion that the Thracian Euchitae in the eleventh century indulged in nocturnal orgies cannot, as we have seen, be applied to the Bogomils.
⁴ Cf. supra, p. 240. Cf. Florinsky, op. cit. pp. 33–40.
⁵ See J. Benoist, *Histoire des Albigeois et des Vaudois*, vol. I, p. 296.
⁶ Op. cit. pp. 65–72.
⁷ *Summa de Catharis et Leonistis*: Martène et Durand, *Thesaurus novus anecdotorum*, vol. V, cols. 1773–4. Cf. Moneta of Cremona (*Adversus Catharos et Valdenses*, ed. T. A. Ricchinius, Rome, 1743, lib. III, cap. II, p. 233): 'Sclavi... dicunt, quod Deus pater justorum tres Angelos misit in mundum: Unus ex eis formam mulieris accepit in mundo isto; et hic dictus est Maria. Alii vero duo viriles formas sumpserunt, scilicet Christus et Johannes Evangelista.'
⁸ Ibid. col. 1775.

association with Nazarius must hence be dated in the first half of the thirteenth century, at the time when he was himself a Patarene teacher in Lombardy. Nazarius's initiation into the Bulgarian Bogomil sect, which, according to Reinerius, occurred about sixty years previously, can therefore be dated in the second half of the twelfth century. The doctrine taught by Nazarius is the familiar Docetism, one of the main articles of the Bogomil faith.

Reinerius thus mentions three titles which, he claims, existed among the Bulgarian Bogomils in the later twelfth and in the thirteenth centuries: 'episcopus', 'antiquissimus' and 'filius major'. The first two are probably synonymous, as they are applied to the same person and both appear to designate the highest rank in a given hierarchy. The title of 'episcopus' was also given in 1167 to Nicetas of Constantinople.[1]

This raises the following questions: did these titles belong in fact to the leaders of the Bulgarian Bogomils and, if so, what were their Slavonic equivalents? It has already been shown that there is no ground for maintaining that a regular hierarchy existed among the Bulgarian Bogomils in the tenth and eleventh centuries. The followers and immediate successors of the *pop* Bogomil appear to have organized the sect on democratic lines. Basil, the leader of the Byzantine Bogomils in the late eleventh and early twelfth centuries, was surrounded by twelve 'apostles', and this same title is attested among the Bulgarian sectarians in the thirteenth century in the *Synodicon*.[2] But the Latin sources of the twelfth and thirteenth centuries apply the term 'Ecclesia' to the Bogomil community in Bulgaria;[3] at the same time, the notion of 'Ordo' appears, apparently synonymous with 'Ecclesia'.[4] It might seem that these terms, applied to the Bogomil communities, are fictitious, as the Latin writers had generally only an indirect acquaintance with the Bulgarian sectarians and tended to judge them from their knowledge of the western Cathars and Patarenes, whose communities possessed a highly developed organization, closely modelled on that of the Roman Church.[5]

[1] Cf. supra, p. 156.
[2] Cf. supra, pp. 199, 238.
[3] Cf. supra, pp. 157 et seq.
[4] See P. Limborch, *Liber Sententiarum Inquisitionis Tholosanae* (Amstelodami, 1692), p. 126: 'ordinem sive sectam.'
[5] See Schmidt, op. cit. vol. II, pp. 139–50.

Yet the existence of an organized hierarchy among the Bulgarian Bogomils from the middle of the twelfth century is *prima facie* a probable supposition. It is not likely that the Bogomil sect could have so successfully survived four centuries of the proselytism, repression and persecution instigated against it by the representatives of the Orthodox Church, had its organization remained weak or indefinite. Moreover, to resist the sectional interests and the danger of schisms, invariably present in all sectarian movements, the acceptance of a strong authority and a hierarchy was clearly necessary. It may be supposed that in order to fight the Orthodox Church the Bogomil sect was reduced to adopt its enemy's own weapons: the most effective of these was the centralized ecclesiastical organization, the model of which was to be found in Byzantium. It is thus very likely that the Bulgarian Bogomils, who in the course of the twelfth century borrowed from Byzantium a number of new features in their doctrine and ritual, also derived from the same source a more rigid organization of their communities and a regular hierarchy, in the same manner as the Cathars and Patarenes borrowed many traits of their own organization from the Roman Church. It is not unreasonable to claim that the notions of 'Ecclesia Bulgariae' and 'Ordo de Bulgaria' appeared among the Bulgarian sectarians as the result of the penetration of Bogomilism into Constantinople.

This suggests that the titles of 'episcopus', 'antiquissimus' and 'filius major', given by Latin writers to several important Bogomils, are not fictitious. Moreover, among the Cathars and Patarenes the 'episcopus' was the holder of the supreme rank in the ecclesiastical hierarchy.[1] In the Bosnian Patarene Church, which in many respects was directly influenced by Bulgarian Bogomilism,[2] and which can consequently serve for the present purpose as a connecting link between the Bogomils and the Cathars, the same title existed, and to the Latin term 'episcopus' corresponded the Slavonic 'dyed'.[3] It seems legitimate to conclude that the 'episcopus Ecclesiae Bulgariae', referred to by Reinerius, and also Nazarius himself held among the Bulgarian Bogomils the title of 'dyed', and that the 'dyed of Sredets' mentioned by the *Synodicon* was the head of the Bogomil community in Sredets;

[1] Schmidt, ibid. p. 142.
[2] See Rački, *Rad*, VII, pp. 163 et seq.   [3] Ibid. p. 184.

judging by the analogy with the Bosnian 'dyed', he may also have been the supreme leader of all the Bulgarian Bogomils, though this cannot be affirmed with any degree of certainty.¹

The 'filius major', on the other hand, occupied the second rank in the hierarchy of the Cathars and Patarenes. In Bosnia he was known as 'gost'.² This Slavonic title is not confirmed by any Bulgarian source, but, by analogy with that of 'dyed', it can reasonably be inferred that the title equivalent among the Bulgarian Bogomils to 'filius major' was in fact 'gost'. As for the title of 'antiquissimus', given to Nazarius by Reinerius, although it is tempting by reason of its etymology to relate it to the 'ancianus' of the Cathars and the 'starats' of the Bosnian Patarenes,³ it probably refers to the Bogomil 'dyed': Reinerius would scarcely have given Nazarius simultaneously two titles corresponding to two different ranks in the Bogomil hierarchy; moreover, 'antiquissimus' and 'dyed' both imply the notion of 'elder', and the former was probably used by Reinerius as a complement to 'episcopus', a Western title presumably unknown among the Bulgarian Bogomils.

The exact relation between the different Bogomil communities in Bulgaria is not very clear. The only information on the subject comes from a Latin source: in 1167, Nicetas, the heretical 'bishop' of Constantinople, presiding over the Council of the Cathars at Saint-Félix de Caraman near Toulouse, was questioned by his Western co-religionists on the organization of the Eastern dualistic 'Churches'. He replied: 'Ecclesiae Romanae et Drogometiae et Melenguiae et Bulgariae et Dalmatiae sunt divisae et terminatae, et una ad alteram non facit aliquam rem ad suam contradictionem, et ita pacem habent inter se: similiter vos facite.'⁴ These

¹ The *Synodicon* was first published by N. Palauzov in 1855 (*Vremennik Imperatorskogo Moskovskogo Obshchestva Istorii i Drevnostey Rossiyskikh*, vol. XXI, Moscow), but in an incomplete form which did not contain the reference to 'the djed of Sredets'. Rački, who knew the *Synodicon* only in this edition, was unacquainted with this valuable piece of evidence, which was first revealed by Florinsky in 1883. (Cf. supra, p. 235, n. 1.)
² See Schmidt, ibid.; Rački, ibid.
³ See Schmidt, ibid, pp. 144–5; Rački, loc. cit. p. 185.
⁴ *Notitia conciliabuli apud S. Felicem de Caraman, sub Papa haereticorum Niquinta celebrati*, in M. Bouquet, *Recueil des historiens des Gaules*, Paris, 1806, vol. XIV, pp. 448–50.

principles of decentralization and mutual collaboration, which apparently existed in the twelfth century among the various sectarian communities in the Balkans, are suggestive of some federalistic basis of organization; but precise evidence on this point is lacking.

A much-debated question, closely connected with the problem of the Bogomil hierarchy, is whether the Cathars, the Patarenes and the Bogomils owed obedience to one supreme heretical 'Pope'. Schmidt has collected all the evidence of medieval Latin sources which appears to assert the existence of such a dignitary.[1] This heretical 'Pope' was always said to reside in south-eastern Europe—in Bulgaria, according to some sources—in Constantinople, according to others. The latter case is that of Nicetas of Constantinople, who in the records of the council of Saint-Félix de Caraman is called 'Papa haereticorum'. The first view was upheld in a letter by Conrad of Marburg, Papal envoy and later inquisitor in Germany, written in 1223: 'ille homo perditus est, qui extollitur super omne quod colitur, aut quod dicitur Deus, jam habet perfidiae suae praeambulum haeresiarcha, quem haeretici Albigenses Papam suum vocant, *habitantem in finibus Burgarorum* [sic], *Croaciae et Dalmatiae juxta Hungarorum nationem*.'[2] The existence of a supreme Bogomil 'Pope' has sometimes been upheld in recent days.[3]

But this theory is rejected by authoritative scholars. Schmidt, after a careful study of all the relevant sources, concluded that the heretical 'Pope' is a purely fictitious character.[4] Rački, who was of the same opinion, pointed out that according to Nicetas himself the Balkan dualistic communities were organized on a federalistic basis, which in itself precludes the possibility of a supreme central authority.[5]

---

[1] Op. cit. vol. II, pp. 146–7.

[2] *Epistolae Gervasii Praemonstratensis Abbatis*, ep. CXXIX: in C. L. Hugo, *Sacrae Antiquitatis Monumenta*, vol. I, p. 116.

[3] In particular by F. Legge ('Western Manichaeism and the Turfan Discoveries', *J.R.A.S.* 1913, p. 73), who states that 'all Southern Europe is said to have been parcelled out into Manichaean dioceses whose bishops paid allegiance to a Manichaean Pope seated in Bulgaria'.

[4] Ibid. pp. 145–50. Cf. J. Guiraud, *Histoire de l'Inquisition au Moyen Âge* (Paris, 1935), vol. I, pp. 232–4.

[5] Op. cit. *Rad*, vol. X, pp. 185–6.

The valuable indication concerning the hierarchy of the thirteenth-century Bogomils, provided by the *Synodicon of the Tsar Boril*, is the only information obtainable on this subject from Bulgarian sources. The considerable value of the *Synodicon* as a historical document lies in the fact that it does not limit itself, like most of the other sources on the Bogomils, to an exposition of their errors in matters of faith and morals, but gives an account of the organization and customs of the sect and of the social behaviour of its members.

Another article of the *Synodicon* which may perhaps be taken as descriptive of the Bogomil customs curses 'those who on the 24th of June, the birth of John the Baptist, practise magic (влъшвенїа) and gather fruits and that night perform foul mysteries like the pagan rites (сквръннаа творять таинства и еллинстѣи службѣ подобнаа)'.[1] The allusion is to the pagan festival of 'Иванъ-день', still celebrated to-day by the southern Slavs.[2] The place occupied by this clause in the *Synodicon* in the section dealing with the Bogomil doctrines suggests, unless it is interpolated,[3] that the Fathers of Trnovo recognized a definite connection between this pagan ceremony and the customs of the sect. This is not the only indication of the connection between paganism and heresy in Bulgaria.[4] The Bogomils in particular, whose contact with the masses was always close, are frequently associated in the sources with everything that has come to be regarded as popular superstition or magic and with the remnants of pre-Christian paganism.[5] One cannot be certain, however, whether the pagan rites were adopted by the Bogomils to their own doctrines or whether their connection with paganism was a tactical one, based on the necessity of fighting the common foe—the Orthodox Church. The latter alternative is perhaps the more probable.

The following articles of the *Synodicon* have sometimes been taken to refer to the Bogomils: 'cursed be those who either by some magic or by herbs, spells, enchantment, devilish witchcraft

---

[1] *Synodicon*, p. 44.
[2] Cf. supra, p. 67, n. 1.
[3] This is Puech's opinion (op. cit. p. 344, n. 1).
[4] Cf. supra, p. 95.
[5] See M. G. Popruzhenko, Синодик царя Бориса, *I.R.A.I.K.* (1900), vol. v, Supplement, pp. 168–9; M. S. Drinov, Южные славяне и Византия, pp. 74–5; P. Kemp, *Healing Ritual*, pp. 159–78.

or poison try to injure the tsar, anointed by God.... Cursed be those who assist thieves, murderers, robbers and other such people.'[1] Popruzhenko, relying on Cosmas's statement that the heretics urged the people to civil disobedience, concluded that these dissident elements anathematized in the *Synodicon* are Bogomils.[2] However, this is extremely hypothetical: even apart from the fact that the last two clauses are not found in the section of the document which deals with Bogomilism, the evidence for Popruzhenko's claim is quite insufficient. Although the Bogomils were in opposition to the central government in Boril's reign, there were many other dissatisfied and rebellious sections of the community which were seeking to bring about the tsar's downfall.[3]

A notable feature of the *Synodicon of the Tsar Boril* is the absence in its articles against the Bogomils of any reference to their ethical teaching. In particular, the Bogomils are accused neither of rejecting marriage nor of condemning the eating of meat. From this fact Popruzhenko has drawn the unwarrantable conclusion that only the doctrines of the Bogomils were condemned at the Council of Trnovo, while their moral behaviour was not only considered innocuous, but even viewed with some favour by the Church.[4] This arbitrary separation of ethics from doctrine shows a misunderstanding of the attitude of the Orthodox Church towards heresy. The asceticism of the Bogomils and that of all dualistic sects, which is based on a hatred of matter and a denial of its sanctification through Grace, was always considered by the Church to be essentially immoral, whatever its outward resemblances to Christian asceticism. The Church never ceased to condemn all dualistic heresies for this very reason. The absence of any condemnation of Bogomil ethics in the *Synodicon* can be explained by the nature of this document. It is not a polemical work against the Bogomils like the *Sermon against the Heretics*, but a doctrinal handbook for the use of the ecclesiastical authorities. In its form—that of concise anathemas suitable to be read on public occasions—it is in the tradition of the Byzantine *Synodica* which provided fixed formulae for the solemn anathematizing of the doctrines of past and contemporary heretics, carried

[1] *Synodicon*, p. 74.
[2] Loc. cit. pp. 169–70.
[3] Cf. supra, p. 234.
[4] Loc. cit. pp. 164–6.

out in Orthodox cathedral churches on the first Sunday in Lent. The prototype of all such manuals is the *Synodicon for the Sunday of Orthodoxy*, which contains a series of anathemas of the principal doctrines of the Bogomils, but also without any reference to their ethical teaching.[1]

The *Synodicon of the Tsar Boril* contains one of the most complete accounts we possess of the Bogomil sect. Its completeness is due to the fact that it combines the results of direct observation of the sect with those of the earlier investigations of Bulgarian and Byzantine Churchmen, particularly, as Popruzhenko has shown, of Cosmas and Euthymius Zigabenus. The *Sermon against the Heretics*, whose author had close contact with the life and customs of the Bogomils, could supply much first-hand information. It became popular very early among the southern Slavs, and the Bulgarian Churchmen at the beginning of the thirteenth century were certainly well acquainted with it.[2] The *Panoplia Dogmatica*, as the most comprehensive account of the Bogomil doctrines and the source of all subsequent anti-Bogomil productions, inevitably influenced the *Synodicon*. Popruzhenko thinks that by 1211 there existed a Bulgarian translation of the *Panoplia*, or at least of its chapter against the Bogomils.[3]

---

[1] The *Synodicon for the Sunday of Orthodoxy* (the first Sunday in Lent) was edited by Th. Uspensky (*Zapiski imperatorskogo novorossiyskogo Universiteta*, Odessa, 1893, vol. LIX, pp. 407–502). This version is based on an original document composed after the Seventh Oecumenical Council and directed against Iconoclasm, but amplified in the course of the eleventh century. The articles against the Bogomils were added by order of Alexius Comnenus. Further clauses were added to the *Synodicon* in the twelfth and fourteenth centuries. See A. Petrovsky's article, 'Анафема' in *Pravoslavnaya Bogoslovskaya Entsiklopediya* (ed. by A. P. Lopukhin; St Petersburg, 1900), vol. I, pp. 679–700.

Another Byzantine document similar in form to the *Synodicon of the Tsar Boril* is the collection of formulae of abjuration to be recited by those Bogomils who were received into the Church. This document, which probably dates from the reign of John Comnenus, was published by L. Thallóczy, 'Beiträge zur Kenntniss der Bogomilenlehre,' *Wiss. Mitt. Bosn. Herz.* Vienna, 1895, vol. III, pp. 360–71) under the title: Τοὺς ἀπὸ τῆς μυσαρᾶς αἱρέσεως τῶν Πογομίλων τῇ ἁγιωτάτῃ τοῦ Θεοῦ μεγάλῃ ἐκκλησίᾳ προσερχομένους ἀπὸ Μανιχαίων καὶ αὐτοὺς καταγομένους καὶ χείρονας τούτων ὄντας, χρὴ προσδέχεσθαι οὕτως.

[2] See Popruzhenko, loc. cit. p. 112.

[3] Popruzhenko's hypothesis is based on the study of a Serbian MS. of the *Panoplia* in the monastery of Khilandar on Mount Athos, which contains, in his opinion, a number of bulgarisms (ibid. pp. 113–15).

250   THE BOGOMILS

The edicts promulgated by the Council of Trnovo did not succeed, however, in destroying Bogomilism in Bulgaria; there is evidence that the heresy was rampant in the reign of Boril's successor, John Asen II (1218–41). During his reign Bulgaria became the strongest power in the Balkans. John Asen II sought to achieve what had been Symeon's idea—the unification of all southern Slavs under the Bulgarian sceptre and within the framework of the Orthodox Church. His crowning success was the establishment in 1235 of the autocephalous Bulgarian patriarchate at Trnovo after formal consent of the four eastern patriarchs.[1] This naturally caused him to abandon the allegiance to the Roman See paid by his predecessors Kaloyan and Boril. Pope Gregory IX, incensed by this defection and by the fact that the Bulgarian tsar had concluded an alliance with the emperor of Nicaea to destroy the Latin Empire of Constantinople, instigated a crusade against Bulgaria. He urged the king of Hungary, Bela IV, to become its leader. As an additional motive for this crusade, Gregory IX, in his letters to Bela and to the Emperor Baldwin II of Constantinople (February 1238), complained that Bulgaria was 'full of heretics' who were apparently under the direct protection of John Asen II.[2] The crusading army assembled in Hungary but never crossed the frontier, as the Bulgarian tsar successfully manœuvred to keep his allies, the Cumans, as a perpetual threat to Hungary and Constantinople.

The final stage in the historical development of Bogomilism, its decline and disappearance in the fourteenth century, now remains to be studied. The decadence of Bogomilism, which followed so closely on its great efflorescence in Bulgaria during the twelfth and thirteenth centuries, was due to several features inherent in this sect as well as to the general characteristics of the time. In spite of its internal coherence and of the external organization which it had borrowed from Byzantium, Bogomilism always remained a somewhat diffuse heresy, eminently changeable and

[1] See Jireček, op. cit. pp. 248–62; Pogodin, op. cit. pp. 85–93; Spinka, op. cit. pp. 109–13.
[2] See A. Theiner, *Vetera monumenta historica Hungariam sacram illustrantia* (Romae, 1859–60), vol. I, p. 160: 'Perfidus...Assanus...receptat in terra sua hereticos et defensat, quibus tota terra ipsa infecta dicitur et repleta.'

adaptable to circumstances. This peculiarity, which rendered the task of fighting it extremely difficult, and hence increased the danger presented by Bogomilism to the Church, became in later times a source of weakness to the sect. Unlike the Paulicians, the Bogomils were unable to retain the purity of their teaching and with time absorbed from kindred sects and movements a number of features which were originally alien to them. Some of these could not fail to have a detrimental effect on the strength and stability of the Bogomil sect. This applies particularly to Massalian elements which, it seems, occupied progressively more and more place in Bogomil doctrine and ethics. The Massalian sect, as we have seen, penetrated into Bulgaria during the eighth and ninth centuries, exerted a strong influence on Bogomilism during the rise of this sect in the tenth and continued to exist alongside of it in the eleventh century. Probably at the end of the twelfth century, a fusion seems to have occurred between Bogomilism and Massalianism, which continued throughout the thirteenth and led to the complete identification of both sects in the fourteenth.[1] Until the fourteenth century, however, an important difference remained between the two: whereas the Massalians were generally accused of the practice of sexual immorality for pseudo-religious motives, the Bogomils were always noted for their moral austerity. There can be no doubt that this trait was a source of great strength to the Bogomil sect, as it lent some justification to its claim to follow the true evangelical life and goes far to explain its great fascination for the masses and its steadfastness in persecution. By the fourteenth century, however, under the increased influence of Massalianism, the Bogomils had entirely lost their reputation of puritanism and had become associated with the most extreme forms of sexual indulgence. This was no doubt partly due to the general moral decline in the

[1] Cf., however, a different interpretation by Puech (op. cit. pp. 292–303), who explains the gradual substitution of Massalianism for Paulicianism as the fundamental source of Bogomilism partly by the subjective impressions and stereotyped notions of the medieval heresiologists, partly by the probable fact that they had no very precise knowledge of Massalianism and often tended to attribute to the Massalians doctrines which they encountered in Bogomilism. Puech's arguments are not without weight, but, it seems to me, do not really refute the cumulative evidence of the sources, which points very strongly to the increasing influence of certain Massalian doctrines on Bogomilism.

fourteenth century, which affected all classes of Bulgarian society and weakened the resistance of the Bogomils to the disruptive influence of Massalian practices. At the same time, by absorbing alien teachings, Bogomilism became more and more syncretic and gradually lost its inner coherence.

We possess no information on the Bogomil sect in the second half of the thirteenth century. But it can be safely assumed that the sect survived the legislation of 1211 and continued to thrive in Bulgaria. The condition of the country between the death of John Asen II (1241) and the accession of John Alexander (1331) was favourable to its growth: the social chaos which succeeded the death of John Asen II, the several dynastic upheavals, the invasions of Greeks, Tatars and Serbs, the impending menace of the Turks,[1] must have encouraged Bogomil proselytism, always successful in troubled times. By a policy of strong centralization John Alexander (1331–71) was able to restore to Bulgaria for a time some measure of stability, prosperity and prestige, but politically he was largely dependent on his brother-in-law, the Serbian Tsar Stephen Dušan, and in the intellectual, social and economic realms Byzantine influence was supreme.[2]

The fourteenth century was a period of severe crisis for the Byzantine Church. The general political instability, the internal divisions in the Church and the intellectual and moral decline of some of the clergy produced confusion and dissatisfaction among the people. The higher clergy was often unable to command obedience and respect. Their flocks, unsettled in mind, vainly seeking for solutions to the pressing spiritual and material problems, were prone, in an atmosphere saturated with demonology and magic, to grasp at any new and strange teaching even of the most unorthodox kind. In this atmosphere of spiritual decadence, the best elements in the Church rallied round those monasteries where the purity of the Orthodox faith was preserved. The great monastic revival in the fourteenth century was expressed in the Hesychast movement, which played a central role in the history of the Byzantine and the Bulgarian Churches in the

[1] See Jireček, op. cit. pp. 263–96; Spinka, op. cit. pp. 113–16.
[2] See K. Radchenko, Религиозное и литературное движение в Болгарии в эпоху перед турецким завоеванием, *Universitetskie Izvestiya* (Kiev, 1898), pp. 29–46.

fourteenth century and is indirectly connected with the history of Bogomilism.

Hesychasm is a general term applied to the mystical trend of Eastern Orthodox monasticism, whose aim is the pursuit of pure contemplation and union with God by means of inner prayer. Inherent in Orthodox monasticism from the third and fourth centuries, Hesychasm was revived in the fourteenth century by St Gregory of Sinai (d. 1346) on Mount Athos and received a doctrinal justification in the theological works of St Gregory Palamas (d. 1359).[1]

The teaching of the Hesychasts aroused violent opposition in certain Byzantine ecclesiastical circles, and led to a bitter controversy which raged for some twenty-five years and ended in 1368 in a solemn vindication of Hesychasm and the canonization of Gregory Palamas at a council in Constantinople.[2] The quarrel was largely a philosophical one, though the opponents of Hesychasm were really attacking the contemplative tradition of Orthodox monasticism. The leaders of this opposition, the Calabrian monk Barlaam and the historian Nicephorus Gregoras, sought to discredit the Hesychasts by accusing them of Massalianism, which was then held to be identical with Bogomilism.[3]

[1] An exposition of the teaching of the fourteenth-century Hesychasts naturally lies outside the scope of this work. Its historical connections with Bogomilism will be indicated in the following pages. There are very few satisfactory accounts of Hesychasm; Western writers, in particular, often repeat biased opinions on the subject. M. Jugie's articles ('Palamas' and 'Palamite (Controversy)' in the D.T.C. vol. XI) paint the best historical background of Hesychasm, but are seriously vitiated by the author's prejudice against St Gregory Palamas. K. Radchenko's Религиозное и литературное движение в Болгарии (loc. cit.) gives a useful historical introduction to Hesychasm, but is most inadequate from the theological point of view. The best accounts of the Hesychast doctrines are by Fr. B. Krivoshein (Аскетическое и богословское учение св. Григория Паламы, Athos, 1935. Seminarium Kondakovianum, Prague, 1936, vol. VIII; Engl. tr. Eastern Churches Quarterly, 1938, vol. III, nos. 1-4) and by the Archimandrite C. Kern ('Les éléments de la théologie de Grégoire Palamas,' Irénikon, Chevetogne, 1947, vol. XX, pts 1-2).

[2] See M. Jugie, 'Palamite (Controversy)', D.T.C. vol. XI.

[3] Barlaam entitled his polemical work against the Hesychasts, c. 1337, Κατὰ Μασσαλιανῶν (see Jugie, loc. cit. cols. 1738, 1779). Gregoras writes of St Gregory Palamas and his followers: κακίαν ἄλλην ἐπὶ κακίᾳ προσεπεδαψιλεύσαντο πολυειδῆ καὶ πολύμορφον. Τίνα δὴ ταύτην; τὴν τῶν Εὐχιτῶν δηλαδὴ καὶ Μασσαλιανῶν. (Historiae Byzantinae, l. XXXII; C.S.H.B. vol. III, p. 396.) Elsewhere he identifies the Massalians with the Bogomils (cf. infra, p. 254).

The absurdity of this accusation is obvious from the slightest acquaintance with the teaching of St Gregory Palamas. Not only does he specifically condemn the 'accursed Massalians [who] think that those among them who are worthy behold the essence of God',[1] but his teaching rests on the principle, fundamentally opposed to any conception of dualism, that the human body in itself is not evil and can be transfigured even in this life by the Spirit.[2] But the accusation is interesting as it illustrates the fear inspired by the Bogomils and the Massalians in the fourteenth century. Moreover, the wilful confusion between the Orthodox mystical movement and these heresies can be explained by certain apparent similarities between them. The great importance attributed to inner prayer by the Hesychasts could easily be taken by their enemies to correspond to the Massalian view of prayer as alone capable of driving out the demon living in man. The distinction made by St Gregory of Sinai between θεωρία (or ἡσυχία), the supreme aim of the contemplative life, and πρᾶξις, or preparation, whose value is only relative, could be falsely taken to imply a rejection of the discipline of the Church, and particularly of the sacraments, as cramping and unnecessary. The Hesychasts also taught that the most efficacious method of spiritual advancement was the constant repetition of the 'Jesus Prayer' ('Lord Jesus Christ, Son of God, have mercy on me, a sinner'); the Bogomils held that all prayers except the Lord's Prayer were 'babblings'. The essentially contemplative nature of Hesychasm could be compared by the anti-Hesychasts with the total rejection of manual labour preached by the Massalians and the Bogomils. Finally, these sects shared the monastic character of Hesychasm: both Massalianism and Bogomilism recruited many adherents in the monasteries, which were also the centres of Hesychasm, and in the fourteenth century, as it will be shown, the Bogomil heresy spread to Mount Athos, the stronghold of Hesychasm.

In these controversies the opponents of Hesychasm, and particularly Nicephorus Gregoras, identify Massalianism with Bogomilism. This identification, it will be seen, occurs in all fourteenth-century sources dealing with these sects. It seems undeniable that all distinction between them had disappeared by then.

[1] *Homilia XXXV, In...Domini transformationem, P.G.* vol. CLI, col. 448.
[2] See Krivoshein, op. cit.

In the first half of the fourteenth century, before the outbreak of the Hesychast controversy, the Massalian or Bogomil heresy penetrated to Mount Athos, the shrine of Orthodoxy. For this we have the evidence of Nicephorus Gregoras[1] and of the author of the Bulgarian *Life of Saint Theodosius of Trnovo*,[2] both of whom identify Bogomilism with Massalianism. According to the latter source, the heresy came to the Holy Mountain from Thessalonica. The spread of Bogomilism from its stronghold in Macedonia to Thessalonica was undoubtedly facilitated by the natural route of the Vardar, the main artery connecting Macedonia with Byzantium. Thessalonica, on the other hand, was the principal link between the monasteries of Mount Athos and the outside world; periodically monks would visit the city to replenish their supplies or transact commercial business. It was probably in such circumstances that, according to the *Life of Saint Theodosius*, a number of monks from Athos, during their stay in Thessalonica, became corrupted by the teachings of a certain nun, Irene, outwardly pious, but a Massalian at heart. Having returned to their monasteries, they spread the heresy over the Holy Mountain, where it became rampant 'for three years or more'.[3] The behaviour of these heretical Athonite monks was somewhat scandalous: 'they offended the local monasteries by begging, and when they lacked bread, drink or fuel, they cut down the olive trees outside the enclosures of the monasteries and did many other vexatious things'.[4] However, apart from the practice of begging, which can be regarded as a Massalian trait, our source gives no information on the doctrines or customs of the heretics on Athos. Nor is Nicephorus Gregoras any more informative on this subject.[5] It

---

[1] *Hist. Byzantinae*, C.S.H.B. vol. II, pp. 714, 718–20, 876.
[2] Житіе и жизнь преподобнаго ѡтца нашегѡ Ѳеодосія, иже въ Терновѣ постничествовавшегося, ed. O. Bodyansky, *Chteniya v imperatorskom obshchestve istorii i drevnostey rossiyskikh pri Moskovskom Universitete* (Moscow, 1860), vol. I.
[3] *Life of Saint Theodosius*, p. 6.   [4] Ibid.
[5] According to Nicephorus Gregoras, Callistus, patriarch of Constantinople, who lived on Mount Athos at that time and was later himself accused of Massalianism, claimed to have discovered some monks on the Holy Mountain about to throw images of Our Lord and of the saints on to dungheaps (op. cit. vol. III, p. 543). But Gregoras's evidence on the whole matter is very suspect, owing to his anti-Hesychast bias and his desire to discredit Callistus, who was a leading Hesychast. Moreover, it is not likely that such treatment of icons could have gained much support on Mount Athos.

seems likely that those Massalian doctrines which spread over Mount Athos were related to prayer and contemplation, features of the heresy which could offer potential points of contact with the views of the Hesychast monks.[1]

The heresy did not flourish long on the Holy Mountain: some three years after its appearance the monks convened a council which anathematized the heretics and expelled the ringleaders from Athos. Some of them went to Constantinople, Thessalonica and Berrhoea, others penetrated into Bulgaria.[2] But the scandal flared up with even greater intensity in Constantinople, where it became centred round the person of the Patriarch Callistus. During his second patriarchate (1355–63) Callistus received a letter from the monks of Athos, accusing the monk Niphon Scorpio of Massalianism. This Niphon, who had formerly lived on the Holy Mountain and was a close friend of Callistus, had already been accused of Bogomilism in 1350 but had succeeded in justifying himself.[3] The monks' suspicions were confirmed by the confession of Niphon's servant Bardarius, who on his deathbed, twelve years after the expulsion of the Massalians from Athos, revealed that his master had actually accepted the heretical doctrine.[4] Seeing that the patriarch protected Niphon, the anti-Hesychast party in Byzantium seized this opportunity to launch a violent campaign against Callistus, accusing him also of Massalianism. The patriarch, however, successfully confuted these attacks and had his opponents condemned.[5] It is not clear from this tendentious account of Nicephorus Gregoras whether Niphon was really a Massalian or whether he was simply a Hesychast like Callistus.

[1] On the sole evidence of Nicephorus Gregoras, one might be tempted to think that these heretical teachings were simply the Hesychast views of the Athonite monks and to ascribe their denunciation as Massalian by Gregoras to his anti-Hesychast bias. However, the evidence of the *Life of Saint Theodosius* precludes such an interpretation: for there these doctrines are explicitly termed Massalian or Bogomil; its author, who was clearly a supporter of Hesychasm, would never have confused it with Massalianism.
[2] *Life of Saint Theodosius*, p. 6; Nicephorus Gregoras, op. cit. vol. II, pp. 718–20; cf. F. Miklosich and J. Müller, *Acta Patriarchatus Constantinopolitani* (Vindobonae, 1860), vol. I, pp. 296–300. Cf. Bishop Porfiry (Uspensky), История Афона. Часть III: Афон монашеский, отд. 2 (St Petersburg, 1892), pp. 274–82.
[3] Miklosich and Müller, ibid.
[4] Nicephorus Gregoras, op. cit. vol. III, pp. 260–1.
[5] Ibid. vol. III, pp. 532–46.

But the accusations levelled against the patriarch were obviously based on the deliberate confusion between Hesychasm and Massalianism: the only way by which the anti-Hesychast party could hope to discredit Callistus was to charge him with Massalianism or Bogomilism; a direct attack on Hesychasm was no longer possible, since the doctrines of St Gregory Palamas had been recognized as Orthodox by a council in Constantinople in 1351.[1]

In view of the extremely close relations between the Byzantine Empire and Bulgaria in the fourteenth century,[2] it is not surprising to find that Hesychasm penetrated into Bulgaria even before its final triumph in Byzantium. The chief protagonist of Bulgarian Hesychasm was St Theodosius of Trnovo. After wandering from monastery to monastery in search of the true ascetic life, Theodosius was initiated into the way of contemplation by St Gregory of Sinai, who was then living in Paraoria, to the north of Adrianople, on the boundaries of the Byzantine Empire and Bulgaria. Theodosius became the favourite disciple of the great master of Hesychasm and, after St Gregory's death (1346), succeeded to the position of teacher to the group of his disciples. He then visited the great centres of Hesychasm, Athos, Thessalonica and Mesembria, and finally settled in Bulgaria, where the Tsar John Alexander gave the group of his disciples, numbering some fifty, a tract of land on the hill of Kiliphar near Trnovo.[3]

St Theodosius occupies an important position in the history of the Bulgarian Church: he was an ardent supporter of the Oecumenical Patriarch Callistus against his own immediate superior, the Bulgarian Patriarch Theodosius,[4] and also the leader of Bulgarian Hesychasm and the chief opponent of heresy in the reign of John Alexander. The two latter aspects of his activity are

[1] See M. Jugie, 'Palamite (Controverse)', loc. cit. cols. 1790–2.
[2] See Radchenko, loc. cit. pp. 169 et seq. Although the Patriarch of Trnovo had been granted nominal autocephality in 1235, he remained in practice under the domination of the Oecumenical See.
[3] *Life of Saint Theodosius*, pp. 3–5.
[4] The Bulgarian patriarch was trying to assert his complete independence of the Patriarch Callistus. St Theodosius, on the other hand, was united to Callistus by their common devotion to the memory of their master St Gregory of Sinai and by their championship of Hesychasm. For the struggle between Callistus and the Bulgarian patriarch, see Radchenko, loc. cit. pp. 180–4; V. N. Zlatarski, *Geschichte der Bulgaren*, pp. 171–2; Spinka, op. cit. pp. 117–18.

described in the *Life of Saint Theodosius*. This work was until recently generally ascribed to the Patriarch Callistus, who appears as the author in the title of the manuscript published by Bodyansky. V. S. Kiselkov, however, has shown fairly conclusively that the document in its present form is not the work of Callistus, but a compilation of a later date, probably of the fifteenth century, based on a shorter Greek version written by Callistus, but now lost.[1]

Behind the outward splendour of John Alexander's Byzantinized court and government, Bulgaria was in a sorry state; never had the economic oppression of the people been so heavy and the gulf between the privileged classes and the peasants so profound; the lack of inner unity in the country, the constant wars between Bulgaria and the Empire, the frequent and terrible devastations by the Ottoman Turks,[2] which were the determining causes of the collapse of the Second Bulgarian Empire at the close of the fourteenth century, could only favour the spread of heretical teachings. These could develop all the easier, as a marked decadence was observable among the Bulgarian clergy, not excluding the monastic *élite*.[3] Hesychasm alone, as in Byzantium, promoted a spiritual revival. The people, thus deprived in many cases of moral guidance, were living in an atmosphere of great mental and material instability, where scepticism and rationalism were combined with excessive credulity and a readiness to accept any extravagant teaching, and where extreme asceticism coexisted with extreme immorality.[4] In these circumstances, it is not surprising that Bogomilism again raised its head in the fourteenth century.

The first direct evidence of Bogomilism in Bulgaria since the Council of Trnovo in 1211 can be found at the beginning of the fourteenth century. A council held in June 1316 under the presidency of the metropolitan of Heracleia in Thrace, who held the title of ἔξαρχος πάσης Θράκης καὶ Μακεδονίας, judged the priest Garianus, accused of having contracted heresy from his association

[1] Житието на св. Теодосий Търновски като исторически паметникъ (Sofia, 1926), pp. i–lii.
[2] See Pogodin, op. cit. pp. 106–13.
[3] Cf. the significant admission of St Theodosius's biographer: Скудни бо оубѡ бяху тогда во странахъ болгарскихъ, иже добродѣтель проходящіи (op. cit. p. 3).   [4] See Radchenko, loc. cit. pp. 205–6.

with the Bogomils.¹ The Patriarch Philotheus also tells us that St Gregory Palamas held *c.* 1317 a victorious discussion with some 'Marcionites or Massalians' (Μαρκιανιστῶν ἢ Μασσαλιανῶν) at the monastery of Mount Papikion on the borders of Thrace and Macedonia.² There can be no doubt that these heretics were Bogomils.³ Macedonia, the home of Bogomilism, probably still remained its stronghold in the fourteenth century, when the confused political state of this region was undoubtedly favourable to the sect.⁴

The rest of our knowledge of fourteenth-century Bogomilism is derived from the *Life of Saint Theodosius*, which gives the saint the credit for personally conducting the struggle against a number of different heresies in Bulgaria. The exclusive role played by St Theodosius in fighting heresy, contrasted with the complete insignificance of the rest of the Bulgarian clergy, appears to have been considerably exaggerated by his biographer.⁵

The 'heretics' fought by St Theodosius were of two kinds: on the one hand isolated teachers of false doctrines and, on the other, members of well-known sects, Bogomils (or Massalians) and Jews. Disciplinary action or persuasion were sufficient to deal with the former; the latter were only defeated, it seems, after their condemnation by two specially convened councils.

The individual heretics dealt with by St Theodosius were two monks, Theodoret and Theodosius. The first is said to have come from Constantinople to Trnovo, to have been an accomplished physician and to have taught an incongruous mixture of anti-Hesychast doctrines, paganism and magic.⁶ He appears to have

---

¹ See F. Miklosich and J. Müller, *Acta Patriarchatus Constantinopolitani*, vol. I, p. 59: ὁ... δὲ παπᾶς Γαριάνος ἦλθεν εἰς τοὺς Πατερίνους, καὶ ἔδωκαν αὐτὸν ὑπέρπυρα πεντήκοντα καὶ ἄλογον, καὶ ἐγένετο εἰς μετ' αὐτούς. Garianus was acquitted.

² *Gregorii Palamae Encomium*, P.G. vol. CLI, col. 562.

³ Cf. M. Jugie, 'Palamas', loc. cit. col. 1736.

⁴ See Radchenko, loc. cit. p. 173.

⁵ The two main studies of the *Life of Saint Theodosius*, by Radchenko and Kiselkov, suffer from opposite defects: the former from an attempt to build up ingenious but unjustifiable hypotheses, the latter from excessive scepticism. Kiselkov's view that the whole struggle of St Theodosius against heresy is apocryphal and based on a confusion between the saint and another—unidentified—monk Theodosius is substantiated by no conclusive argument.

⁶ *Life of Saint Theodosius*, p. 5. Apparently he taught the people 'to worship a certain oak and to receive healing from it'.

been astonishingly successful in Trnovo, not only among the simple folk but also among many high-placed people (множае и въ нарочитыхъ и славныхъ). The scandal ended by the timely intervention of St Theodosius, who confounded Theodoret and had him banished.[1] Though not a Bogomil himself, Theodoret probably appealed to the Bogomils by his views on magic and paganism. In any case, his spectacular success testifies to the religious confusion in Bulgaria at that time and to the readiness of the people to follow any new teacher.

The monk Theodosius behaved even more extravagantly: he wandered from place to place, preaching the dissolution of marriage ties, gathered round him a group of men and women whom he persuaded to walk about naked and indulge in unbridled promiscuity. St Theodosius apparently succeeded in bringing him and his followers to their senses.[2] The behaviour of this Theodosius closely resembles that ascribed to the fourteenth-century Bogomils, whose moral austerity had largely disappeared under the influence of Massalianism. The combination of asceticism with immorality, originally a Massalian feature, is now attributed to the Bogomils in the *Life of Saint Theodosius*.

This document next describes the arrival in Trnovo of Lazarus and Cyril Bosota, who belonged to the group of monks who had been expelled from Mount Athos for their adherence to the Massalian heresy.[3] After a brief period of concealment, they began to preach in the open and corrupted a certain priest, Stephen, who became their leading disciple. The heresy which they taught in Bulgaria was Massalianism, which in two passages in the *Life of Saint Theodosius* is said to be synonymous with Bogomilism.[4] Their behaviour caused a great sensation in Trnovo: Lazarus walked about naked and urged the necessity of castration, Cyril Bosota preached the dissolution of marriage.[5] According to our source, the scandal was so grave that the Bulgarian patriarch himself, being a 'simple man', was baffled and appealed for help to St Theodosius, who urged the convocation of a council to pass judgement on the heretics. His advice

---

[1] *Life of Saint Theodosius*, pp. 5–6.
[2] Ibid. pp. 7–8.   [3] Cf. supra, p. 256.
[4] Богомилскую, сирѣчь масаліанскую, ересь. (Ibid. pp. 8, 11.)
[5] Ibid. p. 6.

BOGOMILISM IN THE SECOND BULGARIAN EMPIRE 261

was followed, and the council met under his presidency, probably c. 1350.[1]

The description of St Theodosius's interrogation of the Bogomils is purely conventional. As in the *Synodicon of the Tsar Boril*, the heretics are described as dumbfounded by the theological skill and eloquence of the prosecutor. Lazarus repented of his errors, but Bosota and Stephen remained obdurate, and were branded on the face and banished from Bulgaria.[2] But the enumeration of the Bogomil doctrines and practices is not stereotyped and sheds some light on the state of the Bogomil sect in the fourteenth century. Some of the doctrines taught by the followers of Lazarus, Cyril and Stephen were already held by the Bogomils in the tenth century and had thus remained unchanged for four hundred years: the dualism between the heavenly God and the evil creator of

[1] St Theodosius's biographer is clearly biased against the Bulgarian patriarch. Kiselkov has shown (op. cit. pp. xlix et seq.) that the latter was anything but a simple-minded man, incapable of taking the most elementary measures to safeguard his flock from heresy. Doubtless St Theodosius's role in fighting heresy was not as exclusive as his biographer would like us to believe.

The Bulgarian patriarch, in whose time the council against the Bogomils was convened, was a contemporary of Callistus who was twice patriarch of Constantinople, from 1350 to 1354 and from 1355 to 1363. The anti-Bogomil council must have met during his first patriarchate. (See Jireček, op. cit. p. 314.) Spinka maintains that this Bulgarian patriarch was called Symeon (op. cit. p. 117). Symeon was, in fact, still patriarch of Trnovo in 1346, as Kiselkov has shown (op. cit. p. xlix). But in 1348 the patriarchal throne of Bulgaria was occupied by Theodosius, which is proved by a note in a Bulgarian Gospel-Book written in that year (in the collection of Bulgarian MSS. belonging to Robert Curzon, 15th Baron Zouche). See P. T. Gudev, Български рѫкописи въ библиотеката на лордъ Zouche, *S.N.U.* (1892), vol. VIII, p. 167. Thus the anti-Bogomil council of c. 1350 must have taken place in the patriarchate of Theodosius.

[2] Kiselkov has levelled against the authenticity of this council all the weight of his considerable critical talent. He attempts to prove (op. cit. pp. xxv–xxix) that in reality there was no anti-Bogomil council c. 1350, that the author of the *Life of Saint Theodosius* was guilty of a chronological confusion and that his evidence applies to the Council of Trnovo of 1211. Kiselkov's most important arguments are: (1) The vagueness of the hagiographer about the date, place and minutes of the council; (2) the reference to the Bulgarian patriarch as 'simple', whereas our knowledge of that personage suggests just the opposite. But Kiselkov's arguments are not conclusive and his theory can scarcely be accepted. The scathing allusion to the Bulgarian patriarch is quite sufficiently explained by the antipathy of the hagiographer towards one who had been a consistent opponent of St Theodosius.

this world,[1] the rejection of images and of the Cross, the denial of the Real Presence in the Eucharist.[2] Others reveal a specifically Massalian origin: following their ideal of evangelical poverty and their insistence on continual prayer, the Bogomils told St Theodosius: 'we embrace poverty and pray unceasingly...for this reason we are "the poor in spirit" blessed by Our Lord'.[3] They also rejected manual labour.[4] Cyril Bosota claimed that dreams were in reality divine visions.[5] When discovered and threatened with punishment, the Bogomils were wont to swear their innocence by the most solemn oaths and 'curse the Massalian heresy', only to return to it at the first opportunity.[6]

The *Life of Saint Theodosius* is the first source to attribute the practice of sexual immorality to the Bogomils. St Theodosius accuses them of submitting to the 'natural passions' on the grounds that 'our nature is a slave to the demons'.[7]

After the condemnation of the Bogomils, the Bulgarian Church was confronted with the aggressive behaviour of the Jews. The Tsar John Alexander had married a Jewess, after forcing his first wife to enter a nunnery. The new tsaritsa became a zealous Christian and generously endowed monasteries and churches. Nevertheless, the Bulgarian Jews apparently hoped to gain her support, but 'they were mistaken in their undertaking'.[8] Ac-

---

[1] Два начала суть, едино оубо благо, другое же зло (op. cit. p. 7). However, this formulation of the cosmological dualism in terms of *two independent principles* is a Paulician and not a Bogomil feature. The evidence of all the previous sources shows that the creator of this world was for the Bogomils not a *principle*, parallel with God, but an inferior creature, generally regarded as a fallen angel. (Cf. supra, pp. 123–5). The belief in two *principles* may have been ascribed to the Bogomils by the author of the *Life of Saint Theodosius* for one of three reasons: (1) a possible influence of Paulicianism on Bogomilism in the fourteenth century; (2) a confusion between the Paulician and Bogomil doctrines; (3) an insufficiently profound acquaintance with Bogomilism itself. The first alternative does not seem very likely, since we possess no evidence of any special influence exerted by the Paulicians on the Bogomils after the tenth century; in spite of frequent contacts, both sects always remained clearly distinct from each other. The second or third alternative is probably the correct one.

[2] Который бѣсъ научи васъ попирати святыя иконы и животворящій крестъ, и прочыя священныя сосуди; еще же и святымъ тайнамъ яко просту причащатися хлѣбу. (Ibid.)

[3] Ibid. p. 6.   [4] Ниже ручнагѡ дѣла дѣлати. (Ibid.)

[5] Сонная же мечтанія боговидѣнія быти оучаше. (Ibid. p. 6.)

[6] Ibid. p. 7.   [7] Ibid. p. 6.   [8] Ibid. p. 8.

cording to the *Life of Saint Theodosius*, the Jews 'blasphemed the images of Our Lord Jesus Christ and of His Most Pure Mother... spurned the churches of God and the sacrifices offered therein' and inveighed against Orthodox priests and monks. Again St Theodosius intervened, and on his advice the tsar, in agreement with the tsaritsa and the patriarch, convened a council in 1360 which was attended by the tsar's son and the leading hierarchs of the Bulgarian Church.[1] Together with Judaism, the council condemned Bogomilism and the anti-Hesychast teachings of Barlaam and Acyndinus. The Bogomils and the anti-Hesychasts were anathematized and banished from Bulgaria; the Jews convicted of blasphemy were sentenced to death, but were reprieved through the mercy of the tsar; one ringleader recanted and was received into the Church; of the other two who remained steadfast in their faith the one was killed by an angry mob, the other was punished by having his tongue, lips and ears cut off.[2]

The Judaizing movement in fourteenth-century Bulgaria appears to have been strong. Melioransky connects the doctrines condemned at the Council of 1360 with the outbreak of Judaism in Macedonia, which came to the notice of the ecclesiastical authorities in Thessalonica between 1324 and 1336.[3] But Judaism in itself could scarcely have called for the convocation of a special council, since the Byzantine Nomocanons were not lacking in articles against the Jews which could easily have been applied in Bulgaria. The reason for this solemn condemnation in 1360 probably lies in the nature of the doctrines attributed to the Jews in the *Life of Saint Theodosius*. Not one of them is specifically Jewish: the rejection of icons, churches, the Eucharist and of Holy Orders are features at least as characteristic of Bogomilism. It is significant that although the Council of 1360 appears to have met primarily in order to deal with the Jews, their doctrines were anathematized together with those of the Bogomils, notwithstanding the fact that

[1] Ibid. The document gives a list of their names and dioceses.
[2] Ibid. pp. 8–9. Spinka erroneously states that 'three leaders of the Judaizing party were put to death' (op. cit. p. 121).
[3] К истории противоцерковных движений в Македонии в XIV веке, p. 72. The Jews of Thessalonica were accused of magic, of relations with evil spirits, of attacking what they considered to be an excessive cult rendered to saints and relics to the detriment of the worship of God, and of denying the Resurrection of the Body.

the latter had been condemned at the Council of Trnovo only a few years previously. The contact between Bogomilism and Judaism in the fourteenth century is undeniable and evidence of it can already be seen in the thirteenth century.[1] In 1360 it was probably not so much Judaism in itself which presented a danger to the Bulgarian Church as its association with the dreaded Bogomil heresy.

The *Life of Saint Theodosius of Trnovo* is the last Bulgarian source containing evidence of Bogomilism, and illustrates the final stage in the evolution of the sect at the end of its history of four hundred years in Bulgaria. Among the main reasons for the decadence of the sect in the fourteenth century were, as we have seen, the strong influence of Massalianism and the general moral decline of the age. What is remarkable in the history of Bogomilism is not that it was eventually undermined by these influences, but that it succeeded in resisting them for so long. The example of many other sects of a kindred nature shows that the boundary between extreme asceticism and unbridled immorality is a narrow one. The fact that the Bogomils were until the fourteenth century perhaps the greatest ascetics and puritans of the Middle Ages testifies to the considerable vigour and independence of the sect.

Another characteristic feature of the sect, present throughout its history but expressed more clearly in the *Life of Saint Theodosius*, is its increasingly syncretic character, due to its versatility and opportunistic adaptability to circumstances. To carry out their proselytism or to elude persecution the Bogomils never scrupled to ally themselves with other religious and secular movements and even to affect conformity with their greatest enemy, the Orthodox Church. This eclectic tendency became more pronounced with time, and from the thirteenth century onwards Bogomilism is associated more and more frequently with paganism, magic, popular superstitions and with the teachings of other sects, such as the Massalians and the Jews. In one sense this diffuseness of Bogomilism greatly facilitated its spread and hampered the task of its persecutors. The Orthodox Churchmen recognized this only too well, as can be seen from their angry attacks on the 'hypocrisy' of the Bogomils. The Orthodox habit of classing many forms of religious nonconformity under the

[1] Cf. supra, p. 241.

heading of Bogomilism is thus partly justifiable. Between the tenth and fourteenth centuries Bogomilism was undoubtedly the most dangerous of all the heresies confronting the Orthodox Church. It is significant that the last words of St Theodosius, that fighter of many heresies in the fourteenth century, spoken on his deathbed to his disciples, urged them to fly above all 'from the Bogomil, that is to say the Massalian heresy'.[1] But from another point of view the increasingly syncretic character of Bogomilism could not fail to further its disintegration, by obscuring among the heretics the consciousness of their own sectarian traditions. It is safe to assert that many of the ugly features which in the fourteenth century passed as Bogomilism would have been disowned by the *pop* Bogomil.

This increasing decadence of Bogomilism in the fourteenth century largely explains the fact that after the capture of Trnovo by the Sultan Bayazid (17 July 1393), when the Second Bulgarian Empire fell under the yoke of the Turks, the sect apparently disintegrated of itself and the Bogomils disappeared for ever from the scene of Bulgarian history. The exact behaviour of the Bulgarian Bogomils towards the Turkish invaders is unknown, but it may be inferred by analogy with that of the Bosnian Patarenes: these openly supported the Turks against their own Catholic rulers, and after the conquest of Bosnia (1463) many of them accepted Islam.[2] It is probable that the Bulgarian Bogomils were also sympathetic to the Turks, who were generally more tolerant than the Christians in matters of religion and who, moreover, in all the Slavonic countries which they conquered, tried at first to win the sympathies of the peasants. It is generally thought that some Bulgarian Bogomils became Moslems,[3] while others accepted

---

[1] Бѣгати...яко же лѣпо есть богомилскія, сирѣчь масаліанскія, ереси. (*Life of Saint Theodosius*, p. 11.)

[2] See Rački, op. cit. *Rad*, vol. vIII, pp. 174–5; D. Prohaska, *Das kroatisch-serbische Schrifttum in Bosnien und der Herzegowina* (Zagreb, 1911), pp. 34 et seq.; J. A. Ilić, *Die Bogomilen in ihrer geschichtlichen Entwicklung*, pp. 83–91.

[3] See Rački, loc. cit. p. 187. Jireček thinks that many Bogomils became Moslem even before the Turkish conquest of Bulgaria. (История Болгар, Odessa, 1878, p. 461.) Indirect evidence of the conversion of the Bogomils to Islam is perhaps provided by the name *torbeshi*, of Bogomil origin, applied to-day to the Moslem Bulgarians, or *pomaks*, of central Macedonia. (Cf. supra, pp. 166–7.)

Orthodox Christianity.[1] The first must have hoped to obtain, in return for the outward recognition of Islam, a degree of freedom and toleration which was refused them by their Christian masters; the second were probably governed by expediency or fear, as Christianity was the only religion to enjoy recognized rights under Turkish rule. This behaviour of the Bogomils is not surprising, since outward conformity to the discipline of the Church was never incompatible with their principles and was indeed frequently practised by them.

The diffuse and syncretic character of Bogomilism in the final period of its history, which resulted in the rapid disintegration of the sect after the Turkish conquest, appears more clearly by contrast with Paulicianism. The Paulicians, from the very moment of their appearance in the Balkan peninsula in the eighth century, remained in self-contained communities, religiously and ethnically distinct from the Bulgarians, and never mixed with the people to the same extent as the Bogomils. Although this isolation rendered the Paulicians less dangerous to the Church, it also permitted them to survive the Turkish invasion. Though converted to Roman Catholicism in the seventeenth century, the descendants of the medieval Paulicians exist in Bulgaria, and particularly round Philippopolis, to the present day.[2]

After the Turkish conquest, we hear no more of the Bogomils in Bulgaria, and it may be presumed that they soon became sub-

[1] See Ivanov, op. cit. p. 36.
[2] See D. E. Takela, Нѣкогашнитѣ павликяни и сегашнитѣ католици въ Пловдивско, *S.N.U.* (1894), vol. XI, pp. 110 et seq.; 'Les anciens Pauliciens et les modernes Bulgares catholiques de la Philippopolitaine', *Le Muséon* (Louvain, 1897), vol. XVI; E. Fermendžin, 'Acta Bulgariae ecclesiastica ab a. 1565 usque ad a. 1799', *Monumenta spectantia historiam Slavorum meridionalium* (Zagreb, 1887), vol. XVIII; I. Dujčev, *Il cattolicesimo in Bulgaria nel sec. XVII* (Rome, 1937). Cf. L. F. Marsigli, *Stato militare dell' Imperio Ottomanno* (Haya, 1732), p. 24; Jireček, *Cesty po Bulharsku*, pp. 101-3, 278; L. Miletich, Заселението на католишкитѣ българи въ Седмиградско и Банатъ, *S.N.U.* (1897), vol. XIV, pp. 284-96. In 1717 Lady Mary Wortley Montagu discovered the descendants of the Paulicians at Philippopolis. She wrote of them: 'I found at Philippopolis a sect of Christians that called themselves Paulines. They shew an old church where, they say, St Paul preached; and he is the favourite saint, after the same manner as St Peter is at Rome; neither do they forget to give him the same preference over the rest of the apostles.' *The Letters and Works of Lady Mary Wortley Montagu* (ed. Lord Wharncliffe; London, 1893), vol. I, p. 290. (Letter dated Adrianople, 1 April 1717.)

merged in the sea of Islam. In the Byzantine Empire, however, which retained a precarious independence until the capture of Constantinople by Mohammed II in 1453, the sect was still extant at the beginning of the fifteenth century. Symeon, archbishop of Thessalonica (1410–29), mentioned the Bogomils (whom he also calls *Kudugeri*) as the most dangerous heretics of his diocese and urged his flock to fly from them on account of their insidious proselytism.[1] But afterwards obscurity descends on the sect and the Bogomils vanish for ever from Bulgaria and the Byzantine Empire. A vague dualistic tradition which has left an imprint on south Slavonic folk-lore and has inspired many Bulgarian popular legends[2] is all that remains to-day of the sway once held over the minds of men by the most powerful sectarian movement in the history of the Balkans.

[1] *Dialogus contra haereses*, P.G. vol. CLV, cols. 65–74, 89–97. Cf. supra, p. 166.
[2] See Ivanov, op. cit. pp. 327–82 and supra, p. 154.

APPENDIX I

THE CHRONOLOGY OF COSMAS

The determination of the precise date at which the *Sermon against the Heretics* was composed is of considerable importance for the question of the origins of Bogomilism. The problem is a complex one. Some chronological evidence can be deduced from the text of the document:

(1) Cosmas refers to the reign of the Tsar Peter as past.[1]

(2) He urges his readers to imitate 'John the New Presbyter and Exarch' and states that many of them know him.[2] This personage is generally thought to be the celebrated Bulgarian Churchman and writer John the Exarch who lived in the reign of the Tsar Symeon (893–927).[3]

(3) Cosmas alludes to widespread misery among the peasants, due to wars and foreign invasions.[4] Although he may be referring to the invasions of Magyars and Pechenegs who, after 934, periodically swept over Bulgaria,[5] it is perhaps more likely that the allusion is to the even greater devastations of 969–72, when the Russians of Svyatoslav and the Byzantine army of John Tzimisces fought a series of fierce battles on Bulgarian soil.[6]

There appears to be some difficulty in reconciling the first two pieces of evidence: John the Exarch was a contemporary of Symeon, who died in 927, and although the date of his death is unknown, it is generally thought that he could not have outlived the Tsar Peter, who died in 969. The different solutions proposed to the problem of Cosmas's chronology depend on the relative degree of importance attached by scholars to each of the three above-mentioned passages of the *Sermon against the Heretics*.

The general opinion is that Cosmas lived in the second half of the tenth century.[7] Some scholars date his work between the

[1] Popruzhenko, op. cit. p. 2. Cf. supra, p. 117.
[2] Подражаите Ивана прозвитера новаго, егоже и.ѡт вас самѣх мнози знаютъ бывшаго пастуха і ексарха иже в земли болгарьстіи. (Ibid. p. 79.) [3] See supra, p. 89, n. 3.
[4] Видяще селикы бѣды ратныа и всѣхь настоящих золъ земли сеи. (Ibid. p. 46.)
[5] See V. N. Zlatarski, История, vol. I, pt 2, pp. 541–4, 567–70.
[6] See Zlatarski, ibid. pp. 600–32; Puech, op. cit. p. 24.
[7] Zlatarski, ibid. p. 566; S. Runciman, *A History of the First Bulgarian Empire*, p. 191; I. Ivanov, Богомилски книги и легенди, p. 21.

death of Peter and the fall of the First Bulgarian Empire (1018),[1] others, more precisely, in the reign of Samuel.[2]

Two recent works on Cosmas, by Blagoev[3] and Trifonov,[4] propose fresh solutions to the problem. Blagoev suggests that Cosmas wrote his work during the first years of Peter's reign, i.e. soon after 927, and thus attempts to overcome the chronological difficulty by making Cosmas practically a contemporary of John the Exarch. But Blagoev's arguments are unconvincing, and the improbability of his theory is easily pointed out by Trifonov. Trifonov is the first to have studied the problem really critically; he shows fairly conclusively that Cosmas must have written his *Sermon* after Peter's death. But he attempts to explain away the difficulty connected with John the Exarch by maintaining that 'John the New Presbyter' is not the celebrated collaborator of Symeon, but a certain John the Presbyter who lived in Bulgaria at the beginning of the eleventh century, and thus concludes that Cosmas wrote his work soon after 1026.[5] This is undoubtedly the weakest point in Trifonov's thesis and reveals his obvious desire to find a substitute for John the Exarch to support his own theory. There is no evidence, in particular, that this John the Presbyter ever held the position of Exarch. A new attempt to overcome the chronological difficulties in Cosmas's evidence has been made by A. Vaillant, who offers the following solution to the problem of Cosmas's chronology.[6] He accepts as true both conclusions suggested by a straightforward reading of the *Sermon*, i.e. that at the time it was composed the Tsar Peter was already dead and John the Exarch still alive. Cosmas's allusions to the evil times he takes to be a reference to the misery and disorder consequent on the conquest of Bulgaria by the

[1] E. Golubinsky, Краткий очерк истории православных церквей, p. 109; M. Genov, Пресвитеръ Козма и неговата Беседа противъ богомилството, *B.I.B.* (1929), vol. III, p. 70.

[2] A. Gilferding, Собрание сочинений, vol. I, p. 228; N. Osokin, История Альбигойцев (Kazan, 1869), p. 141; Rački, op. cit. *Rad*, vol. VII, p. 108; K. Jireček, *Geschichte der Bulgaren*, pp. 434–5; A. N. Pypin and V. D. Spasovich, История славянских литературъ (2nd ed.), vol. I, p. 66; M. S. Drinov, Исторически прѣгледъ на българската църква, p. 50, n. 23; M. G. Popruzhenko, Св. Козмы Пресвитера Слово на Еретики (1907), p. xiv; J. A. Ilić, *Die Bogomilen in ihrer geschichtlichen Entwicklung*, p. 18.

[3] Беседата на Презвитер Козма против богомилите, *G.S.U.* (1923), vol. XVIII.

[4] Бесѣдата на Козма Пресвитера и нейниятъ авторъ, *S.B.A.N.* (1923), vol. XXIX.

[5] Ibid. p. 74. See the criticism of Trifonov's theory in Puech, op. cit. p. 21.

[6] Puech, op. cit. pp. 19–24.

Byzantines in 972. Vaillant supposes that John the Exarch, who must have been born *c.* 890, was still alive in 972, and, as an old man of more than eighty years of age, was living in retirement in some Bulgarian monastery. This quite plausible hypothesis permits him to come to the following conclusion: 'Quand Cosmas écrit, la paix n'est pas encore rétablie, mais la ruine de la Bulgarie est consommée: son traité a paru aussitôt après 972, non pas dans les années, mais dans les mois qui ont suivi l'occupation de la Bulgarie par les Grecs.' This solution seems quite acceptable.

APPENDIX II

THE *POP* JEREMIAH

The Bulgarian priest (*pop*) Jeremiah is frequently associated in historical sources with the *pop* Bogomil. He is mentioned for the first time in the tenth century by Sisinnius II, patriarch of Constantinople (995–8), as an author of heretical writings.[1] The monk Athanasius of Jerusalem, who lived not later than the middle of the thirteenth century, denounced in a letter to a certain Pank the 'lying fables' of 'Jeremiah the Presbyter', particularly one concerning the Holy Cross.[2] The oldest known version of the Slavonic Index of forbidden books, in the so-called 'Pogodin's Nomocanon' which dates from the fourteenth century, quotes the titles of several legends and fables whose authorship is attributed to 'Jeremiah, the Bulgarian *pop*'.[3] On the evidence that Jeremiah held heretical views, that he lived in the tenth century (since he is mentioned by Sisinnius), and that a seventeenth-century document brands him with the very same epithet 'not beloved of God' (Богу не милъ) which Cosmas applied to the *pop* Bogomil, many scholars have thought that Bogomil and Jeremiah were one and the same person. Šafařík suggested that Bogomil received the name Jeremiah when he took the monastic vows or when he entered the Church.[4] Rački, on the contrary, supposed that the original name of the heresiarch was Jeremiah, and that, following the example of the Paulician elders who were given to taking other names in their capacity of sectarian leaders, he assumed the leadership of the Bulgarian sect under the name of Theophilus, of which the Slavonic equivalent is Bogomil.[5] That Bogomil and Jeremiah

---

[1] See V. Jagić, История сербско-хорватской литературы, p. 101.

[2] А иже то почелъ еси слово Іеремѣя прозвитера, еже ѡ древѣ честнѣмъ...ѿт него же навыкъ звяжеши, то басни лживыя челъ еси. See M. Sokolov, Материалы и заметки по старинной славянской литературе (Moscow, 1888), vol. I, pp. 108–9; cf. Pypin and Spasovich, op. cit. p. 67.

[3] See Jagić, op. cit. pp. 102 et seq.; Sokolov, op. cit. pp. 109 et seq.; Pypin and Spasovich, op. cit. p. 72; Ivanov, op. cit. p. 53.

[4] *Památky hlaholského písemnictví*, p. lx. According to the rule of the Orthodox Church, a change of name is obligatory after the assumption of the cowl.

[5] Op. cit. *Rad*, vol. VII, p. 94.

are identical was considered probable or certain by a number of scholars.[1]

M. Sokolov was the first to examine the problem by a critical analysis of the sources. He convincingly showed that the arguments adduced in favour of this identification are not only indecisive, but are proved to be false by reference to the historical documents.[2] The index of forbidden books, issued by the Russian metropolitan Zosima (1490–4), mentions 'Jeremiah, *son and disciple of Bogomil*'.[3] A passage in a seventeenth-century manuscript of the Russian Solovki monastery denounces the lies of 'Jeremiah the *pop, disciple of Bogomil*'.[4] Finally, a sixteenth-century index of the Moscow Synodal Library states that 'the authors of heretical books in Bulgaria were the *pop* Jeremiah, the *pop* Bogomil and Sidor the Frank'.[5] There can no longer be any doubt that Bogomil and Jeremiah were two different persons.

Was Jeremiah really a Bogomil, as the Russian sources suggest? The only information we possess about his views is that supplied by the letter of the monk Athanasius and, on the other hand, by the group of writings attributed to Jeremiah, particularly *The Legend of the Cross* and *Falsehoods about Fever and Other Illnesses*.

Athanasius's denunciation of the 'lying fables' of Jeremiah contains no allusions to any Bogomil teaching.[6] *Falsehoods about Fever and Other Illnesses* (the story of St Sisinnius and the twelve daughters of Herod)[7] is a mixture of Christian apocryphal legends and pagan magical lore and, likewise, shows nothing specifically Bogomil.[8] It is in any case doubtful whether Jeremiah was the author of this work.[9]

---

[1] V. Jagić, История сербско-хорватской литературы, loc. cit. p. 101; E. Golubinsky, Краткий очерк истории православных церквей, p. 156; K. Jireček, *Geschichte der Bulgaren*, p. 175; A. N. Pypin and V. D. Spasovich, История славянских литератур, vol. I, p. 65; A. N. Veselovsky, Славянские сказания о Соломоне и Китоврасе, p. 173; J. A. Ilić, *Die Bogomilen*, p. 18; and N. P. Kondakov, О манихействе и богумилах, *Seminarium Kondakovianum* (1927), vol. I, p. 290.

[2] Op. cit. pp. 113–19, 141–2.

[3] Sokolov, ibid. p. 115. Cf. A. N. Pypin, История русской литературы, vol. I, p. 451, n. 2. The term 'son' must be understood in a spiritual sense.

[4] Sokolov, ibid. p. 116.

[5] Творци быша еретическимъ книгамъ въ болгарскои земли попъ Еремѣи, да попъ Богумилъ и Сидорь Фрязинъ (ibid.); cf. Ivanov, op. cit. p. 50.

[6] See Sokolov, op. cit. pp. 119–22.

[7] Cf. S. Runciman, *The Medieval Manichee*, p. 83.

[8] See Pypin and Spasovich, op. cit. vol. I, pp. 86–8.

[9] See Runciman, ibid.

The Legend of the Cross is a complex compilation of diverse material. It consists of a number of separate episodes, linked somewhat loosely by the history of the wood from which the Holy Cross was made, from the time of Moses (or, in some cases, from the beginning of the world) to the Crucifixion. Some of these stories are apocryphal episodes from the life of Jesus, e.g. *How Christ was made a priest*; *How Christ ploughed with the plough*; *How Christ called Probus His friend*. They occur in all versions of the *Legend*, and are certainly not Bogomil; we even find in them such essentially anti-Bogomil traits as the glorification of the Cross (which is the main theme of the *Legend*), a respectful attitude to the Old Testament, the veneration of the Blessed Virgin, the recognition of the sanctity of the priesthood and of images, and even an indirect vindication of manual work and civil obedience.[1] Most of these stories are old Christian apocryphal legends, closely related to passages of the *Palea*,[2] and form the basis of the south Slavonic versions of the *Legend*.[3] How then are we to explain the fact that the Russian Churchmen established a definite connection between Jeremiah and Bogomil and the impression which we derive from the repeated condemnation of Jeremiah's 'fables' by ecclesiastical authorities that they were probably popular among the Bulgarian Bogomils? The explanation probably lies in the Russian versions of *The Legend of the Cross*, which are considerably less innocuous than the Balkan ones, and which begin the story of the Wood of the Cross with the creation of the world. In one Russian version we read that the tree, whose wood was later used for making the Cross upon which Christ was crucified, was planted in Paradise by Satanael, who then existed alone with God.[4] And another version begins with these typically Bogomil words: 'when God created the world, only He and Satanael were in existence'.[5]

We do not know how these Bogomil ideas came to figure in the Russian version of a work so incompatible with Bogomil teaching in other respects, and why, on the other hand, they are absent

---

[1] See Sokolov, ibid. pp. 123–8.
[2] For the *Palea*, see infra, pp. 281–2.
[3] The Bulgarian version of *The Legend of the Cross* was published by V. Jagić, 'Opisi i izvodi iz nekoliko južnoslovinskih rukopisa', *Starine*, 1873, vol. v, pp. 79–95), and the Serbian one by Sokolov (op. cit. pp. 84–107).
[4] N. S. Tikhonravov, Памятники отреченной русской литературы (St Petersburg, 1863), vol. I, pp. 305–13; I. Porfiriev, Апокрифические сказания о новозаветных лицах и событиях, *Sbornik otdel. russk. yazyka i slovesnosti Imper. Akad. Nauk* (1890), vol. LII, no. 4, pp. 55–61.
[5] M. Gaster, *Lectures on Greeko-Slavonic Literature*, p. 36.

from the south Slavonic versions, where one might have expected to find them. Probably those versions of the *Legend* which have come down to us in Russian manuscripts had, before their spread to Russia, been amplified by a Bulgarian Bogomil. But it is highly doubtful whether the Bogomil elements which they contain are due to the pen of Jeremiah.

There does not seem to be sufficiently strong evidence for maintaining that Jeremiah was in any real sense a Bogomil. He appears to have been above all a compiler of Christian apocryphal legends, and it is doubtful whether even in this field he displayed any creative originality. But the compilations associated with his name (above all *The Legend of the Cross*) and which, owing to their unauthoritative and apocryphal origin, were pronounced heretical by the Church, may have been used by the Bogomils who, as the Russian versions of the *Legend* show, were not averse to embroidering them with their dualistic cosmology.[1]

Apart from the fact that Jeremiah probably lived in the second half of the tenth century, nothing is known about his life. It seems probable either that he was a contemporary of Cosmas, in which case the latter's silence concerning him may be due to the fact that Jeremiah's writings were pronounced heretical only later, or that he lived at the very end of the tenth century, when Cosmas was already dead.[2]

[1] I cannot, for reasons outlined above, agree with Runciman's view that 'Jeremiah was, with Bogomil, co-founder of the Bogomils' (op. cit. p. 91). Surely what little we know of Jeremiah does not warrant this assertion. As for the 'Dragovitsan school', which, according to Runciman, was founded by Jeremiah (ibid.), I have already stated my reasons for believing that it must be taken to refer not to the Bogomils, but to the Paulicians of Thrace (see supra, pp. 158–62).

[2] See D. Tsukhlev, История на българската църква, p. 678.

APPENDIX III

THE DATE OF THE BOGOMIL TRIAL IN CONSTANTINOPLE

The precise date of the arrest, trial and condemnation of Basil the Bogomil and of his followers in Constantinople at the beginning of the twelfth century cannot be directly obtained from historical sources. Anna Comnena, our principal informant on the matter, left a blank space for the date in her manuscript of the *Alexiad*, doubtless with the intention of filling it in later after precise verification (μετὰ δὲ ταῦτα τοῦ ἔτους * διϊππεύοντος τῆς βασιλείας αὐτοῦ):[1] such lacunae are not uncommon in the *Alexiad*.[2] However, her narrative clearly shows that the measures against the Bogomils were taken at the very end of the reign of Alexius Comnenus; moreover, she refers to these measures as ὕστατον ἔργον καὶ ἆθλον τῶν μακρῶν ἐκείνων πόνων καὶ κατορθωμάτων τοῦ αὐτοκράτορος.[3] Some scholars have concluded from this that the event occurred in the last year of Alexius's reign, i.e. in 1118.[4]

But this view contradicts Anna's own statement that the trial of the Bogomils occurred in the patriarchate of Nicholas III the Grammarian, who held office from 1084 to 1111.[5] Spinka explains away this difficulty by ascribing Anna's reference to Nicholas to 'one of her rather frequent slips of memory'.[6]

On the other hand, a number of scholars have accepted Anna's testimony that Nicholas the Grammarian took part in the prosecution of the Bogomils and consequently place this event in the last years of his patriarchate: the year 1110 is suggested by K. Paparrhegopoulos[7] and M. I. Gedeon,[8] the year 1111 by Jireček,[9] Rački[10] and Pogodin.[11]

[1] *Alexiad*, lib. xv, cap. 8, p. 350.
[2] See G. Buckler, *Anna Comnena*, p. 251.
[3] *Alexiad*, lib. xv, cap. 10, p. 364.
[4] F. Chalandon, *Essai sur le règne d'Alexis Ier Comnène*, p. 319; M. Spinka, *A History of Christianity in the Balkans*, p. 98; I. Klincharov, Попъ Богомилъ и неговото време, p. 74.
[5] See C. D. Cobham, *The Patriarchs of Constantinople* (Cambridge, 1911), p. 94.　　　　[6] Op. cit. p. 98, n. 7.
[7] Ἱστορία τοῦ ἑλληνικοῦ ἔθνους (Athens, 1871), vol. IV, p. 546.
[8] Πατριαρχικοὶ Πίνακες (Constantinople, 1891), pp. 338–47.
[9] Op. cit. p. 212.　　　　[10] Op. cit. *Rad*, vol. VII, p. 116.
[11] История Болгарии, p. 51.

The truth of their opinion and the reliability of Anna's statement are borne out by circumstantial historical evidence. It is highly improbable that Alexius's prosecution of the Bogomils took place in 1118: the emperor's last illness, from which he never recovered, overcame him in January or February of that year,[1] and it is obvious from his daughter's account that the trial of the heretics stretched over a period of at least several months. Moreover, the period between 1112 and 1118 was almost entirely occupied with wars against the Seljuq Turks and the Cumans—with the exception of part of 1114, when Alexius busied himself with the conversion of the Paulicians of Philippopolis—and during his brief stays in Constantinople the emperor was weary and sick[2] and hence not likely to have shown the energy and persistence with which his daughter credits him in interrogating and arguing with the Bogomils. The most satisfactory period for dating the Bogomil trial is between 1109 and 1111. Alexius was then in Constantinople and actively applying himself to the internal affairs of the State.[3]

There is thus no reason to doubt Anna's statement that the trial of the Bogomils in Byzantium occurred in the patriarchate of Nicholas the Grammarian, i.e. not later than 1111. Her assurance that 'this was the last and crowning act of the emperor's long labours and successes' is, no doubt, somewhat incorrect, as his conversion of the Paulicians took place a few years later, in 1114,[4] but this inaccuracy is easier to explain than her alleged confusion of the two patriarchs, Nicholas the Grammarian and John IX (1111–34), both of whom she doubtless knew personally.

[1] See Chalandon, op. cit. p. 275.
[2] Ibid. pp. 265–76.
[3] Ibid. pp. 254–7.
[4] Cf. supra, pp. 193–5.

APPENDIX IV

BOGOMILISM IN RUSSIA, SERBIA, BOSNIA AND HUM

1. RUSSIA

The question whether Bogomilism ever spread to Russia is very insufficiently known, as the extant sources are practically silent on this matter. However, a few scattered hints can be gleaned which suggest that individual Bogomils may have proselytized in Russia between the eleventh and the fifteenth centuries.

Nikon's Chronicle, composed in the sixteenth century, states that in 1004 the authorities of the Russian Church arrested a heretic by the name of Adrian for reviling 'the laws of the Church, the bishops, the priests and the monks'. In prison Adrian soon repented of his errors.[1] Russian historians, with some plausibility, have generally taken him to be a Bogomil from Bulgaria.[2] There is no evidence, however, that he had any appreciable following in Russia.[3]

The Russian Primary Chronicle in the Laurentian redaction describes under the year 1071 the appearance in Beloozero in northern Russia of 'two magicians' (два волхва), guilty of the murder of a number of women whom they accused of causing a famine in the region of Rostov. When brought before the civil authorities the magicians made an interesting confession of faith: to the question 'how was man created?' they replied: 'God washed Himself in the bath, and after perspiring, dried Himself with straw and threw it out of heaven upon the earth. Then Satan quarrelled with God over which of them should create man out of it. But the Devil made man, and God set a soul in him. As a result, whenever man dies, his body goes to the earth and his

[1] Патриаршая или Никоновская Летопись: Полное собрание русских летописей, изд. Археографическою Коммиссиею (St Petersburg, 1862), t. IX, p. 68.

[2] Metropolitan Makary, История русской церкви (2nd ed., St Petersburg, 1868), vol. I, pp. 227–8; E. Golubinsky, История русской церкви (Moscow, 1904), vol. I, pt 2, pp. 791–3.

[3] The brief reference in *Nikon's Chronicle* under the year 1123 to the heretic Dmitr (Dimitri), who was imprisoned by the metropolitan of Kiev (ibid. p. 152), has also been regarded as an allusion to Bogomilism in Russia (Makary, op. cit. vol. II, pp. 316–17). But the evidence here is really insufficient, as we are told nothing further about this heretic.

soul to God.' Asked then 'in what God do you believe?', they answered: 'in Antichrist', who 'dwells in the abyss'.[1] The first part of the magicians' confession refers to a popular legend which still existed among the Finnish mordva in the nineteenth century.[2] But the contents and terminology of the second part are unmistakably Bogomil.

The *Kiev Paterik*, the composition of which was begun in the early thirteenth century, contains perhaps a hint that some Orthodox circles in Kievan Russia were not entirely impervious to the influence of dualism. In the story of the temptation of St Nikita, hermit of the Kiev monastery of the Caves and later bishop of Novgorod (1096–1108), we are told how Nikita, smitten with pride, decided against the advice of his abbot to exchange the community life for one of complete seclusion. In his dangerous solitude he was ensnared by the Devil, who appeared in the shape of an angel and persuaded him to abandon prayer for study and teaching. With the aid of the Devil, Nikita gained a reputation for great wisdom, owing particularly to his unequalled knowledge of the Old Testament. 'He knew all the Jewish books well', but refused to read the New Testament. Nikita was eventually saved by the prayers and exorcisms of his fellow-monks and, on coming to his senses, vigorously disclaimed any knowledge of the Old Testament.[3] It is possible that we have here an echo of the Bogomil rejection of the Old Testament.[4]

The heresies of the 'Strigolniki' and of the 'Judaizers' which flourished in north-western Russia in the fourteenth and fifteenth centuries respectively have sometimes been at least indirectly related to Bogomilism by historians. The 'Strigolniki' appeared in Novgorod and Pskov in the last quarter of the fourteenth century. The temporary success enjoyed by this sect, of which we know but little, is usually explained by the ravages caused in Russia by the Black Death, by the penetration of rationalist ideas from western Europe and by the protest of the heretics against the laxity of the

---

[1] Лаврентьевская Летопись. Вып. I: Повесть Временных Лет, изд. второе (Leningrad, 1926): Полное собрание русских летописей, изд. Историко-Археографической Комиссиею, t. I, cols. 175–7; Engl. tr. by S. H. Cross, 'The Russian Primary Chronicle', *Harvard Studies and Notes in Philology and Literature* (Cambridge, Mass. 1930), vol. XII, pp. 240–1. Cf. N. K. Chadwick, *The Beginnings of Russian History* (Cambridge, 1946), pp. 112–13.

[2] See История русской литературы, publ. by the Academy of Sciences of the U.S.S.R., t. I (ed. by A. S. Orlov and others; Moscow, 1941), p. 84.

[3] Києво-Печерський Патерик (ed. D. Abramovich; Kiev, 1931), pp. 124–7.

[4] История русской литературы, ibid. p. 65.

clergy.[1] It is possible, though as yet far from certain, that in their rejection of the hierarchy of the Church, which they accused of simony, and of the sacraments, and in their emphasis on moral rigorism, the 'Strigolniki' were influenced by the dualist tradition.[2] But we cannot speak of any direct influence of Balkan Bogomilism on this movement.

The 'Judaizers' are of greater interest. The precise nature of this movement is still somewhat mysterious and controversial. It was undoubtedly connected with the heresy of the 'Strigolniki', arose in Novgorod *c.* 1470, spread in 1479–80 to Moscow, where it gained some popularity in high ecclesiastical and political circles, was condemned by a council of the Russian Church in 1490 and was finally extinguished by persecution at the beginning of the sixteenth century. The two foremost opponents of the 'Judaizers', Gennady, archbishop of Novgorod, and St Joseph of Volokolamsk, seem to have considered them in the main as adherents of the Jewish faith. They were said to disbelieve in the Trinity and in the divinity of Christ, to prefer the Old Testament to the New, to have adopted some Jewish practices, to reject the cult of saints, icons and relics, to shun the use of churches, to revile the monks and the whole ecclesiastical hierarchy and to claim the right of free interpretation of the Holy Scripture. Some at least of these doctrines are certainly suggestive of Judaism, and as some Jews in Novgorod are known to have been active in spreading them, several Russian historians regarded the movement as essentially Jewish in character.[3] Others, however, have challenged this view. I. Panov, in his searching study of the 'Judaizers', considered the movement to have been essentially a compromise between a form of Christian rationalism and a liberal and philosophical interpretation of Judaism, and thought that the Jewish elements were gradually relegated to a secondary role.[4] More recently M. N. Speransky expressed the opinion that the mainstrings of the 'Judaizing' movement were the growing influence of western rationalist ideas in north-west Russia and the opposition of the clergy and people

[1] For the 'Strigolniki', see Golubinsky, op. cit. vol. II, pt 1, pp. 396–407; M. N. Speransky, История древней русской литературы (3rd ed.; Moscow, 1921), vol. II, pp. 51–3.

[2] Speransky, op. cit. p. 52. Cf. A. S. Orlov (Древняя русская литература, Moscow, Leningrad, 1945, p. 239), who compares the name of 'Strigolniki', which is derived from a verb meaning 'to cut' (cloth), with that of *tisserands*, *texerants*, often given to the French Cathars.

[3] Makary, op. cit. vol. VI, pp. 80 et seq.; Golubinsky, op. cit. vol. II, pt 1, pp. 560–607.

[4] Ересь жидовствующих, *Zh.M.N.P.* (Jan.-March 1877), vols. CLXXXVIII–CXC.

of Novgorod to the centralizing autocracy of Moscow, which culminated in the conquest of Novgorod by Ivan III in 1475 and in the final extinction of the proud city's independence in 1480. The Novgorod Jews, in his view, tried to turn to their advantage the growing rationalism of the city, with the result that the teaching of the sect became infused with Jewish elements. The heretics regarded themselves not as Jews, but as authentic Christians; moreover, it is unlikely that a basically Jewish movement would have gained much success in Russia at that time.[1]

It is probable that the Russian 'Judaizing' heresy was not simply a revival of Judaism and that its origin and character are more complex. Western rationalism, which came to Russia through the close connections of Novgorod with northern and western Europe, and the separatist tendencies of the anti-Moscow party in the city undoubtedly played some part in the movement. But we may go further and ask whether any other influences were exerted on the Novgorod heretics at the end of the fifteenth century.

Both Gennady and St Joseph drop some significant hints. In a letter to Ioasaf, archbishop of Rostov and Yaroslavl, written on 25 February 1489, Gennady states that the doctrines of the Novgorod sect are 'Judaism, mixed with the Massalian heresy'.[2] The 'Judaizers' are identified with or related to the Massalians five times in the same letter. Unfortunately Gennady does not give his reasons for relating the two sects to each other, beyond stating that the followers of each forswear themselves fearlessly and 'celebrate the Divine Liturgy unworthily'. St Joseph likewise says that the Novgorod heretics 'hold secretly to the Massalian heresy'.[3] If these references to the ancient sect are not simply heresiological clichés,[4] the 'Massalian' doctrines ascribed to the Russian 'Judaizers' may well be elements of Bogomilism;[5] in the Balkans Bogomilism and Massalianism were considered to be

---

[1] Speransky, op. cit. vol. II, pp. 53–9. Golubinsky attempted to answer this last argument by supposing that the Jews attracted the Russians by their clever proselytism and by exploiting the popular interest in astrology (loc. cit. pp. 595–6), but this does not seem very convincing.

[2] In *Chteniya v imperat. obshch. istorii i drevnostey rossiyskikh pri Moskov. Universit.* (Moscow, 1847), vol. VIII.

[3] Просвѣтитель, или ѡбличеніе ереси жидовствующихъ (3rd ed.; Kazan, 1896), p. 44.

[4] St Joseph, at least, seems well acquainted with the doctrine of Massalianism (cf. ibid. p. 456) and is hence not likely to have used the term in a loose sense.

[5] This is Panov's opinion (loc. cit. vol. CXC, pp. 12–13).

identical after the thirteenth century.[1] We cannot, however, be certain whether the 'Massalian' features of the Russian heresy included more than a rejection of the material objects of the Christian cult and of the hierarchy of the Church; and several doctrines attributed to the Novgorod heretics (especially the preference for the Old Testament) are scarcely compatible with the tenets of Bogomilism. Yet the suspicion that the Russian 'Judaizers' were not impervious to the influence of the Balkan sect is confirmed by Gennady's statement that the Novgorod heretics made use of Cosmas's treatise against the Bulgarian Bogomils.[2]

In the absence of further documentary evidence, we can only surmise that the indirect references to Bogomilism which we find in contemporary works concerning the heresy of the Russian 'Judaizers' may be due to a spread of Bogomil doctrines to Russia, presumably in the fourteenth century, when Russian culture once more became subject to a strong influence of the Balkan Slavonic countries. It is perhaps significant that the connection which might be traced between Judaism and Bogomilism in Novgorod in the late fifteenth century existed in the thirteenth and fourteenth centuries in Bulgaria.[3]

Some influence of Bogomilism is noticeable in the large apocryphal literature which circulated widely in Russia during the Middle Ages. Most of these works were Byzantine in origin and were brought to Russia in Bulgarian translations; in some of them the dualistic bent of a Bogomil intermediate is clearly discernible, and the very term 'Bulgarian fables' by which they were known in Russia is suggestive of their Balkan provenance.

The *Palea*, the famous and partly apocryphal Old Testament Bible which enjoyed great popularity in medieval Russia, shows evidence of having been remodelled on its way from Byzantium by the Bulgarian Bogomils. In its account of the creation and fall of the angels, the chief of the rebellious heavenly host is given the name of Satanael.[4] The defeat of Satanael is attributed to the Archangel Michael, who inherits the divine particle *el*, while his opponent, deprived of his divinity, becomes the Devil.[5] The same

[1] See supra, pp. 222, 254, 260.
[2] *Chteniya*, loc. cit. p. 5. Cf. Popruzhenko, Козма Пресвитер, Sofia, 1936 (Български Старини), vol. XII, pp. xxvii–xlii.
[3] See supra, pp. 241, 263–4.
[4] See M. Gaster, *Lectures on Greeko-Slavonic Literature*, pp. 153–65; S. Runciman, *The Medieval Manichee*, p. 85.
[5] I. Porfiriev, Апокрифические сказания о ветхозаветных лицах и событиях, *Sbornik otdel. russk. yazyka i slovesnosti Imper. Akad. Nauk* (1877), vol. XVII, no. 1, p. 86.

teaching is ascribed to the Byzantine Bogomils by Euthymius Zigabenus.[1]

A prominent part in Russian apocryphal literature is played by the *Legend of the Cross*, attributed to the heretical Bulgarian *pop* Jeremiah, and probably popular among the Bogomils.[2] In one Russian version it is stated that the tree which was later used for making the Cross was planted in Paradise by Satanael, who then existed alone with God.[3] Another version of the *Legend* begins with these words, which are obviously Bogomil: 'when God created the world, only He and Satanael were in existence.'[4]

Passages showing a remarkable similarity to parts of the Bogomil *Secret Book* are to be found in the Slavonic version of the apocryphal Apocalypse of St John[5] and in a Russian medieval manuscript in the library of the Solovki monastery.[6]

The legend of *The Sea of Tiberias*, which is regarded as Bogomil in origin, must have been popular in medieval Russia, for it has come down to us in several Russian manuscripts. It describes the collaboration of God and Satanael in the creation of the earth.[7]

The apocryphal *Gospel of Thomas* (known in its Slavonic versions as *The Childhood of Jesus*), *The Vision of Isaiah*, *The Book of Enoch*, which were adapted and used by the Bulgarian Bogomils, were already known in Russia in the Kiev period.[8]

Finally, unmistakable traces of Bogomil doctrines, myths and terminology (mixed with Christian apocryphal stories) can be found in later south Russian folk-lore, particularly in a number of Ukrainian cosmogonical legends, which describe the dual rule of God and the Devil (sometimes called Satanael) over the universe, their collaboration in the creation of man and their rivalry in the present world.[9]

[1] See supra, p. 210.   [2] Cf. supra, pp. 271–4.

[3] N. S. Tikhonravov, Памятники отреченной русской литературы, vol. I, pp. 305–13; I. Porfiriev, Апокрифические сказания о новозаветных лицах и событиях, *Sbornik* (1890), vol. LII, no. 4, pp. 55–61.

[4] Gaster, op. cit. p. 36.

[5] V. Jagić, 'Opisi i izvodi iz nekoliko južnoslovinskih rukopisa', *Starine* (1873), vol. V, p. 77.

[6] I. Porfiriev, Апокрифические сказания о ветхозаветных лицах, *Sbornik* (1877), vol. XVII, no 1, p. 86.

[7] I. Ivanov, Богомилски книги, pp. 287–311; История русской литературы, vol. I, p. 84.

[8] See Ivanov, op. cit. pp. 227–48; История русской литературы, ibid. p. 85; Speransky, op. cit. vol. I, pp. 264–6. Cf. supra, p. 154, n. 2.

[9] See the study of some of these Ukrainian legends by K. Radchenko, Этюды по богомильству, *Izvestiya otdel. russk. yazyka i slovesnosti Imperat. Akad. Nauk* (St Petersburg, 1910), vol. XV, pt 4, pp. 73–131.

There is an apparent contradiction between the complete absence of any direct and explicit references to native Bogomilism in Russian sources and the comparative abundance of Bogomil elements in Russian apocryphal literature. There would seem to be two reasons for this. It is probable that Bogomilism, as a sectarian movement, never struck deep roots in medieval Russia, and that its success there was limited to the sporadic and intermittent proselytism of individuals who were either Russian heretics or, perhaps, missionaries from Bulgaria. Moreover, the cosmogonical legends and apocryphal Old and New Testament stories which contain Bogomil elements probably captured the fancy of the medieval Russian mostly for their narrative interest and literary merit, and their dualist background either passed largely unnoticed, or was not consciously associated with any formally heretical set of doctrines.[1]

2. SERBIA

The geographical position of medieval Serbia was favourable to the penetration of Bogomilism into that country from neighbouring Macedonia. The range of the Shar Planina, the sole barrier between Serbia and Macedonia, could be traversed without difficulty at its two extremities: to the north-east, immediately north of Skoplje, the valley of the southern Morava connected Macedonia with Niš and the Danube;[2] moreover, the valley of the Sitnica led past Kosovo Polje to Raška and the valley of the Ibar. On the other hand, at the western extremity of the Shar Planina, the road was open from Macedonia up the valley of the Black Drina to Dioclea and Zeta.[3] Hence it is not surprising to find evidence of connections between Serbia and Macedonia already in the second half of the tenth century, at the time of the great development of Bogomilism in the latter country.[4] In

[1] See Golubinsky, op. cit. vol. I, pt 2, p. 794; Speransky, op. cit. vol. I, pp. 262-3.
[2] This was the medieval trade-route linking Thessalonica with Belgrade. The town of Raš, the centre of the Serbian medieval kingdom of Raška, lay along this route and was also connected with Sardica (Sofia) by an ancient road passing through Niš. See K. Jireček, 'Die Handelsstrassen und Bergwerke von Serbien und Bosnien während des Mittelalters', *Abh. böhm. Ges. Wiss.* (Prague, 1879), VI. Folge, 10. Bd, p. 32.
[3] See Jireček, ibid. pp. 62-8.
[4] Examples of these connections are Samuel's invasion of Dioclea in 998 and his relations with Vladimir, prince of Dioclea. See V. N. Zlatarski, История, vol. I, pt 2, pp. 706-13; K. Jireček, *Geschichte der Serben*, vol. I, pp. 205-6.

the eleventh and twelfth centuries the link between Macedonia and Serbia was strengthened through the influence of the archbishopric of Ochrida.[1]

The penetration of Bogomilism from Macedonia into Serbia was undoubtedly facilitated by these natural waterways. The Bogomils are first heard of in Serbia in the reign of Stephen Nemanja, Grand Župan of Raška (1168–96). His *Life*, written by his son, Stephen the First-Crowned, tells us that on becoming aware of the prevalence of the heretics in his realm, Nemanja summoned a general assembly of the land (*sabor*), which inflicted dire punishment on the Bogomils: their leader had his tongue cut out, his followers were either executed or banished, their property confiscated and their heretical books burnt.[2]

Nemanja's measures appear to have been most successful: for 150 years we hear no more of the Bogomils in Serbia. The main credit for this must be given to his younger son Rastko, who became the first archbishop and great organizer of the autocephalous Church of Serbia, and is known in history under the name of St Sava. By firmly establishing the Serbian Church on truly popular and national foundations St Sava dealt the greatest possible blow to Bogomilism, by depriving the heretics of one of their most potent weapons of proselytism, namely the reaction against an excessive Byzantine influence in Church and State. The medieval Church of Bulgaria was never able to achieve this, for, owing to historical, geographical and political reasons, Bulgaria, unlike Serbia, was incapable of resisting the stream of Byzantinism. It was undoubtedly largely due to St Sava's great apostolic work that, while Bogomilism flourished for four centuries in Bulgaria, its influence in Serbia was arrested at the end of the twelfth century. Bogomilism is mentioned for the last time in Serbia in the famous code of laws of the Tsar Stephen Dušan, promulgated in 1349–54. The *Zakonik* imposes fines, flogging, branding or exile, according to the social status or the

[1] See Jireček, op. cit. pp. 219–22.
[2] *Život Sv. Symeona od Králé Štěpána*: P. J. Šafařík, *Památky Dřevního Písemnictví Jihoslovanův*, pp. 6 et seq. H. W. V. Temperley, describing Nemanja's persecution of the Bogomils, suggests that their doctrines 'have been greatly misrepresented by Orthodox opponents. Its main principle [i.e. that of Bogomilism] does not appear to have been the dualism or equality of good and evil, as is often asserted.' (*History of Serbia*, London, 1917, p. 43.) But apart from the fact that the terms 'dualism' and 'equality of good and evil' are by no means synonymous (which can be seen in the so-called 'mitigated dualism' of the Bogomils), Prof. Temperley's statement flatly contradicts the evidence of all Bulgarian and Byzantine sources on the Bogomils, as well as that of the written monuments of the Bogomils themselves. The fact that the Bogomil doctrine was dualistic is to-day established beyond any doubt.

obstinacy of the offenders, on the followers of the 'Babun faith'.[1]

3. BOSNIA AND HUM

The problem of the penetration of Bogomilism into Bosnia and Hum (present-day Herzegovina) is a vast and complex one and cannot be even briefly outlined here. In no other country did the heresy (known in Bosnia as the Patarene faith) have such widespread repercussions on the internal and external history of the people. The notable differences in doctrine between the Bogomils and the Bosnian Patarenes make it impossible to treat the latter simply as part of the 'Bogomil question'. The most comprehensive study of the history of the Patarene sect in Bosnia and Hum is Rački's 'Bogomili i Patareni' (*Rad*, vol. VII, pp. 126–79; vol. VIII, pp. 121–75). The general histories of V. Klaić (*Geschichte Bosniens von den ältesten Zeiten bis zum Verfalle des Königreiches*, Leipzig, 1885) and of Spinka (*A History of Christianity in the Balkans*, pp. 157–83) reiterate in the main the results of Rački's investigations. The principal sources used by Rački are: (1) *Monumenta Serbica spectantia historiam Serbiae, Bosnae, Ragusii*, ed. F. Miklosich, Vienna, 1858. (2) 'The State Documents of the Republic of Dubrovnik', ed. M. Pučić (Споменици Сръбски, Belgrade, 1858). (3) The MSS. of the Vatican archives published by A. Theiner (*Vetera monumenta historica Hungariam sacram illustrantia*, 1859–60, and *Vetera monumenta Slavorum meridionalium*, 1863). Rački's work on Paterenism in Bosnia is still unsurpassed, though some of his opinions need revision in the light of more recent historical research, in particular that of Ć. Truhelka ('Testamenat gosta Radina', *Glasnik Zemaljskog Muzeja u Bosni i Hercegovini*, Sarajevo, July–September, 1911, pp. 355–75).

A brief history of the Bosnian Patarenes is given by Runciman (*The Medieval Manichee*, pp. 100–15), who, however, scarcely mentions their doctrines.

Incomplete studies of the problem can also be found in the works of Petranović (Црьква Босаньска и крьстяни), J. Asbóth (*An Official Tour through Bosnia and Herzegovina*, London, 1890), D. Prohaska (*Das kroatisch-serbische Schrifttum in Bosnien und der Herzegowina*, Zagreb, 1911, pp. 18–55), P. Rovinsky (Материал для истории богумилов в сербских землях, *Zh.M.N.P.* vol. CCXX, March, 1882, pp. 32–51).

A survey of the studies devoted to the Bosnian Patarene movement can be found in the article of J. Šidak, 'Problem "bosanske crkve" u našoj historiografiji od Petranovića do Glušca (Prilog rješenju t.zv. bogumilskog pitanja)', *Rad* (1937), vol. CCLIX, pp. 37–182.

[1] *Zakonik Stefana Dušana, cara Srpskog* (ed. S. Novaković; Belgrade, 1898), pp. 14, 67 (articles nos. 10, 85).

APPENDIX V

BOGOMILS, CATHARS AND PATARENES

The problem of the influence of Bogomilism on the development of dualistic heresy in western Europe still awaits a definitive study. Western medievalists for the most part have not investigated the Slavonic Bogomil sources, while Slavonic historians have generally taken the filiation of the Cathars and Patarenes from the Bogomils for granted but have not attempted a detailed study of Western dualism from the point of view of its connection with Bogomilism.[1]

A full examination of this question does not fall within the scope of the present work, but the following notes indicate the position reached by modern scholarship on the subject and posit certain aspects of the problem for future research.

The first modern scholar to have examined the links between Eastern and Western dualism in some detail was Schmidt. He acknowledged the strong influence exerted by the Bogomils on the Cathars, but denied that the latter derived their teaching from the former. His main arguments are as follows: (1) The view that the Bogomils and the Cathars are linked by their common Gnostic inheritance is refuted by the fact that the doctrines of the Cathars reveal no great influence of Gnosticism. (2) Catharism, at least in the twelfth and thirteenth centuries, was a unified and harmonious system; whereas 's'il y a une doctrine incomplète, c'est bien plutôt celle des Bogomiles'. (3) Traces of Catharism can be found in France already at the end of the tenth century, while, in Schmidt's opinion, Bogomilism only appeared in the second half of the eleventh century. (4) He regards Catharism and Bogomilism as parallel branches of a 'dualisme cathare primitif'.[2] But these arguments of Schmidt, which have not so far been refuted, are not conclusive and are based on erroneous notions of Bogomilism. As we have seen, there is no evidence of any direct connection between Gnosticism and Bogomilism; Bogomilism, after its penetration into Byzantium at the end of the eleventh century, was very far from being 'une doctrine incomplète'; moreover, it arose in Bulgaria a century

---

[1] A good summary of the available evidence will be found in S. Runciman, *The Medieval Manichee*, especially pp. 163–70.
[2] *Histoire et doctrine de la secte des Cathares ou Albigeois*, vol. II, pp. 263–6.

earlier than Schmidt imagined; finally, the existence of a 'dualisme cathare primitif' can be substantiated by no reliable evidence.

On the other hand, eminent Slavists like Rački and Ivanov regard the Patarenes and the Cathars as offshoots of the Bulgarian Bogomils. An interesting attempt to disprove their view was made by the Russian scholar E. Anichkov.[1] In his opinion Catharism developed not as the result of the penetration of Slavonic Bogomilism into Italy and France, but from the remnants of ancient Manichaean traditions, preserved for many centuries in western Europe.

Such is the position of the problem to-day. The view commonly held by Slavonic scholars that the development of dualism in the twelfth and thirteenth centuries throughout all southern Europe, from the Black Sea to the Atlantic, was simply due to a gradual spread of Bogomilism from Bulgaria to Serbia, Bosnia, northern Italy and southern France, is over-simplified and should be reconsidered after a careful study of the origins of Catharism. It is significant in this respect that positive evidence of the influence exerted by Bogomilism on the Patarenes and the Cathars only exists from the second half of the twelfth century, when dualism was already widespread in western Europe.

The principal arguments adduced in support of the theory that Bogomilism exerted a direct influence on the development of Catharism can be summarized as follows:

I. INTERNAL EVIDENCE

(1) A name frequently given to Catharism by its Catholic opponents in France in the thirteenth century was *Bulgarorum haeresis*; the Cathars themselves were often called *Bulgari, Bolgari, Bogri, Bugres*.[2] It is a well-known fact that the French word 'bougre', which since then has been synonymous with 'sodomite' and was originally applied by the Catholics to the Cathars, is derived from 'Bulgarus'.[3]

(2) The name 'Cathars' (καθαροί) is of Greek origin.

(3) From the eleventh century place-names and family names in northern Italy, such as Bulgaro, Bulgari, Bulgarello, Bulgarini indicate that relations existed from early times between Italy and Bulgaria.[4]

---

[1] 'Les survivances manichéennes en pays Slaves et en Occident', *R.E.S.* (1928), vol. VIII, pp. 203–25.
[2] See Schmidt, op. cit. vol. II, p. 282.
[3] See Ivanov, op. cit. p. 41, n. 1.
[4] See Schmidt, ibid. p. 286; Rački, op. cit. *Rad*, vol. VII, p. 106.

(4) Certain Cathars, condemned to the stake in 1146 in Cologne, admitted: 'hanc haeresim usque ad haec tempora occultatam fuisse a temporibus martyrum *et permansisse in Graecia et quibusdam aliis terris*'. These 'other lands' may refer to 'Bulgaria.[1]

(5) The version of the New Testament used by the French Cathars and the Italian Patarenes was not the Vulgate but another translation from the Greek. Schmidt suggests that this was a translation into Latin of the Slavonic version of St Cyril and St Methodius.[2]

(6) The ritual of the Cathars, connected with initiatory prayer-meetings, was undoubtedly influenced by that of the Bogomils.[3]

(7) Finally, a number of doctrines held by the Cathars are to be found in Bogomilism: the Docetic Christology, the opposition to the instituted Church, the repudiation of marriage, the emphasis on asceticism, the exclusive preference for the Lord's Prayer, and, above all, the belief that the Devil was the son of God, and also the unjust steward and the lord of this world.[4]

II. Historical connections between the Bogomils and the Western Dualists

Evidence of Latin sources of the twelfth century suggests that at that time the dualists of western Europe regarded the Balkans as the fount of their teaching.

It has been shown that the 'Ecclesia Bulgariae' and the 'Ecclesia Dugunthiae', situated respectively in Macedonia and in Thrace, were considered by the Cathars as the origin of all the dualistic communities in Italy and France.[5] Towards the middle of the twelfth century Marcus, the leader of the Italian Patarenes, belonged to the 'Ecclesia Bulgariae'.[6] In 1167 Nicetas, the chief of the heretics of Constantinople, journeyed to Lombardy with the aim of converting Marcus to the 'absolute dualism' professed by the 'Ecclesia Dugunthiae'. He was successful in his task.[7] In the same year Nicetas travelled from Lombardy to France, where he presided over the council of the Cathars at Saint-Félix de Caraman. Largely owing to his personal authority Nicetas succeeded in imposing the doctrines of the 'Ecclesia Dugunthiae' on the French and Italian dualists. So great were Nicetas's prestige and reputation that his Catholic opponents, in the face of this union of all

[1] See Schmidt, op. cit. vol. I, p. 2, n. 1; Rački, op. cit. *Rad*, vol. VII, p. 92.
[2] Op. cit. vol. II, p. 274.
[3] See Ivanov, op. cit. pp. 113 et seq.; Runciman, op. cit. pp. 163–6.
[4] See Puech, op. cit. pp. 340–1.
[5] Cf. supra, pp. 157 et seq.   [6] Schmidt, op. cit. vol. I, p. 61.
[7] Schmidt, ibid. pp. 58–9, 61; Rački, ibid. pp. 121–2.

dualistic heretics from the Bosporus to the Pyrenees, ascribed to him the title of 'Pope of the heretics'.[1] However, his triumph was short-lived. Soon after the council of 1167, a certain Petracus arrived in Lombardy *des parties d'outremer* and by discrediting Nicetas secured the adherence of a large number of Patarenes to the 'Ecclesia Bulgariae'.[2] It is generally thought that Petracus came from Bulgaria and that he was a Bogomil.[3]

In the second half of the twelfth century the Bogomil 'dyed' Nazarius brought to the Italian Patarenes the Bogomil *Liber Sancti Johannis*.[4] In the middle of the thirteenth century Reinerius Sacchoni stated that the origin of the Cathar heresy was in the Balkans.[5]

Finally, the close connection between the Cathar heresy and the cloth industry in western Europe is partly due to the fact that many Cathar missionaries were cloth merchants who, together with the woven fabrics of Byzantium, brought the dualist doctrines to southern France.[6]

The author of a recent study of the Cathar doctrines, while admitting their close resemblance to the teachings of Mani, speaks of a 'hiatus historique' between the two.[7] It can no longer be doubted that Bogomilism, at least in several respects, is this 'missing link'.

[1] See M. Bouquet, *Recueil des historiens des Gaules*, vol. XIV, pp. 448–9; this title has been proved to be fictitious (cf. supra, p. 246).
[2] See N. Vignier, *Recueil de l'histoire de l'Église*, p. 268.
[3] Schmidt, ibid. p. 61; Rački, ibid. p. 122.
[4] Cf. supra, pp. 226–8, 242.   [5] Cf. supra, pp. 157 et seq.
[6] See Runciman, op. cit. p. 169.
[7] A. Dondaine, *Un traité néo-manichéen du XIIIe siècle: Le 'Liber de Duobus Principibus', suivi d'un fragment de rituel Cathare*, Rome, 1939, pp. 52–7.

BIBLIOGRAPHY

I. Sources

Acta et diplomata res Albaniae mediae aetatis illustrantia, vol. I, collegerunt et digesserunt L. de Thallóczy, K. Jireček and E. de Sufflay. Vindobonae, 1913.

Albericus, monk of Trois-Fontaines. *Chronicon. M.G.H. Ss.* vol. XXIII.

Allatius, L. *De Ecclesiae occidentalis atque orientalis perpetua consensione.* Coloniae Agrippinae, 1648.

Anastasius Bibliothecarius. *Historia de Vitis Romanorum Pontificum. P.L.* CXXVIII.

Anna Comnena. *Alexiad*, ed. L. Schopenus, *C.S.H.B.* 1839–78. Translated by E. Dawes, *The Alexiad of the Princess Anna Comnena.* London, 1928.

Annales Fuldenses. M.G.H. Ss. vol. I.

Annalista Saxo. *Annales. M.G.H. Ss.* vol. VI.

Anonymi Gesta Francorum et aliorum Hierosolymitanorum, ed. B. A. Lees, Oxford, 1924; ed. L. Bréhier (*Histoire anonyme de la Première Croisade*), Paris, 1924.

Attaliates, Michael. *Historia*, ed. I. Bekker. *C.S.H.B.* 1853.

Balsamon, Theodore. *In Canones Ss. Apostolorum, Conciliorum...Commentaria. P.G.* vol. CXXXVII.

—— *Photii Patriarchae Constantinopolitani Nomocanon*, G. Voellus and H. Justellus, *Bibliotheca Juris Canonici Veteris*, vol. II, Lutetiae, 1661.

Bonacursus. *Contra Catharos.* S. Baluzius, *Miscellanea*, ed. Mansi, vol. II, Lucae, 1761.

Bouquet, M. *Recueil des historiens des Gaules et de la France.* Paris, 1806.

Cedrenus, Georgius. *Historiarum Compendium*, ed. I. Bekker. *C.S.H.B.* 1838–9.

Cinnamus, Joannes. *Historiae*, ed. A. Meineke. *C.S.H.B.* 1836.

Conybeare, F. C. *The Key of Truth. A Manual of the Paulician Church of Armenia.* Oxford, 1898.

Cosmas. Слово на Еретики, ed. by M. G. Popruzhenko, Odessa, 1907; later edit. by Popruzhenko, Козма Пресвитер, болгарский писатель X века. Sofia, 1936. (Български Старини, vol. XII.)

Definitio sanctae et oecumenicae synodi Ephesinae contra impios Messalianitas. Mansi, vol. IV.

Demetrius Chomatianus. *Epistolae*, in Pitra, *Analecta*, vol. VII.

Enoch. *Book of the Secrets of Enoch*, trans. by W. R. Morfill. Oxford, 1896.

Ephraim, St. *St Ephraim's Prose Refutations of Mani, Marcion and Bardaisan.* Ed. C. W. Mitchell, London, 1912–21.

Epiphanius, St. *Adversus Haereses. P.G.* vols. XLI, XLII.

Eustathius of Thessalonica. *Manuelis Comneni laudatio funebris. P.G.* vol. CXXXV.

Euthymius of Acmonia. Ἐπιστολὴ Εὐθυμίου μοναχοῦ τῆς περιβλέπτου μονῆς σταλεῖσα ἀπὸ Κωνσταντινουπόλεως· πρὸς τὴν αὐτοῦ πατρίδα στηλιτεύουσα τὰς αἱρέσεις τῶν ἀθεωτάτων καὶ ἀσεβῶν πλάνων τῶν Φουνδαγιαγιτῶν ἤτοι Βογομίλων. See Ficker.

—— *Liber invectivus contra haeresim exsecrabilium et impiorum haereticorum qui Phundagiatae dicuntur. P.G.* vol. CXXXI. [In the *Patrologia Graeca* this work is falsely attributed to Euthymius Zigabenus.]

Euthymius, Patriarch of Trnovo. *See* Kałużniacki.

Euthymius Zigabenus. *See* Zigabenus.

Eznik of Kolb. *Against the sects*, ed. by S. Weber. *Ausgewählte Schriften der armenischen Kirchenväter*, Bd I. München, 1927.
Fermendžin, E. 'Acta Bulgariae ecclesiastica ab a. 1565 usque ad a. 1799.' *Monumenta spectantia historiam Slavorum meridionalium*, vol. XVIII. Zagreb, 1887.
Ficker, G. *Die Phundagiagiten: Ein Beitrag zur Ketzergeschichte des byzantinischen Mittelalters*. Leipzig, 1908. [This work contains the text of the letter of Euthymius of Acmonia concerning the Bogomils.]
Formula of abjuration for Paulicians: Quo modo haeresim suam scriptis oporteat anathematizare eos qui e Manichaeis accedunt ad sanctam Dei catholicam et apostolicam Ecclesiam. *P.G.* vol. I, cols. 1461–72.
Genesius. *Regum* libri IV, ed. C. Lachmann. *C.S.H.B.* 1834.
Gennadius Scholarius. *Œuvres complètes de Gennade Scholarios*, publiées par L. Petit, X. Siderides, M. Jugie, t. IV. Paris, 1935.
Gennady, Archbishop of Novgorod. 'Letter to Ioasaf, Archbishop of Rostov and Yaroslavl.' *Chteniya v imperatorskom obshchestve istorii i drevnostey rossiyskikh pri Moskovskom Universitete*, vol. VIII. Moscow, 1847.
Georgius (Hamartolus) Monachus. *Chronicon*, ed. de Boor. Leipzig, 1904. Also ed. Muralt, St Petersburg, 1859.
Germanos II, Patriarch of Nicaea. *Epistula ad Constantinopolitanos contra Bogomilos*. Ficker, *Die Phundagiagiten*.
—— *In exaltationem venerandae crucis et contra Bogomilos*. *P.G.* vol. CXL.
Gervase of Chichester. *Epistolae*. C. L. Hugo, *Sacrae Antiquitatis Monumenta*, vol. I. Stivagii, 1725.
Gesta Roberti Wiscardi. M.G.H. Ss. vol. IX.
Glycas, Michael. *Annales*, ed. I. Bekker. *C.S.H.B.* 1836.
Gregoras, Nicephorus. *Historiae Byzantinae*, ed. L. Schopenus and I. Bekker. *C.S.H.B.* 1829–55.
Gregorius Barhebraeus. *Chronicon ecclesiasticum*, ed. J. B. Abbeloos and T. J. Lamy. Lovanii, 1872.
Grumel, V. *Les Regestes des Actes du Patriarcat de Constantinople*, vol. I, fasc. 2. Constantinople, 1932.
Guillelmus Bibliothecarius. *Vita Hadriani II*. J. S. Assemanus, *Kalendaria Ecclesiae Universae*, vol. II. Romae, 1755.
Helmoldus. *Chronica Slavorum. M.G.H. Ss.* vol. XXI. English tr. by F. J. Tschan. New York, 1935.
Hilarion, St, Bishop of Moglena. See *Life of St Hilarion*.
Hincmar, Archbishop of Rheims. *Annales Bertiniani. M.G.H. Ss.* vol. I.
Histoire anonyme de la Première Croisade, éd. et trad. par L. Bréhier. Paris, 1924.
Incerti panegyricus Constantio Caesari, ed. by G. Baehrens. Leipzig: Teubner, 1911.
Interrogatio Iohannis et apostoli et evangelistae in cena secreta regni celorum de ordinatione mundi istius et de principe et de Adam. R. Reitzenstein, *Die Vorgeschichte der christlichen Taufe*. Leipzig und Berlin, 1929.
Isidore of Seville. *Etymologiae*. *P.L.* vol. LXXXII.
Ivanov, I. Произходъ на Павликянитѣ споредъ два български ръкописа *S.B.A.N.* (Dept. of hist. and philol.) vol. XXIV, 1922.
—— Богомилски книги и легенди. Sofia, 1925.
Jagić, V. 'Opisi i izvodi iz nekoliko južnoslovinskih rukopisa', *Starine*, vol. V, 1873. (*Jugoslavenska Akademija Znanosti i Umjetnosti*, Zagreb.)
John Damascene, St. *De haeresibus*. *P.G.* vol. XCIV.

John the Exarch. Шестодневъ, ed. O. Bodyansky and N. Popov, *Chteniya v imperatorskom obshchestve istorii i drevnostey rossiyskikh pri Moskovskom Universitete*, vol. III. Moscow, 1879.
Joseph, St, Abbot of Volokolamsk. Просвѣтитель, или ѡбличеніе ереси жидовствующихъ, 3rd ed. Kazan, 1896.
Kałużniacki, E. *Werke des Patriarchen von Bulgarien Euthymius*. Vienna, 1901.
Key of Truth, The. See Conybeare.
Kiev Paterik. Києво-Печерський Патерик, ed. D. Abramovich. Пам'ятки Мови та Письменства Давньої України, t. IV. Kiev, 1931.
Kmosko, M. *Antiquorum testimonia de historia et doctrina Messalianorum sectae. Patrologia Syriaca*, pars I, t. 3. Paris, 1926.
Krmčaja Ilovička, ed. V. Jagić. *Starine*, vol. VI, 1874. (*Jugoslavenska Akademija Znanosti i Umjetnosti*, Zagreb.)
Langlois, V. *Collection des historiens anciens et modernes de l'Arménie*. Paris, 1867–9.
Лаврентьевская Лѣтопись. *See Russian Primary Chronicle*.
Liber Sancti Johannis. See Interrogatio Iohannis.
Life of Saint Hilarion of Moglena. Житіе и жизнь прѣподобнаго ѡтца нашего Иларіѡна, епископа мегленскаго, въ немже и како прѣнесень бысть въ прѣславный градъ Трънѡвъ, съписано Евѳиміемъ патріархомъ Трьновьскыимь. *See* Kałużniacki. Also ed. Dj. Daničić, *Starine*, vol. I, 1869.
Life of Saint Theodosius of Trnovo. Житіе и жизнь преподобнаго ѡтца нашегѡ Ѳеодосія, иже въ Терновѣ постничествовавшегося, ed. O. Bodyansky. *Chteniya v imperatorskom obshchestve istorii i drevnostey rossiyskikh pri Moskovskom Universitete*, vol. I. Moscow, 1860.
Macarius, St. *Fifty spiritual homilies*, ed. by A. J. Mason. London, 1921.
Mani. 'The so-called Injunctions of Mani, translated from the Pahlavi of Denkart', 3, 200, ed. by A. V. W. Jackson, *J.R.A.S.* 1924.
Manichaean Psalm-Book, A, ed. by C. R. C. Allberry. Stuttgart, 1938.
Manichäische Handschriften der Staatlichen Museen Berlin, herausgegeben in Auftrage der pr. Akad. der Wissensch., unter Leitung von C. Schmidt, Bd I, *Kephalaia*. Stuttgart, 1935–7.
Manichäische Homilien (*Manichäische Handschriften der Sammlung A. Chester Beatty*, Bd I), herausgegeben von H. J. Polotsky. Stuttgart, 1934.
Mansi, J. D. *Sacrorum conciliorum nova et amplissima collectio*. Florence, Venice, 1759–98; new volumes ed. by L. Petit, Paris, 1901– .
Mas'ūdī (Al-). *Les Prairies d'Or*. Texte et traduction par C. Barbier de Meynard et Pavet de Courteille, vol. VIII. Paris, 1874.
—— *Le Livre de l'Avertissement et de la Révision*. Traduction par B. Carra de Vaux. Paris, 1896.
Matthew of Edessa. 'Chronique de Matthieu d'Édesse', French tr. by F. Dulaurier, *Bibliothèque historique arménienne*. Paris, 1858.
Migne, J. P. *Patrologiae Cursus Completus*. Series Graeco-Latina. Paris, 1857–66.
—— *Patrologiae Cursus Completus*. Series Latina. Paris, 1844–55.
Miklosich, F. and Müller, J. *Acta Patriarchatus Constantinopolitani*, vol. I. Vindobonae, 1860.
Moneta of Cremona. *Adversus Catharos et Valdenses*, ed. T. A. Ricchinius. Rome, 1743.
Monumenta Germaniae Historica, ed. by G. H. Pertz and others. Hanover, 1826–

Monumenta Serbica spectantia historiam Serbiae, Bosnae, Ragusii, ed. F. Miklosich. Vienna, 1858.
Monumenta spectantia historiam Slavorum Meridionalium. Zagreb, Jugoslavenska Akademija Znanosti i Umjetnosti, 1868-1918.
Moses Chorensis. *Histoire d'Arménie.* French tr. by P. E. Le Vaillant de Florival. Venice, 1841.
Nicephorus, Patriarch of Constantinople. *Opuscula Historica,* ed. C. de Boor. Leipzig, 1880.
Nicetas Choniates. *De Manuele Comneno* in *Historia,* ed. I. Bekker. *C.S.H.B.* 1835.
Nicholas of Methone. *Orationes,* ed. A. K. Demetrakopoulos. 'Εκκλ. Βιβλιοθήκη. Leipzig, 1866.
Nicholas I, Pope. *Epistolae. P.L.* vol. cxix.
—— *Responsa ad Consulta Bulgarorum. P.L.* vol. cxix.
Nikon's Chronicle. See Патриаршая или Никоновская Летопись.
Palamas, St Gregory. *Homiliae, P.G.* vol. cli.
Патриаршая или Никоновская Летопись: Полное собрание русских летописей, изданное Археографическою Коммиссиею, tt. ix-xiii. St Petersburg, 1862.
Peter, Abbot. Πέτρου ἐλαχίστου μοναχοῦ 'Ηγουμένου περὶ Παυλικιανῶν τῶν καὶ Μανιχαίων, ed. by J. K. L. Gieseler. Göttingen, 1849.
Petrus Siculus. *Historia Manichaeorum qui et Pauliciani dicuntur. P.G.* vol. civ.
Philotheus, Patriarch of Constantinople. *Gregorii Palamae Encomium. P.G.* vol. cli.
Photius, Patriarch of Constantinople. *Bibliotheca. P.G.* vol. ciii.
—— *Contra Manichaeos. P.G.* vol. cii.
—— *Epistolae. P.G.* vol. cii. Also I. N. Valetta, Φωτίου τοῦ σοφωτάτου καὶ ἁγιωτάτου πατριάρχου Κωνσταντινουπόλεως ἐπιστολαί, 'Εν Λονδίνῳ, 1864.
Pitra, J. B. *Analecta sacra spicilegio Solesmensi parata.* Paris, 1876-91.
Procopius of Caesarea. *De Bello Gothico.* Leipzig: Teubner, 1905.
Psellus, Michael. *Dialogus de daemonum operatione.* Ed. by J. Boissonade, Nuremberg, 1838. Also in *P.G.* vol. cxxii. A French translation by P. Moreau is given by E. Renauld: 'Une traduction française du Περὶ ἐνεργείας δαιμόνων de Michel Psellos'. *Revue des Études Grecques,* vol. xxxiii. Paris, 1920.
Regino of Prüm. *Chronicon. M.G.H. Ss.* vol. i.
Rhalles, G. and Potles, M. Σύνταγμα τῶν θείων καὶ ἱερῶν κανόνων. Athens, 1852-9.
Russian Primary Chronicle, The. Лаврентьевская Летопись. Вып. I: Повесть Временных Лет, изд. второе (ed. by E. F. Karsky), Leningrad, 1926: Полное собрание русских летописей, изд. Историко-Археографической Коммиссиею Академии Наук СССР, t. i. English tr. by S. H. Cross, 'The Russian Primary Chronicle', *Harvard Studies and Notes in Philology and Literature,* vol. xii. Cambridge, Mass. 1930.
Sacchoni, Reinerius. *Summa de Catharis et Leonistis: Maxima Bibliotheca veterum Patrum,* ed. M. de La Bigne, vol. xxv. Lugduni, 1677. Also in E. Martène et U. Durand, *Thesaurus novus anecdotorum,* vol. v. Paris, 1717. Also in A. Dondaine, *Un traité néo-manichéen du XIIIe siècle, suivi d'un fragment de rituel Cathare.* Rome, 1939.
Šafařík, P. J. *Památky Dřevního Písemnictví Jihoslovanův.* Prague, 1873.
Samuel of Ani. *Ausgewählte Schriften der armenischen Kirchenväter,* vol. i. München, 1927.

Scylitzes, Joannes. *Historia*, ed. I. Bekker. *C.S.H.B.* 1839.
Sebêos, Bishop. *Histoire d'Héraclius*. Paris, 1904.
Sigebertus Gemblacensis. *Chronicon. M.G.H. Ss.* vol. VI.
Споменици Сръбски, 1395–1423, ed. M. Pučić. Belgrade, 1858.
Stephen Dušan, Tsar of Serbia. *Zakonik Stefana Dušana, cara Srpskog*, ed. S. Novaković. Belgrade, 1898.
Stephen Nemanja (St Symeon), Grand Župan of Serbia. *Život Sv. Symeona od Kréle Štěpane*: in P. J. Šafařík, *Památky Dřevního Písemnictví Jihoslovanův*.
Stephen of Taron. *Des Stephanos von Taron armenische Geschichte*, ed. H. Gelzer and A. Burckhardt. Leipzig, 1907.
Suidas. *Lexicon*, ed. A. Adler, vol. I. Leipzig, 1928.
Symeon Magister. *Annales*, ed. I. Bekker. *C.S.H.B.* 1838.
Symeon, Archbishop of Thessalonica. *Dialogus contra haereses. P.G.* vol. CLV.
Synodicon for the Sunday of Orthodoxy, ed. Th. Uspensky. *Zapiski imperatorskogo novorossiyskogo Universiteta*, vol. LIX. Odessa, 1893.
Synodicon of the Tsar Boril, ed. M. G. Popruzhenko, Синодик царя Борила (Български Старини, vol. VIII). Sofia, 1928.
Thallóczy, L. 'Beiträge zur Kenntniss der Bogomilenlehre' (Formulae for the abjuration of converted Bogomils). *Wiss. Mitt. Bosn. Herz.* vol. III. Vienna, 1895.
Theiner, A. *Vetera monumenta historica Hungariam sacram illustrantia*. Romae, 1859–60.
—— *Vetera monumenta Slavorum meridionalium historiam illustrantia*. Romae, 1863.
Theodoret. *Haereticarum Fabularum Compendium. P.G.* vol. LXXXIII.
Theodosius, St, of Trnovo. See *Life of St Theodosius*.
Theophanes. *Chronographia*, ed. I. Classenus. *C.S.H.B.* 1839–41 (revised text, ed. de Boor; Leipzig, 1883).
Theophanes Continuatus. *Chronographia*, ed. I. Bekker. *C.S.H.B.* 1838.
Theophylact, Archbishop of Ochrida. *Epistolae. P.G.* vol. CXXVI.
—— *Historia Martyrii XV Martyrum. P.G.* vol. CXXVI.
—— *Vita Sancti Clementis Bulgarorum Archiepiscopi. P.G.* vol. CXXVI.
Theophylact, Patriarch of Constantinople. Письмо патриарха Константинопольского Феофилакта царю Болгарии Петру, ed. N. M. Petrovsky, *Izvestiya otdeleniya russkogo yazyka i slovesnosti Imperatorskoy Akademii Nauk*, vol. XVIII, tom. 3, 1913.
Tikhonravov, N. S. Памятники отреченной русской литературы. St Petersburg, 1863.
Timotheus Presbyter. *De receptione haereticorum. P.G.* vol. LXXXVI (1).
Tollius, J. *Insignia itinerarii italici*: 'Victoria et Triumphus de impia et multiplici exsecrabilium Massalianorum secta, qui et Phundaïtae et Bogomili, nec non Euchitae, Enthusiastae, Encratitae, et Marcionitae appellantur'. (Formula of abjuration for converted Bogomils.) Trajecti ad Rhenum, 1696.
Truhelka, Ć. 'Testamenat gosta Radina', *Glasnik Zemaljskog Muzeja u Bosni i Hercegovini*, Sarajevo. July–September, 1911.
Villehardouin, Geoffroy de. *La Conquête de Constantinople*, éditée et traduite par E. Faral. Paris, 1938–9.
Zigabenus, Euthymius. *De haeresi Bogomilorum narratio*. Ficker, *Die Phundagiagiten*.
—— *Panoplia Dogmatica. P.G.* vol. CXXX.
Zonaras, Joannes. *Annales* and *Epitome Historiarum*, ed. M. Pinder, T. Büttner-Wobst. *C.S.H.B.* 1841–97.

II. Modern Works

Aboba-Pliska. Материалы для болгарских древностей. *I.R.A.I.K.* vol. x. Sofia, 1905.
ADONTZ, N. 'Samuel l'Arménien, roi des Bulgares.' *Mém. Acad. Belg.* (cl. des lettres) (série in 8°), t. xxxix. Bruxelles, 1938.
ALFARIC, P. *Les Écritures manichéennes.* Paris, 1918.
—— *L'Évolution intellectuelle de Saint Augustin.* Paris, 1918.
AMANN, E. 'Novatien et Novatianisme.' *D.T.C.* vol. xi.
ANDRÉADÈS, A. 'Deux livres récents sur les finances byzantines.' *B.Z.* vol. xxviii, 1928.
ANICHKOV, E. Язычество и древняя Русь. St Petersburg, 1914.
—— 'Les survivances manichéennes en pays Slaves et en Occident.' *R.E.S.* vol. viii. 1928.
ARNOLD, G. *Unpartheyische Kirchen- und Ketzer-Historien.* Schaffhausen, 1740–2.
ASBÓTH, J. *An Official Tour through Bosnia and Herzegovina.* London, 1890.
BALASCHEV, G. Климентъ епископъ словѣнски и службата му по старъ словѣнски прѣводъ. Sofia, 1898.
BANDURI, A. *Imperium Orientale.* Venetiis, 1729.
BARDY, G. 'Montanisme.' *D.T.C.* vol. x.
—— *Paul de Samosate.* Louvain, 1923.
BAREILLE, G. 'Borboriens.' *D.T.C.* vol. ii.
—— 'Encratites.' *D.T.C.* vol. v.
—— 'Euchites.' *D.T.C.* vol. v.
—— 'Gnosticisme.' *D.T.C.* vol. vi.
BENOIST, J. *Histoire des Albigeois et des Vaudois.* Paris, 1691.
BENVENISTE, E. *The Persian Religion according to the chief Greek Texts.* Paris, 1929.
BESHEVLIEV, V. Първобългарски надписи. *G.S.U.* (Faculty of philology and history), vol. xxxi, pt 1. 1934.
BIDEZ, J. 'Michel Psellus.' *Catalogue des manuscrits alchimiques grecs*, vol. vi. Bruxelles, 1928.
BLAGOEV, N. P. История на старото българско държавно право. Sofia, 1906.
—— Беседата на Презвитер Козма против богомилите. *G.S.U.* (Faculty of Law), vol. xviii. 1923.
—— Произходъ и характеръ на царь Самуиловата държава. *G.S.U.* (Faculty of Law), vol. xx. 1925.
BOBCHEV, S. S. Римско и византийско право в старовремска България. *G.S.U.* (Faculty of Law), vol. xxi, 1925.
—— Симеонова България отъ държавно-правно гледище. *G.S.U.* (Faculty of Law), vol. xxii. 1926–7.
BONWETSCH, G. N. 'Paulicianer.' *R.E.* vol. xv.
BOUSSET, W. *Hauptprobleme der Gnosis.* Göttingen, 1907.
BROWNE, E. G. *A Literary History of Persia.* Cambridge, 1928.
BROZ, I. *Crtice iz hrvatske književnosti.* Zagreb, 1886, 1888.
BUCKLER, G. *Anna Comnena.* Oxford, 1929.
BURKITT, F. C. *The Religion of the Manichees.* Cambridge, 1925.
BURY, J. B. *A History of the Eastern Roman Empire.* London, 1912.
BUTLER, Dom E. C. *Monasticism. C.M.H.* vol. i.

Cambridge Economic History (The), vol. I, ed. by J. H. Clapham and E. Power. Cambridge, 1941.
Cambridge Medieval History (The), vol. I, ed. by H. M. Gwatkin and J. P. Whitney; vol. IV, ed. by J. R. Tanner, C. W. Previté-Orton and Z. N. Brooke. Cambridge, 1911, 1927.
CASARTELLI, L. C. *La philosophie religieuse du Mazdéisme sous les Sassanides.* Paris, 1884.
CHADWICK, N. K. *The Beginnings of Russian History.* Cambridge, 1946.
CHALANDON, F. *Essai sur le règne d'Alexis Ier Comnène.* Paris, 1900.
—— *Jean II Comnène et Manuel Ier Comnène.* Paris, 1912.
CHARLESWORTH, M. P. *Trade-Routes and Commerce of the Roman Empire.* 2nd ed. Cambridge, 1926.
CHAVANNES, E. and PELLIOT, P. 'Un traité manichéen retrouvé en Chine, traduit et annoté.' *J.A.* 1911, 1913.
CHRISTENSEN, A. 'A-t-il existé une religion zurvanite?' *Le Monde Oriental*, vol. XXV. Uppsala, 1931.
—— *L'Iran sous les Sassanides.* Copenhagen, 1936.
CLÉDAT, L. *Le Nouveau Testament traduit au XIIIe siècle en langue provençale, suivi d'un Rituel Cathare.* Paris, 1887.
COBHAM, C. D. *The Patriarchs of Constantinople.* Cambridge, 1911.
ĆOROVIĆ, V. Богомили. Народна енциклопедија српско-хрватско-словеначка, ed. S. Stanojević, vol. I. Zagreb, 1925.
CUMONT, F. 'La date et le lieu de la naissance d'Euthymios Zigabénos.' *B.Z.* vol. XII. 1903.
—— *La Cosmogonie manichéenne d'après Théodore bar Khôni.* (*Recherches sur le Manichéisme*, vol. I.) Bruxelles, 1908.
—— 'La propagation du manichéisme dans l'Empire romain.' *Revue d'histoire et de littérature religieuses*, vol. I. Paris, 1910.
—— *Les religions orientales dans le paganisme romain*, 4th ed. Paris, 1929.
DANIČIĆ, DJ. Рјечник из књижевних старина српских, vol. I. Belgrade, 1863.
—— *Rječnik hrvatskoga ili srpskoga jezika*, vol. I. Zagreb, 1880.
DARMESTETER, J. *Ormazd et Ahriman, leurs origines et leur histoire.* Paris, 1877.
DHALLA, M. N. *History of Zoroastrianism.* New York, 1938.
Dictionnaire de Théologie Catholique, ed. A. Vacant and E. Mangenot. Paris, 1909– .
DIEHL, C. *Byzance, grandeur et décadence.* Paris, 1919.
DÖLLINGER, I. VON. *Beiträge zur Sektengeschichte des Mittelalters.* München, 1890.
DONDAINE, A. *Un traité néo-manichéen du XIIIe siècle: Le 'Liber de Duobus Principibus', suivi d'un fragment de rituel Cathare.* Rome, 1939.
DOROSIEV, L. I. Българскитѣ колонии въ Мала Азия. *S.B.A.N.* (Dep. of hist. and philol.) vol. XXIV. 1922.
DRINOV, M. S. Южные славяне и Византия въ X векe. Moscow, 1876.
—— Исторически прѣгледъ на българската църква. Sofia, 1911.
DU CANGE, C. *Glossarium ad scriptores mediae et infimae Graecitatis*, vol. I. Lugduni, 1688.
DUCHESNE, L. *Histoire ancienne de l'Église.* Paris, 1906–10.
DUFOURCQ, A. *Étude sur les Gesta Martyrum romains.* Paris, 1900.
DUJČEV, I. 'Il cattolicesimo in Bulgaria nel sec. XVII' (*Orientalia Christiana Analecta*, no. 111). Rome, 1937.
DVORNÍK, F. *Les Slaves, Byzance et Rome au IXe siècle.* Paris, 1926.

DVORNÍK, F. 'Deux inscriptions gréco-bulgares de Philippes.' *Bulletin de Correspondance Hellénique de l'École Française d'Athènes*. Paris, 1928.
—— 'Les Légendes de Constantin et Méthode vues de Byzance.' *Byzantinoslavica*, Supplementa I. Prague, 1933.
—— 'The Authority of the State in the Œcumenical Councils.' *The Christian East*, vol. XIV, no. 3. London, 1933.
—— *National Churches and the Church Universal*. London, 1944.
ENGELHARDT, J. G. V. *Kirchengeschichtliche Abhandlungen*. Erlangen, 1832.
FARLATUS, D. *Illyricum sacrum*. Venetiis, 1751–1819.
FAYE, E. DE. *Gnostiques et Gnosticisme*. Paris, 1913.
FILIPOV, N. Произходъ и сѫщность на богомилството. *B.I.B.* vol. III, Sofia, 1929.
FLICHE, A. and MARTIN, V. *Histoire de l'Église*, vol. VII. Paris, 1940.
FLORINSKY, T. К вопросу о богомилах. *S.L.* St Petersburg, 1883.
FLÜGEL, G. *Mani, seine Lehre und seine Schriften*. Leipzig, 1862.
FRACASSINI, U. 'I nuovi studi sul manicheismo.' *G. Soc. Asiat. Ital.* (n.s.) vol. I. Florence, 1925.
FREEMAN, E. A. *The History of the Norman Conquest of England*. Oxford, 1867–71.
FRESHFIELD, E. H. *Roman Law in the later Roman Empire*. Cambridge, 1932.
FRIEDRICH, J. 'Der ursprüngliche bei Georgios Monachos nur theilweise erhaltene Bericht über die Paulikianer.' *S.B. bayer. Akad. Wiss.* (philos.-phil. hist. Kl.). München, 1896.
GASTER, M. 'Bogomils.' *Encyclopaedia Britannica*, 11th ed. vol. IV.
—— *Ilchester Lectures on Greeko-Slavonic Literature*. London, 1887.
GEDEON, M. I. Πατριαρχικοὶ Πίνακες: Εἰδήσεις ἱστορικαὶ βιογραφικαὶ περὶ τῶν πατριαρχῶν Κωνσταντινουπόλεως, 36–1884. Constantinople, 1891.
GENOV, M. Пресвитеръ Козма и неговата Беседа противъ богомилството. *B.I.B.* vol. III. 1929.
GIBBON, E. *The History of the Decline and Fall of the Roman Empire*, ed. J. B. Bury. London, 1909–14.
GIESELER, J. K. L. 'Untersuchungen über die Geschichte der Paulicianer.' *Theologische Studien und Kritiken*. Hamburg, 1829, pt 1.
—— *Lehrbuch der Kirchengeschichte*. Bonn, 1835–49.
—— 'Über die Verbreitung christlich-dualistischer Lehrbegriffe unter den Slaven.' *Theologische Studien und Kritiken*. Hamburg, 1837, pt 2.
GILFERDING, A. История сербов и болгар: Собрание сочинений, vol. I. St Petersburg, 1868.
GOLUBINSKY, E. История русской церкви. Moscow, 1901–11.
—— Краткий очерк истории православных церквей болгарской, сербской и румынской. Moscow, 1871.
GRÉGOIRE, H. 'Autour des Pauliciens.' *Byzantion*, vol. XI. Paris, 1936.
—— 'Les sources de l'histoire des Pauliciens: Pierre de Sicile est authentique et "Photius" un faux.' *Bull. Acad. Belg.* (classe des lettres), vol. XXII. Bruxelles, 1936.
—— 'Sur l'histoire des Pauliciens.' *Bull. Acad. Belg.* (classe des lettres), ibid.
GUDEV, P. T. Български рѫкописи въ библиотеката на лордъ Zouche. *S.N.U.* vol. VIII. 1892.
GUIRAUD, J. 'Le Consolamentum Cathare.' *R.Q.H.* vol. LXXV. 1904.
—— *Histoire de l'Inquisition au Moyen Âge*. Paris, 1935–8.

HAGEMANN, H. *Die Römische Kirche und ihr Einfluss auf Disciplin und Dogma in den ersten drei Jahrhunderten*. Freiburg, 1864.
HARNACK, A. *Lehrbuch der Dogmengeschichte*, 4th ed. vol. I. Tübingen, 1909.
—— *Marcion: das Evangelium vom Fremden Gott*, 2nd ed. Leipzig, 1924.
—— 'Monarchianismus.' *R.E.* vol. XIII.
—— *Die Mission und Ausbreitung des Christentums in den ersten drei Jahrhunderten*, 4th ed. Leipzig, 1924.
HAUG, M. *Essays on the sacred language, writings and religion of the Parsis*, 3rd ed. by E. W. West. London, 1884.
HEFELE, C. and LECLERCQ, H. *Histoires des Conciles*. Paris, 1907– .
HERGENRÖTHER, J. *Photius, Patriarch von Constantinopel*. Regensburg, 1867–9.
HEUSSI, K. *Der Ursprung des Mönchtums*. Tübingen, 1936.
HOLLARD, A. *Deux hérétiques: Marcion et Montan*. Paris, 1935.
HUSSEY, J. M. *Church and Learning in the Byzantine Empire* (867–1185). Oxford, 1937.
ILIĆ, J. A. *Die Bogomilen in ihrer geschichtlichen Entwicklung*. Sr. Karlovci, 1923.
История русской литературы. Published by the Institute of Literature of the Academy of Sciences of the U.S.S.R.; ed. by P. I. Lebedev-Polyansky and others; vol. I: Литература XI–начала XIII века; ed. by A. S. Orlov and others. Moscow, 1941.
IVANOV, I. Сѣверна Македония. Sofia, 1906.
—— Българитѣ въ Македония, 2nd ed. Sofia, 1917.
—— Св. Иванъ Рилски и неговиятъ монастиръ. Sofia, 1917.
—— Богомилски книги и легенди. Sofia, 1925.
—— Български старини изъ Македония, 2nd ed. Sofia, 1931.
JACKSON, A. V. W. *Zoroastrian Studies*. New York, 1928.
—— *Researches in Manichaeism*. New York, 1932.
JAGIĆ, V. История сербско-хорватской литературы. *Uchenye zapiski imperatorskogo Kazanskogo Universiteta*. Kazan, 1871.
JANIN, R. 'Pauliciens.' *D.T.C.* vol. XII.
JIREČEK, K. *Geschichte der Bulgaren*. Prague, 1876.
—— *Die Heerstrasse von Belgrad nach Constantinopel und die Balkanpässe*. Prague, 1877.
—— История болгар. Odessa, 1878.
—— 'Die Handelsstrassen und Bergwerke von Serbien und Bosnien während des Mittelalters.' *Abh. böhm. Ges. Wiss.* VI. Folge, 10. Band. Prague, 1879.
—— *Cesty po Bulharsku*. Prague, 1888.
—— *Das Fürstentum Bulgarien*. Prague, 1891.
—— *Geschichte der Serben*, vol. I. Gotha, 1911.
—— 'Staat und Gesellschaft im mittelalterlichen Serbien.' *Denkschr. Akad. Wiss. Wien*, vol. LVI, pt 2. Vienna, 1912.
—— *La Civilisation Serbe au Moyen Âge*. Paris, 1920.
JUGIE, M. 'Phoundagiagites et Bogomiles.' *Échos d'Orient*, t. XII. Paris, 1909.
—— 'Palamas' and 'Palamite (Controverse)'. *D.T.C.* vol. XI.
KALAIDOVICH, K. Иоанн, Ексарх болгарский. Moscow, 1824.
KALOGERAS, N. 'Αλέξιος Α' ὁ Κομνηνός, Εὐθύμιος ὁ Ζιγαβηνός καὶ οἱ αἱρετικοὶ Βογομίλοι. *Athenaion*, vol. IX. Athens, 1880.
KAROLEV, R. За Богомилството. *P.S.* vol. III. Braila, 1871.

KAZAROW, G. 'Die Gesetzgebung des bulgarischen Fürsten Krum.' *B.Z.* vol. XVI. 1907.
KEMP, P. *Healing Ritual: Studies in the Technique and Tradition of the Southern Slavs.* London, 1935.
KERN, Archimandrite C. 'Les éléments de la théologie de Grégoire Palamas.' *Irénikon*, vol. XX, pts 1–2. Chevetogne, 1947.
KESSLER, K. 'Mani.' *R.E.* vol. XII.
KIPRIANOVICH, G. Жизнь и учение богомилов по Паноплии Евфимия Зигабена и другим источникам. *P.O.* vol. II. July 1875.
KISELKOV, V. S. Патриархъ Евтимий (животъ и обществена дейность). *B.I.B.* vol. III. 1929.
—— Житието на св. Теодосий Търновски като исторически паметникъ. Sofia, 1926.
KLAIĆ, V. *Geschichte Bosniens von den ältesten Zeiten bis zum Verfalle des Königreiches.* Leipzig, 1885.
KLINCHAROV, I. Попъ Богомилъ и неговото време. Sofia, 1927.
KONDAKOV, N. P. О манихействе и богумилах. *Seminarium Kondakovianum*, vol. I. Prague, 1927.
KRIVOSHEIN, B. Аскетическое и богословское учение св. Григория Паламы, Athos, 1935. *Seminarium Kondakovianum*, vol. VIII, Prague, 1936. English tr., 'The Ascetic and Theological Teachings of Gregory Palamas.' *The Eastern Churches Quarterly*, vol. III, nos. 1–4. London, 1938.
KRUMBACHER, K. *Geschichte der byzantinischen Litteratur*, 2nd ed. München, 1897.
LABRIOLLE, P. DE. *La crise montaniste.* Paris, 1913.
LAMANSKY, V. I. О славянах в Малой Азии, в Африке и в Испании. *Uchenye zapiski Vtorogo Otdeleniya Imperatorskoy Akademii Nauk*, vol. V. St Petersburg, 1859.
LAMBECIUS, P. *Commentaria de augustissima Bibliotheca Caesarea Vindobonensi*, 2nd ed. vol. V. Vindobonae, 1778.
LAURENT, J. 'L'Arménie entre Byzance et l'Islam depuis la conquête arabe jusqu'en 886.' *Bibliothèque des Écoles Françaises d'Athènes et de Rome*, fasc. 117. Paris, 1919.
LAVRIN, J. 'The Bogomils and Bogomilism.' *S.R.* vol. VIII. 1929.
LE COQ, A. VON. 'A short account of the origin, journey and results of the first Royal Prussian expedition to Turfan in Chinese Turkestan.' *J.R.A.S.* 1909.
LÉGER, L. 'L'Hérésie des Bogomiles.' *R.Q.H.* vol. VIII. 1870.
—— *La Mythologie Slave.* Paris, 1901.
LEGGE, F. 'Western Manichaeism and the Turfan Discoveries.' *J.R.A.S.* 1913.
LE QUIEN, M. *Oriens Christianus.* Paris, 1740.
LEVITSKY, V. Богомильство—болгарская ересь. *Kh.Ch.* 1870, pt 1.
LIMBORCH, P. *Liber Sententiarum Inquisitionis Tholosanae.* Amstelodami, 1692.
LOMBARD, A. *Pauliciens, Bulgares et Bons-Hommes en Orient et en Occident.* Geneva, 1879.
—— *Constantin V, empereur des Romains.* Paris, 1902.
LUCHAIRE, A. *Innocent III et la Croisade des Albigeois.* Paris, 1905.
MAKARY, Metropolitan of Moscow. История русской церкви, 2nd ed. St Petersburg, 1866–83.
MANSIKKA, V. *Die Religion der Ostslaven*, FF Communications, no. 43. Helsinki, 1922.
MARSIGLI, L. F. *Stato militare dell' Imperio Ottomanno.* Haya, 1732.

MARTIN, E. J. *A History of the Iconoclastic Controversy*. London, 1930.
MELIORANSKY, B. К истории противоцерковных движений в Македонии в XIV веке: Στέφανος: *Sbornik statey v chest'* F. F. Sokolova. St Petersburg, 1895.
MILETICH, L. Заселението на католишкитѣ българи въ Седмиградско и Банатъ. *S.N.U.* vol. XIV. 1897.
MILLER, K. *Itineraria Romana: Römische Reisewege an der Hand der Tabula Peutingeriana*. Stuttgart, 1916.
MILLET, G. 'La Religion Orthodoxe et les hérésies chez les Yougoslaves.' *R.H.R.* vol. LXXV. 1917.
MISHEW, D. *The Bulgarians in the Past*. Lausanne, 1919.
MKRTTSCHIAN. *See* Ter-Mkrttschian.
MONTAGU, LADY MARY WORTLEY. *Letters and Works*, ed. by Lord Wharncliffe. London, 1893.
MONTFAUCON, B. DE. *Palaeographia Graeca*. Parisiis, 1708.
MÜLLER, F. W. K. 'Handschriften-Reste in Estrangelo-Schrift aus Turfan, Chinesisch-Turkistan.' (I) *S.B. preuss. Akad. Wiss.* 1904; (II) *Abh. preuss. Akad. Wiss.* 1904. Berlin.
MURKO, M. *Geschichte der älteren südslawischen Litteraturen*. Leipzig, 1908.
NEANDER, A. *Allgemeine Geschichte der christlichen Religion und Kirche*. 3rd ed. Gotha, 1856.
NERSOYAN, T. 'The Paulicians.' *The Eastern Churches Quarterly*, vol. V, no. 12. London, 1944.
NIEDERLE, L. *Slovanské Starožitnosti: Život starých Slovanů*. Prague, 1911–25.
—— *Manuel de l'Antiquité Slave*. Paris, 1923, 1926.
NOVAKOVIĆ, S. Охридска Архиепископија у почетку XI века. *Glas Srpske Kraljevske Akademije*, vol. LXXVI. Belgrade, 1908.
NYBERG, H. S. 'Questions de cosmogonie et de cosmologie mazdéennes.' *J.A.* vol. CCXIV, 1929; vol. CCXIX, 1931.
—— 'Forschungen über den Manichäismus.' *Z. Neutestamentliche Wiss. und die Kunde der älteren Kirche*, vol. XXXIV. Giessen, 1935.
—— *Die Religionen des alten Iran*. Leipzig, 1938.
OBOLENSKY, D. 'The Bogomils.' *The Eastern Churches Quarterly*, vol. VI, no. 4. London, 1945.
OECONOMOS, L. *La Vie religieuse dans l'Empire Byzantin au temps des Comnènes et des Anges*. Paris, 1918.
OEDER, J. L. *Dissertatio inauguralis prodromum historiae Bogomilorum criticae exhibens*. Gottingae, 1743.
ORLOV, A. S. Древняя русская литература. Moscow, Leningrad, 1945.
OSOKIN, N. A. История Альбигойцев. Kazan, 1869.
OSTROGORSKY, G. *Studien zur Geschichte des byzantinischen Bilderstreites*. Breslau, 1929.
—— 'Agrarian conditions in the Byzantine Empire in the Middle Ages.' *C.E.H.* vol. I.
PALAUZOV, S. N. Век болгарского царя Симеона. St Petersburg, 1852.
PANCHENKO, B. A. Памятник славян в Вифинии VII века. *I.R.A.I.K.* vol. VIII. 1902.
PANOV, I. Ересь жидовствующих. *Zh.M.N.P.* vols. CLXXXVIII, CLXXXIX, CXC. Jan.-March, 1877.

PAPADOPOULOS-KERAMEUS, A. Βογομιλικά. *V.V.* vol. II. 1895.
PAPARRHEGOPOULOS, K. Ἱστορία τοῦ ἑλληνικοῦ ἔθνους. Athens, 1860–72.
PETIT, L. 'Le Monastère de N.-D. de Pitié.' *I.R.A.I.K.* vol. VI. Sofia, 1900.
—— 'Typicon de Grégoire Pacurianos pour le monastère de Pétritzos (Bačkovo) en Bulgarie.' *V.V.* vol. XI, Supplement no. 1. 1904.
—— 'Arménie.' *D.T.C.* vol. I.
PETRANOVIĆ, B. Богомили, Црьква Босаньска и крьстяни. Zara, 1867.
PETROVSKY, A. Анафема. *Pravoslavnaya Bogoslovskaya Entsiklopediya*, ed. A. P. Lopukhin, vol. I. St Petersburg, 1900.
POGODIN, A. История Болгарии. St Petersburg, 1910.
POLOTSKY, H. J. 'Manichäismus.' Pauly-Wissowa, *Real-Encyclopädie der classischen Altertumswissenschaft*, Supplementband VI. Stuttgart, 1935.
POPOWITSCH, M. 'Bogomilen und Patarener. Ein Beitrag zur Geschichte des Sozialismus.' *Die Neue Zeit*, 24. Jahrg. 1. Bd. Stuttgart, 1905.
POPRUZHENKO, M. G. Козма Пресвитер. *I.R.A.I.K.* vol. XV. Sofia, 1911.
—— Синодик царя Бориса. *I.R.A.I.K.* vol. V, Supplement. 1900.
—— Козма Пресвитер, болгарский писатель X века. Sofia, 1936. (Български Старини, vol. XII.)
PORFIRIEV, I. Уа., Апокрифические сказания о ветхозаветных лицах и событиях по рукописям Соловецкой библиотеки. *Sbornik otdeleniya russkogo yazyka i slovesnosti Imperatorskoy Akademii Nauk*, vol. XVII, no. 1. St Petersburg, 1877.
—— Апокрифические сказания о новозаветных лицах и событиях по рукописям Соловецкой библиотеки. Ibid. vol. LII, no. 4. St Petersburg, 1890.
PORFIRY (USPENSKY), Bishop. История Афона. Часть III. Афон монашеский. Отд. 2. St Petersburg, 1892.
Православная богословская энциклопедия, ed. A. P. Lopukhin. St Petersburg, 1900–11.
PROHASKA, D. *Das kroatisch-serbische Schrifttum in Bosnien und der Herzegowina.* Zagreb, 1911.
PUECH, H.-C. and VAILLANT, A. *Le traité contre les Bogomiles de Cosmas le prêtre.* Paris, 1945.
PYPIN, A. N. История русской литературы, 3rd ed. vol. I. St Petersburg, 1907.
PYPIN, A. N. and SPASOVICH, V. D. История славянских литератур, 2nd ed. vol. I. St Petersburg, 1879.
RAČKI, F. 'Bogomili i Patareni.' *Rad Jugoslavenske Akademije Znanosti i Umjetnosti*, vols. VII, VIII, X. Zagreb, 1869–70.
RADCHENKO, K. Религиозное и литературное движение в Болгарии в эпоху перед турецким завоеванием. *Universitetskie Izvestiya*. Kiev, 1898.
—— Этюды по богомильству. Народные космогонические легенды славян в их отношении к богомильству. *Izvestiya otdel. russk. yazyka i slovesnosti Imperat. Akad. Nauk*, vol. XV, pt 4. St Petersburg, 1910.
RADLOFF, W. *Chuastuanit, das Bussgebet der Manichäer.* St Petersburg, 1909.
RAMBAUD, A. *L'Empire Grec au Xme siècle.* Paris, 1870.
RAMSAY, W. M. *The Historical Geography of Asia Minor.* London, 1890.
—— *The Cities and Bishoprics of Phrygia.* Oxford, 1897.
ROUILLARD, G. 'Une étymologie (?) de Michel Attaliate.' *Rev. de philol. litt. et d'hist. anciennes*, 3ᵉ série, vol. XVI. Paris, 1942.

Rovinsky, P. Материал для истории богумилов в сербских землях. *Zh.M.N.P.* vol. ccxx. March, 1882.
Runciman, S. *The Emperor Romanus Lecapenus and his Reign.* Cambridge, 1929.
—— *A History of the First Bulgarian Empire.* London, 1930.
—— *The Medieval Manichee. A Study of the Christian Dualist Heresy.* Cambridge, 1946.
Šafařík, P. J. *Slovanské Starožitnosti.* Prague, 1837.
—— *Památky hlaholského písemnictví.* Prague, 1853.
Salemann, C. 'Ein Bruchstük manichaeischen Schrifttums im asiatischen Museum.' *Zapiski imperatorskoy akademii nauk* (ist.-fil. otd.), vol. vi. St Petersburg, 1904.
Sathas, C. and Legrand, E. *Les Exploits de Digénis Akritas.* Paris, 1875.
Schaeder, H. H. 'Bardesanes von Edessa in der Überlieferung der griechischen und der syrischen Kirche.' *Zeitschrift für Kirchengeschichte,* vol. li. Gotha, 1932.
—— 'Der Manichäismus nach neuen Funden und Forschungen.' *Morgenland. Darstellungen aus Geschichte und Kultur des Ostens,* Heft xxviii. Berlin, 1936.
—— 'Urform und Fortbildungen des manichäischen Systems.' *Vorträge der Bibliothek Warburg.* Leipzig, Berlin, 1924-5.
Schlumberger, G. *Un Empereur Byzantin au Xe siècle, Nicéphore Phocas,* 2nd ed. Paris, 1923.
—— *L'Épopée Byzantine à la fin du Xe siècle.* Paris, 1896.
Schmidt, C. *Histoire et doctrine de la secte des Cathares ou Albigeois.* Paris, 1849.
Schmidt, C. and Polotsky, H. J. 'Ein Mani-Fund in Ägypten. Originalschriften des Mani und seiner Schüler.' *S.B. preuss. Akad. Wiss.* (philos.-hist. Kl.). Berlin, 1933.
Schnitzer. 'Die Euchiten im 11. Jahrhundert.' *Studien der evangelischen Geistlichkeit Württembergs,* vol. xi, pt 1. Stuttgart, 1839.
Sharenkoff, V. N. *A study of Manichaeism in Bulgaria with special reference to the Bogomils.* New York, 1927.
Šidak, J. 'Problem "bosanske crkve" u našoj historiografiji od Petranovića do Glušca (Prilog rješenju t. zv. bogumilskog pitanja)', *Rad,* vol. cclix. 1937.
Sokolov, I. Состояние монашества въ византийской церкви с половины IX до начала XIII века. Kazan, 1894.
Sokolov, M. Материалы и заметки по старинной славянской литературе, vol. i. Moscow, 1888.
Speransky, M. N. История древней русской литературы. 3rd ed. Moscow, 1921.
Spiegel, F. *Erânische Alterthumskunde.* Leipzig, 1871-8.
Spinka, M. *A History of Christianity in the Balkans.* Chicago, 1933.
Spulber, C. A. *L'Éclogue des Isauriens.* Cernautzi, 1929.
Stanimirov, S. История на българската църква. Sofia, 1925.
Stoop, E. de. 'Essai sur la diffusion du manichéisme dans l'Empire romain.' *Rec. Univ. Gand,* 38e fasc. 1909.
Svoboda, K. 'La Démonologie de Michel Psellos.' *Spisy Filosofické Fakulty Masarykovy University v Brně,* no. 22. Brno, 1927.
Tafel, G. L. F. *De Via Romanorum militari Egnatia, qua Illyricum, Macedonia et Thracia iungebantur.* Tübingen, 1837.
—— *De Thessalonica ejusque agro.* Berlin, 1839.

Tafel, G. L. F. 'Symbolarum criticarum, geographiam Byzantinam spectantium, partes duae.' *Abh. bayer. Akad. Wiss.* (hist. Kl.), vol. v, Abt. 2, pars 1. München, 1849.
Takela, D. E. Нѣкогашнитѣ павликяни и сегашнитѣ католици въ Пловдивско. *S.N.U.* vol. xi. 1894.
—— 'Les anciens Pauliciens et les modernes Bulgares catholiques de la Philippopolitaine.' *Le Muséon*, vol. xvi. Louvain, 1897.
Temperley, H. W. V. *History of Serbia*. London, 1917.
Ter-Mkrttschian, K. *Die Paulikianer im byzantinischen Kaiserreiche und verwandte ketzerische Erscheinungen in Armenien.* Leipzig, 1893.
Tournebize, F. *Histoire politique et religieuse de l'Arménie.* Paris, 1900.
Trifonov, Yu. Бесѣдата на Козма Пресвитера и нейниятъ авторъ. *S.B.A.N.* (Dep. of hist. and philol.) vol. xxix. 1923.
Tsukhlev, D. История на българската църква, vol. i. Sofia, 1910.
Tunitsky, N. L. Св. Климент, епископ словенский. Sergiev Posad, 1913.
Uspensky, Th. Значение византийской и южнославянской пронии. *S.L.* St Petersburg, 1883.
—— К истории землевладения в Византии. *Zh.M.N.P.* vol. ccxxv. Feb. 1883.
—— Богословское и философское движение в Византии XI и XII веков. *Zh.M.N.P.* vol. cclxxvii. Sept. 1891.
Vaillant, A. and Lascaris, M. 'Date de la conversion des Bulgares.' *R.E.S.* vol. xiii, fasc. 1, 2. 1933.
Vasiliev, A. A. Византия и Арабы. *Zapiski ist.-fil. fakulteta imperatorskogo S.-Peterburgskogo Universiteta*, vol. lxvi. 1902.
—— 'On the Question of Byzantine Feudalism.' *Byzantion*, vol. viii. 1933.
—— *Byzance et les Arabes*: vol. i, *La Dynastie d'Amorium 820–67* (French ed. by H. Grégoire and M. Canard); vol. iii, *Die Ostgrenze des byzantinischen Reiches von 363 bis 1071* (by E. Honigmann). Bruxelles, 1935.
Vasilievsky, V. Византия и печенеги (1048–94). *Zh.M.N.P.* vol. clxiv. 1872.
Veselovsky, A. N. Славянскіе сказания о Соломоне и Китоврасе и западные легенды о Морольфе и Мерлине: Собрание Сочиненій, vol. viii, pt i. Petrograd, 1921.
Vignier, N. *Recueil de l'histoire de l'Église.* Leyden, 1601.
Villecourt, Dom L. 'La date et l'origine des "Homélies spirituelles" attribuées à Macaire.' *C.R. Acad. Inscriptions Belles-Lettres.* Paris, 1920.
Vogt, A. *Basile Ier, empereur de Byzance (867–86), et la civilisation Byzantine à la fin du IXe siècle.* Paris, 1908.
Waddell, H. *The Wandering Scholars*, 7th ed. London, 1942.
Waldschmidt, E. and Lentz, W. 'Die Stellung Jesu im Manichäismus.' *Abh. preuss. Akad. Wissensch.* (phil.-hist. Kl.). Berlin, 1926.
Wellnhofer, M. *Die Thrakischen Euchiten und ihr Satanskult im Dialoge des Psellos*: Τιμόθεος ἢ περὶ τῶν δαιμόνων. *B.Z.* vol. xxx. 1929–30.
Wesendonk, O. G. von. 'Bardesanes und Mani.' *Acta Orientalia*, vol. x. Leyden, 1932.
Wolf, J. C. *Historia Bogomilorum.* Vitembergae, 1712.
Zachariae von Lingenthal, K. E. *Jus Graeco-Romanum.* Lipsiae, 1856–84.
Zervos, C. *Un philosophe néoplatonicien du XIe siècle: Michel Psellos.* Paris, 1919.

ZLATARSKI, V. N. Какви канонически книги и граждански закони Борисъ е получилъ отъ Византия. *Letopis na Bŭlgarskata Akademiya na Naukite*, vol. 1. Sofia, 1914.
—— *Geschichte der Bulgaren.* Leipzig, 1918.
—— История на българската държава прѣзъ срѣднитѣ вѣкове, vol. 1, pts 1, 2. Sofia, 1918, 1927.
—— Сколько бесед написал Козма Пресвитер? *Sbornik statey v chest' M. S. Drinova.* Kharkov, 1904.

INDEX

Abel, son of Adam, 208
Abstinentia, 216
Achaia (in Greece), 36 n.
'Achaia', Paulician Church of, 36
Achilleus, St, 150
Acmonia, 174; *see also* Euthymius of Acmonia
Acyndinus, Gregory, Byzantine philosopher, 263
Adam, 138, 180, 208, 227–8, 239; see also *Story of Adam and Eve*
Adoptionism, 41 n., 53, 55–6
Adrian, Russian heretic, 277
Adrian Comnenus, 196 n.
Adrianople, 61, 65, 231–2, 257
Adriatic Sea, 153
Aegean Sea, 38
Africa, 8
Agapius, Manichaean, 25–6, 44 n.; *see also* Aristocritus
Ahriman, 11–12, 14
Albania, 63
Albigenses, Albigensian Crusade, *see* Cathars
Alexander the Smith, Bulgarian heretic, 240
Alexandria, 22
Alexiopolis, *see* Neocastrum
Alexius I, Comnenus, Emperor, 30 n., 31 n., 174 n., 190–6, 199, 203–5, 206 n., 219, 232, 236, 249 n., 275–6
Allegory, use of, 26, 40, 42, 130–1, 212, 217–18
Anchialus (in Thrace), 82
Angelarius, St, disciple of St Methodius, 88
Anna Comnena, 146, 161, 189, 191, 193–206, 222, 233, 238, 275–6
Antalya, Gulf of, 177
Anthony, St, 24
Antichrist, 129, 133 n., 278
Antioch, 17, 50, 55–7, 116, 147, 203 n., 236 n.
Aphrigiy, Bulgarian heretic, 240
Apocalypse of Baruch, 128 n., 154 n., 186 n.
Apocrypha, 155, 156 n., 186, 272–4, 281–3
Arabs, 13 n., 15 n., 29, 35, 37, 56 n., 59, 83, 193 n.

Ararat, Mount, 52
Arcadiopolis, 61
Argaoun (Argovan), 36–7
Arianism, *see* Arius
Aristocritus, Manichaean, 26 n., 43 n., 44 n.; *see also* Agapius
Arius, Arianism, 22, 26 n.
Armenia, Armenians, 9, 17, 22, 27–30, 32–7, 41–3, 45–6, 48, 50–3, 55–6, 57 n., 59–62, 69 n., 70 n., 79–83, 94, 146–7, 149 n., 150 n., 153, 167, 189, 223, 224 n., 232
Arnulf, King of Germany, 87 n.
Arsenius, Higumen, 102 n.
Asceticism, 4, 13, 19, 20–5, 44, 47, 50, 95, 105, 114, 127–9, 141, 144–5, 187, 199, 201, 214–17, 219, 221, 248, 251, 258, 260, 264, 288
Asen, family of, 231
Ashot, Prince of Taron, 150 n.
Asia Minor, 3, 8, 17, 19–22, 26 n., 27–8, 30, 35–7, 39, 45, 48, 50, 70 n., 82, 94, 139 n., 160, 167, 174–7, 181, 183, 187, 196, 219–20, 222
Asoghic, Armenian historian, 147, 149 n.
Asperukh, Khan of Bulgaria, 63–4, 67
Athanasius of Jerusalem, monk, 271–2
Athens, 36 n.
Athos, Mount, 163 n., 249 n., 253–7, 260
Atlantic Ocean, 5, 8, 27, 287
Attaliates, Michael, Byzantine historian, 190, 192
Atticus, Patriarch of Constantinople, 52
Aturpat, Magian High Priest, 13–14
Avars, 59
Avdin, Bulgarian heretic, 240
Azov, Sea of, 83 n.

Baanes, Paulician, 114 n.
Babuna, mountain (in Macedonia), 165
Babuna, river (in Macedonia), 165
Babuni, heretics, 164–5, 167, 285
Babylonia, 5, 7, 16
Bachkovo, monastery of, 103 n.
Balbissa, 220, 221 n.
Baldwin I, Latin Emperor of the East, 232
Baldwin II, Latin Emperor of the East, 250
Balsamon, Theodore, Patriarch of Antioch, 116 n., 128 n., 229

INDEX

Baptism:
  Bogomil view of, 129–30, 179, 181, 213–16, 228, 239–40
  Constantine Chrysomalus's view of, 219–20
  Marcionite view of, 47
  Massalian view of, 49, 240
  Paulician view of, 40
  recognized by *The Key of Truth*, 53
Barburiani, Armenian and Syrian heretics, 27
Bardaisan of Edessa, heretic, 16
Bardarius, servant of Niphon Scorpio, 256
Bardy, G., on Paul of Samosata, 56–7
Barhebraeus, Gregory, Syrian writer, 27
Barlaam, Byzantine theologian, 253, 263
Baruch, see *Apocalypse*
Basil I, the Macedonian, Emperor, 77 n., 90 n.
Basil II, Bulgaroctonus, Emperor, 147–8, 152 n., 160, 169–72, 174, 231
Basil, Bogomil, 145, 146 n., 240
Basil, leader of the Byzantine Bogomils, 146 n., 199–205, 206 n., 214, 217, 219, 222, 240, 243, 275–6
Basil, St, the Great, Archbishop of Caesarea, 17, 22, 103 n.
Basilides, Gnostic, 33
Bayazid I, Ottoman Sultan, 265
Bela IV, King of Hungary, 250
Belgrade, 153, 283 n.
Beloozero, 277
Belyatovo, 192
Berrhoea, 256
'Bethlehem', name of Bogomil community, 216
Bithynia, 175, 176 n.
Bitolj, 102, 153, 163 n., 196 n.
Blachernites, Byzantine heretic, 202
Black Drina, river, 167, 283 '
Black Sea, 8, 27, 83 n., 146, 234, 287
Bogomil, *pop*, heresiarch, 104 n., 117–20, 124–6, 133, 138–9, 145–6, 155 n., 182, 235, 238, 240, 243, 265, 271–3, 274 n.
Bogomil writings, viii, 128–9, 154–5, 156 n., 180 n., 186, 208–9, 210 n., 213 n., 226–8, 242, 282, 284 n., 289
Bogomili (in Macedonia), 165
Bogomilsko Polje, valley in Macedonia, 165
Bohemond I, of Taranto, Prince of Antioch, 163 n., 164 n., 190
Borborites, Armenian heretics, 52

Boril, Tsar of Bulgaria, 234–7, 248, 250;
  see also *Synodicon of the Tsar Boril*
Boris I, Khan of Bulgaria, 70–7, 79–80, 83–5, 87–8, 90–3, 96, 100, 102 n., 109, 151, 161
Bosnia, Bosnian, 8, 134 n., 158, 166, 229, 244–5, 265, 285, 287
Bosporus, the, 29, 176, 289
Bregalnitsa, bishopric of, 84 n.
Buddhism, 139 n.
Bulgaria, Bulgarians, viii, 8–9, 30, 38, Chapters III and IV *passim*, 168–73, 175–7, 181, 183, 186, 195–201, 203 n., 206, 207 n., 210 n., 212, 214, 217–20, 222–4, 226, 227 n., 229 n., Chapter VI *passim*, 268–74, 277, 281–4, 286–9
Bulgarian (Orthodox) Church, see Church, Bulgarian
Bulgarian literature, viii, 151, 154
Bulgars of the Volga, 83 n.
Byzantine (Orthodox) Church, see Church, Byzantine
Byzantine Empire, viii, 27–30, 37, 56, 59–65, 67, 69–74, 77, 79 n., 80, 82, 89, 90 n., 91–2, 94, 96–101, 104, 106, 108–10, 141, 146–8, 150–1, 153, 160, 163 n., Chapter V *passim*, 230–4, 252, 257–8, 267–70, 281, 284, 286
Byzantium, see Constantinople

Caesarea (in Cappadocia), 55
Cain, son of Adam, 208, 240
Callinice, of Samosata, Manichaean woman, 32, 43, 54, 57, 114 n.
Callistus, Patriarch of Constantinople, 255 n., 256–8, 261 n.
Calomena, daughter of Adam, 208
Cappadocia, 17, 19, 32, 50–1, 81, 221–2, 240
Carbeas, Paulician, 29
Cathars, Albigenses, vii, 9, 121 n., 134 n., 156–7, 162, 183 n., 215–16, 233 n., 234, 242–6, 279 n., 286–9
Cedrenus, George, Byzantine chronicler, 94, 147, 150, 184 n.
Chaldaean teaching, 187 n.
Chalybes, land of the, 146
Chernebog, pagan god of the Polabians, 68 n.
Chersonese, the (in Thrace), 184 n.
Childhood of Jesus, see *Gospel of St Thomas*
China, Chinese, 7, 16
Choreutes, see Massalians
Christ, see Jesus Christ

INDEX 307

Christopher Lecapenus, Emperor, 112 n.
Chrysochir, Paulician, 29–30, 38
Church, Bulgarian (Orthodox), 30, 71–9, 83–97, 103 n., 104 n., 109–10, 125, 132, 136, 138, 139 n., 141–2, 148–9, 151–2, 169–70, 234–50, 252, 257–64, 284
Church, Byzantine (Orthodox), vii, 28, 40, 62, 64–5, 72–3, 76–9, 82, 84, 90–2, 96, 111–17, 134 n., 135 n., 149, 151, 168–70, 172, 173 n., 188, 195, 219–22, 225–6, 233, 244, 248–50, 252–7, 263
Church, Roman (Catholic), Roman Catholicism, Roman Catholics, Western Church, 74, 76–9, 82, 86, 144 n., 233 n., 243–4, 250, 265–6, 287–8; see also Papacy
Church, Russian (Orthodox), 86, 272–4, 277–81
Church, Serbian (Orthodox), 284
Churches:
  attitude of Bogomils to, 122, 130, 142, 181, 202, 214, 239
  attitude of Bulgarian Jews to, 263
  attitude of Constantine Chrysomalus to, 220
  attitude of Russian 'Judaizers' to, 279
  attitude of Thonraki to, 53
Cibossa, 33–4, 36, 70 n.
Cibyrrhaeots, Theme of, 177
Cilicia, 36
Cimbalongus, pass (in Bulgaria), 152 n.
Cinamon, Greek slave, 65 n.
Civil disobedience, 137–8, 140–1, 220, 248
Clement, St, disciple of St Methodius, 88–94, 110, 154–6
Clement of Sosandra, heretic, 220–2
Cologne, 288
Colonea (in Armenia Minor), 33–6, 70 n.
Colonia (Staria) (in Macedonia), 70 n.
'Colossae', Paulician Church of, 36–7
'Comitopuli', 148, 152
Comnena, see Anna
Confession of sins, 133, 135, 220
Conrad of Marburg, Papal inquisitor in Germany, 246
Consolamentum, 216
Constans II, Emperor, 32
Constantine the Great, Emperor, 194
Constantine IV, Pogonatus, Emperor, 28, 34

Constantine V, Copronymus, Emperor, 60–2, 64 n., 80, 94, 131 n., 146, 176 n., 214 n.
Constantine IX, Monomachus, Emperor, 174
Constantine, Bishop, Bulgarian writer, 89 n.
Constantine, St, see Cyril, St
Constantine Chrysomalus, Byzantine heretic, 137 n., 219–20
Constantine ('Silvanus') of Mananali, Paulician, 32–5, 43, 45, 48, 114 n.
Constantinople (Byzantium), 8, 28, 30, 35, 40, 52, 61–2, 64, 67, 69 n., 70, 72, 73 n., 82, 89, 90–2, 102 n., 153, 156, 158, 162, 166, 173–4, 176, 183, 184 n., 190–1, 195, 197 *et seq.*, 214 n., 217, 219–22, 225–6, 231, 243–6, 250, 253, 255–7, 259, 267, 271, 275–6, 288–9
Constantinople, Councils of, see Councils
Contemplation, 44, 50, 95, 144, 253–4, 256
Conybeare, F. C., 30 n., 36 n., 41 n., 53 n., 55
Corinth, 36
'Corinth', Paulician Church of, 36
Cosmas II, Atticus, Patriarch of Constantinople, 221–2
Cosmas, Bulgarian priest, 99 n., 104–8, 110, 117–45, 153 n., 154 n., 156, 178, 179 n., 180, 182, 185, 199, 202, 206 n., 212, 218–20, 227 n., 237–9, 241, 248–9, 268–70, 274, 281
Council of Ephesus (Third Oecumenical Council, 431), 51
Council of Gangra (*c.* 330), 22–3
Council of Trnovo (1211), 235–7, 240, 247–8, 250, 258, 261 n., 264
Councils of Constantinople (869–70), 65 n., 77; (879), 158 n.
Councils of Nicaea (First Oecumenical Council, 325), 115–16; (Seventh Oecumenical Council, 787), 249 n.
Croatia, Croatian, 164, 246
Cross (the):
  attitude of Bogomils to, 122, 130, 135, 142, 150–1, 179, 181, 211, 214, 221, 239, 262, 282
  attitude of Jeremiah to, 271–4
  attitude of Paulicians to, 40, 224 n.
  attitude of Thonraki to, 53
  See also *Legend of the Cross*
Crusades, Crusaders, vii, 9, 158, 163 n., 164 n., 197, 231–3, 250

20-2

Culeon, Paulician, 190 n., 194–5
Cumans (Polovtsi), 172, 189–90, 192–3, 197, 231, 250, 276
Cusinus, Paulician, 194–5
Cynochorion (in the Pontus), 36
Cyril, St (Constantine), Apostle of the Slavs, 79 n., 89–90, 93 n., 288
Cyril Bosota, Bogomil, 260–2
Cyrillic alphabet, 90
Cyzicus (in Asia Minor), 26 n.

Dalmatia, Dalmatian, 159 n., 245–6
Danube, river, 61, 63, 90 n., 189, 192–3, 231, 283
Dawkins, Prof. R. M., 35 n.
Debar (in Macedonia), 167
Debritsa and Velitsa, bishopric of, 88 n., 156
Demetrius Chomatianus, archbishop of Ochrida, 165 n., 166 n.
Demiurge, the, 4–5, 19, 209, 213
Demonology, demons, 49–50, 66, 129 n., 130, 145, 186–7, 202, 213–14, 221, 240, 252, 254, 262; *see also* Devil
Denkart ('Acts of the Zoroastrian Religion'), 13–14
Develtus (in Thrace), 82
Devil, the (Satan), 53, 64 n., 68 n., 95, 118 n., 122–31, 135, 138, 142, 145, 162, 180, 186, 207–8, 209 n., 220, 227–8, 239–40, 277–8, 281–2, 288; *see also* Demonology, Satanael
Devol (in Macedonia), 88
Diblatius, Bogomil, 199
Dioclea, 93, 149 n., 150 n., 283
Diocletian, Roman Emperor, 60
Dmitr, Russian heretic, 277 n.
Dobromir, Paulician, 190
Dobry, Bogomil, 145, 240
Docetism, 6 n., 25 n., 38, 46, 56, 113, 126, 139 n., 210–11, 224 n., 238, 242–3, 288
Dorylaeum (in Asia Minor), 175 n.
Dragovichi, Dregovichi, 158–60, 165 n., 166 n.
Dragovitsa, river, 158
'Dragovitsan Church', *see* 'Ecclesia Dugunthiae'
Dregovichi, *see* Dragovichi
Dristra, archbishop of, 84 n.
Dualism, dualists, vii, 3–6, 8–14, 16–21, 23–7, 38, 40–1, 44, 46–8, 51–3, 56, 58, 68, 69 n., 78 n., 95, 105, 113–15, 122–5, 127–9, 134 n., 139–40, 141 n., 143–6, 154, 155 n., 156, 161–2, 179 n., 180, 184, 192, 193 n., 202, 207, 208 n., 209 n., 211, 214, 224 n., 227–8, 234, 245–6, 248, 254, 261, 262 n., 267, 274, 279, 281, 283, 284 n., 286–7, 288–9
Dugunthia, Dugrutia, Drogometia, *see* 'Ecclesia Dugunthiae'
Duks, Bulgarian monk, writer, 89 n., 96 n.
Dušan, Stephen, *see* Stephen Dušan
Dyrrhachium (Durazzo), 153

Eastern Orthodox Church, *see* Church (Bulgarian, Byzantine, Russian, Serbian)
'Ecclesia Bulgariae', 157–62, 242–5, 288–9
'Ecclesia Dugunthiae', 157–63, 166 n., 245, 274 n., 288
'Ecclesia Latinorum de Constantinopoli', 157–8
'Ecclesia Melenguiae', 156–7, 245
'Ecclesia Philadelphiae in Romania', 157–60
'Ecclesiae Sclavoniae', 158
Ecloga, code of Leo III and Constantine V, 73
Edessa, 16 n., 17, 50
Egypt, 7–9, 17, 24, 26 n.
Elijah, Prophet, 228
Elucidarium, 154 n., 186 n., 209 n.
Encratites, Encratism, 19–22, 51
England, English, 190 n., 233 n.
Enoch, 228
Enoch, Book of, 154 n., 186 n., 228 n., 282
Enravotas, Bulgarian prince, 65 n.
Enthusiasts, *see* Massalians
Epaphroditus, companion of St Paul, 36 n.
Ephesus, 29
Ephesus, Council of, *see* Council
'Ephesus', Paulician Church of, 36
Ephraim, St, of Edessa, 16 n., 17
Epirus, 63
Ermenrich, bishop of Passau, 76
Eschatology:
Bogomil, 181–2, 215, 228, 229 n., 241
Manichaean, 5
Zoroastrian, 11–12, 14–15
See also Last Judgement, Resurrection of the body
Esoteric teaching, 33, 44, 215 n.

INDEX 309

Eucharist:
  attitude of Bogomils to, 130–1, 179, 181, 203, 214, 239, 262
  attitude of Marcionites to, 47
  attitude of Massalians to, 49
  attitude of Paulicians to, 40, 113
  recognized by *The Key of Truth*, 53
  See also Liturgy
Euchitae, *see* Massalians
Eunomius, bishop of Cyzicus, 26 n.
Euphrates, river, 32, 36, 55
Eustathius of Sebaste, 22–4
Eustratius, metropolitan of Nicaea, 194
Euthymius, Patriarch of Trnovo, 164, 223
Euthymius of Acmonia, monk of Peribleptos, 119, 174–83, 188, 196, 208 n., 222, 228 n., 241
Euthymius Zigabenus, *see* Zigabenus
Eve, 208, 227–8, 239–40; see also *Story of Adam and Eve*
Eznik of Kolb, Armenian bishop, 15 n., 17, 46 n.

Feudalism, 97–101, 138, 172
Finns, Finnish, 278
Flavian, Patriarch of Antioch, 50, 94, 203 n., 236 n.
Folklore, vii, 66 n., 155, 267, 282
Formosus, bishop of Porto, later Pope, 76
France, French, vii, 8–9, 134 n., 162, 163 n., 183 n., 216, 233 n., 234, 242, 286–9
Frankish Empire, Franks, 70–1, 79 n., 231, 233, 272
Frederick I, Barbarossa, Western Emperor, 232
Free will, 3–4, 12
French, *see* France

Gabars, 14 n.
Gabriel-Radomir, Tsar of Bulgaria, 149 n.
Gangra, Council of, *see* Council
Garianus, priest, accused of Bogomilism, 258–9
Gaul, 8
Gegnesius ('Timothy'), Paulician, 36, 40–1, 114 n.
Genesis, Book of, 179 n., 208–9, 227
Gennadius Scholarius, Patriarch of Constantinople, 166
Gennady, archbishop of Novgorod, 279–81
George, St, of Iberia, 110 n.

George Monachus, Byzantine chronicler, 31, 42
Germanicea, 61
Germanos II, Patriarch at Nicaea, 222 n., 241
Germany, Germans, vii, 70, 76–7, 88, 92 n., 99 n., 163 n., 232, 246
Gibbon on the Bogomils, vii–viii
Glycas, Michael, Byzantine chronicler, 200
Gnostics, Gnosticism, viii, 3–4, 11–12, 13 n., 15–16, 19–21, 52, 139 n., 209 n., 286
Gordoserba (in Asia Minor), 175 n.
Goritsa (in Thessaly), 163 n.
Gospel of St Thomas, 154 n., 282
Greece, 63 *et passim*
Grégoire, Prof. H., on the Paulicians, 30–1, 36, 41 n., 43 n.
Gregoras, Nicephorus, Byzantine historian and theologian, 253–6
Gregory IX, Pope, 250
Gregory, Bulgarian monk, writer, 89 n.
Gregory, St, of Sinai, 253–4, 257
Gregory of Narek, Armenian theologian, 51
Gregory Pacurianus, *see* Pacurianus
Gregory Palamas, St, *see* Palamas

Hadrian II, Pope, 77 n.
Haemus, passes of the, 147
Harmenopulus, Constantine, Byzantine writer, 206 n.
Harnack, A., 21, 45, 47
Hebrew language, 91
Hellespont, 184 n.
Helmold of Lübeck, chronicler, 68, 69 n.
Heracleia (in Thrace), metropolitan of, 258
Heraclius, Emperor, 60, 65
Herod, King of Judaea, 216–17
Herzegovina, *see* Hum
Hesychasm, hesychasts, 252–9, 263
Hilarion, St, bishop of Moglena, 164, 223–6, 229, 238 n.
Hincmar, archbishop of Rheims, 74
Historia Manichaeorum, *see* Peter of Sicily
Hum (Herzegovina), 229, 285
Hungary, 246, 250; *see also* Magyars
Huns, 63

Ibar, river, 283
Iconoclasm, Iconoclasts, 28, 41 n., 102 n., 131 n., 140, 166, 214 n., 249 n.

310   INDEX

Icons, images:
  attitude of Bogomils to, 130-1, 142, 181, 214, 221, 239, 262-3
  attitude of Bulgarian Jews to, 263
  attitude of Jeremiah to, 273
  attitude of Massalians to, 255 n.
  attitude of Paulicians to, 41
  attitude of Russian 'Judaizers' to, 279
  attitude of Thonraki to, 53
Ignatius, St, Patriarch of Constantinople, 73 n., 77
Illyricum, 76
Images, see Icons
Incarnation, doctrine of the, 2-3, 26, 38, 44, 46, 56, 113, 140, 211, 215; see also Docetism, Jesus Christ
India, 16
Innocent III, Pope, 231, 233 n., 234
Ioasaf, archbishop of Rostov and Yaroslavl, 280
Iran, Iranian, see Persia
Irene of Thessalonica, Bogomil nun, 255
Isaac II, Angelus, Emperor, 231 n.
Isaac Comnenus, the Sebastocrator, 203
Isaiah, Prophet, 208, 210; see also *Vision of Isaiah*
Isauria, 160 n.
Isidore, bishop of Seville, 92 n.
Islam, see Moslems
Italus, John, 202 n.
Italy, Italians, vii, 8, 134 n., 154 n., 162, 163 n., 216, 242, 287-9
Ivan III, Grand Prince of Moscow, 280

Jacobites, Syrian Monophysites, 61, 189
Janina, 163 n.
Jeremiah, *pop*, Bulgarian heretic, 163 n., 271-4, 282
Jerusalem, 103 n., 107, 176, 214 n.
Jesus Christ:
  Bogomil conception of, 122-3, 126, 130, 134-5, 138, 139 n., 207 n., 210-13, 215-17, 227-8, 238, 242, 288
  Gnostic conception of, 139 n.
  Jeremiah's conception of, 273
  Manichaean conception of, 6, 25
  Marcionite conception of, 139 n.
  Paul of Samosata's conception of, 56
  Paulician conception of, 38, 40, 56, 113
  Russian 'Judaizers'' conception of, 279
  See also Docetism, Incarnation

Jews, Jewish, 41, 47, 83, 90 n., 130, 186 n., 209, 221, 241, 259, 262-4, 278-80; see also Judaism, 'Judaizers'
John I, Tzimisces, Emperor, 146, 148, 151, 172, 188, 268
John II, Comnenus, Emperor, 219, 249 n.
John Alexander, Tsar of Bulgaria, 252, 257-8, 262-3
John Asen II, Tsar of Bulgaria, 234, 250, 252
John, bishop of Otzun, Catholicos of Armenia, 51
John, Bulgarian priest, 269
John, Bulgarian prince, 107, 109
John IX, Patriarch of Constantinople, 276
John, St, the Baptist, 67, 129, 215, 218, 228, 239, 247
John, St, the Evangelist, 227, 242 n.
John, son of Callinice, Manichaean, 32, 43, 54, 57, 114 n.
John Chrysostom, St, 17, 57, 81 n.
John Damascene, St, 51
John of Rila, St, 103
John the Exarch, Bulgarian writer, 89 n., 95, 96 n., 110 n., 118 n., 122, 185, 268-70
John-Vladimir of Dioclea, St, see Vladimir, St, Prince of Dioclea
John-Vladislav, Tsar of Bulgaria, 149 n., 150 n.
Joseph, archbishop of Bulgaria, 87 n.
Joseph ('Epaphroditus'), Paulician, 36, 114 n.
Joseph, St, abbot of Volokolamsk, 279-80
Judaea, 8, 176 n.
Judaism, 2, 13, 44 n., 56, 208 n., 241, 262-4, 278-81; see also Jews
'Judaizers', Russian sect of, 278-81
Justinian II, Emperor, 27-8, 34-5, 176 n.
Justus, Paulician, 35

Kaloyan, Tsar of Bulgaria, 231-4, 250
Kardam, Khan of Bulgaria, 61
Kastoria (in Macedonia), 70 n., 163 n., 164 n., 191
Key of Truth, The, 41 n., 53
Khazars, 83 n., 84 n.
Khilandar, monastery of, 249 n.
Khorasan, 193 n.
Khrabr, Bulgarian monk, writer, 89 n., 92
Kiev, 277 n., 278, 282
Kiev monastery of the Caves, 278

INDEX 311

Kiev (Kievo-Pechersky) Paterik, 278
Kiliphar (near Trnovo), 257
Killing, Bogomil attitude to, 182, 190, 218
Kishevo (in Macedonia), 165 n., 167
Kormisosh, Khan of Bulgaria, 63
Kosara (Theodora ?), Bulgarian princess, 149 n., 150 n.
Kosovo Polje, 283
Kotugeri (in Macedonia), 166
Krum, Khan of Bulgaria, 61, 64, 67, 71, 82, 128 n.
Kudugeri, heretics, 166–7, 267
Kustendil (in Macedonia), 166
Kutugertsi (in Macedonia), 166

Laodicea (in Phrygia), 36
'Laodicea', Paulician Church of, 36
Larissa (in Thessaly), 150
Last Judgement, 26, 181; see also Eschatology
Latifundia, 97–8, 171
Latin Empire of Constantinople, 158, 160, 232, 250
Lazarus, Bogomil, 260–2
Lazarus, opponent of Theophylact of Ochrida, 196 n.
Lecapenus, see Romanus I, Emperor
Lecus, Paulician, 189–90, 192
Legend of the Cross, 271–4, 282
Leo III, the Isaurian, Emperor, 40
Leo IV, the Khazar, Emperor, 61, 146
Leo, 'the Montanist', 39
Leo Stypes, Patriarch of Constantinople, 219
Leontius of Balbissa, heretic, 220–2
Liber Sancti Johannis (*Liber Secretus*, Secret Book), 154 n., 180 n., 208–9, 213 n., 226–8, 242–5, 282, 289
Liturgy:
 attitude of Bogomils to the, 130, 134, 142, 239
 Eustathius of Sebaste and the, 22
 Russian 'Judaizers' and the, 280
 See also Eucharist
Lombardy, 157, 226, 243, 288–9
Louis I, the Pious, Emperor of the West, 70 n.
Louis II, the German, King of Germany, 70, 74–6
Luke, Bulgarian heretic, 240
Lycaonia, 19, 28, 50–1
Lycia, 19, 50
Lydia, 19, 160 n.

Macarius the Egyptian, St, 48 n.
Macedonia, 33, 62–3, 65, 70 n., 82, 83 n., 84 n., 85, 88, 102–3, 109, 110 n., 118 n., 147–8, 150–70, 173, 176–7, 178 n., 189, 193, 196–7, 200 n., 222–6, 229, 234, 241, 255, 259, 263, 265 n., 283–4, 288
'Macedonia', Paulician Church of, 34, 36, 70 n.
Magnaura, school of the, in Constantinople, 89
Magyars, 104, 106, 268; see also Hungary
Malamir, Khan of Bulgaria, 65 n.
Maliunaie, Armenian and Syrian heretics, 27
Mananali (in Armenia), 32, 36
Mandeley of Radobol, Bulgarian heretic, 240
Mani, Manichaeism, Manichaeans, vii, 5–27, 31–3, 37, 40, 42–8, 50–1, 53, 57–8, 69 n., 78 n., 80–1, 95, 112, 114–15, 119, 125, 129, 134 n., 136 n., 178 n., 184–5, 189, 193 n., 198, 200–201, 215, 223–4, 238, 246 n., 287, 289
Manichaean writings, 17–18, 25–6, 32, 43, 44 n.
Manual labour:
 attitude of Bogomils to, 136–7, 143, 254, 262
 attitude of Jeremiah to, 273
 attitude of Massalians to, 50, 254
Manuel I, Comnenus, Emperor, 220, 223, 225, 229
Manuel, archbishop of Adrianople, 65
Marcion, Marcionism, Marcionites, 16, 19, 39 n., 45–8, 50, 53, 58, 139 n., 209 n., 259
Marcus, Italian Patarene, 288
Maria (Irene) Lecapena, Tsaritsa of Bulgaria, 97, 111–12
Marriage, sexual intercourse, 4, 6 n., 13, 19–20, 22, 26, 44, 47, 78 n., 105, 114, 126 n., 127–9, 130 n., 139 n., 140, 141 n., 144, 214, 218, 221, 228, 248, 260, 288
Martyrs, 23 n., 134, 214 n.
Mary, Virgin, Mother of God:
 attitude of Bulgarian Jews to, 263
 Bogomil view of, 126–7, 131 n., 140, 210–11, 215, 228, 238, 242
 Jeremiah's view of, 273
 Paulician view of, 38, 40, 53, 113, 126, 224 n.

Massalians (Choreutes, Enthusiasts, Euchitae, Messalians), Massalianism, ix, 17, 19, 21, 48–52, 58, 93–5, 102, 105–6, 110–11, 114–18, 120 n., 125, 128–9, 137–8, 139 n., 143–5, 149 n., 150 n., 164, 178, 182–8, 198, 201 n., 202, 203 n., 206, 213, 218–22, 238, 240–1, 242 n., 251–7, 259–60, 262, 264–5, 280–1

Matter, doctrine of, 3–6, 20–1, 40–1, 46 n., 114–15, 127, 129, 134 n., 140, 182, 211, 214, 239, 241, 248

Meat, eating of, 6 n., 19–20, 22, 26, 44, 47, 127–8, 129 n., 144, 214, 218, 221, 248

Melitene, 29, 36–7, 50, 60, 94

Melnik (in Macedonia), 156–7; *see also* 'Ecclesia Melenguiae'

Mesembria, 190, 257

Mesopotamia, vii, 9, 17, 27, 32, 50

Messalians, *see* Massalians

Methodius, St, archbishop of Pannonia, Apostle of the Slavs, 79 n., 88–91, 93, 288

Methuselah, 228 n.

Michael I, Rhangabe, Emperor, 28, 37

Michael III, the Drunkard, Emperor, 29 n., 71

Michael IV, the Paphlagonian, Emperor, 171

Michael VII, Ducas, Emperor, 170 n.

Michael, Archangel, 210, 281

Michael, Bogomil, 145–6, 240

Michael, Bulgarian prince, 107, 109

Michael II, Kurkuas, Patriarch of Constantinople, 220

Millet, G., on the Bogomils, 138

Miracles, Bogomil view of, 131, 221

Miroslava, Bulgarian princess, 150 n.

Modalist Monarchianism, *see* Sabellius

Moesia, 63, 65, 160–1

Moglena, 164, 193, 223–6

Mohammed II, Ottoman Sultan, 267

Mojmir, Prince of Moravia, 70

Monasteries, monasticism, monks, 13, 22–5, 42, 44, 50–1, 95, 101–8, 115, 128–9, 132, 137, 148, 156, 174–5, 182, 183 n., 199, 221, 252–6, 258, 263, 279

Monophysites, 60, 61 n., 80, 82–3, 189 n., 223, 225; *see also* Jacobites

Montanus, Montanism, Montanists, 20–21, 39, 50–1

Mopsuestia (in Cilicia), 36

Morava, bishopric of, 84 n.

Morava, river, 63, 283

Moravia, Moravians, 70–1, 79 n., 82, 88–9, 90 n., 92 n.

Mosaic Law, 26, 47, 113, 127, 179 n., 209, 212, 218, 224 n., 239; *see also* Moses

Moscow, 272, 279–80

Moses, Bogomil, 240

Moses, Prophet, 53, 209, 228, 273; *see also* Mosaic Law

Moslems, Islam, 83, 84 n., 167, 193 n., 265–7

Mysticism, 49, 202, 215, 220, 252–4; *see also* Hesychasm, Massalians

Nadim (An-), Arab writer, 7 n., 17

Naum, St, disciple of St Methodius, 88–90, 93, 110 n., 156 n.

Nazarius, Bogomil, 226, 242–5, 289

Nemanja, *see* Stephen Nemanja

Neocaesarea, 36, 80

Neocastrum (Alexiopolis), 195, 232

Neo-Manichaeism, 8–10, 15–16, 18 n., 25–7, 44 n., 48 *et passim*

Neoplatonists, Neoplatonic, 26, 187 n.

Nestorians, 18 n.

New Testament, *see* Testaments

Nicaea, 29, 175 n., 194, 222 n., 250

Nicaea, Councils of, *see* Councils

Nicephorus I, Emperor, 28

Nicephorus II, Phocas, Emperor, 106 n.

Nicephorus, Patriarch of Constantinople, 60

Nicephorus Bryennius, husband of Anna Comnena, 194, 196 n.

Nicetas Choniates, Byzantine historian, 206 n., 232

Nicetas of Constantinople, heretical 'bishop', 156, 158, 159 n., 162, 243, 245–6, 288–9

Nicholas, father of the 'Comitopuli', 148

Nicholas, St, monastery of, in Constantinople, 219

Nicholas Mysticus, Patriarch of Constantinople, 97 n.

Nicholas III, the Grammarian, Patriarch of Constantinople, 204, 275–6

Nicholas I, Pope, 75–7, 78 n., 79 n., 80, 83, 85

Nicomedia, 29

Nikita, St, bishop of Novgorod, 278

Nikon's Chronicle, 277

Nile, river, 24

INDEX 313

Nilus, monk, 202 n.
Niphon, Bogomil, 221–2
Niphon Scorpio, accused of Bogomilism, 256
Niš (Nish), 153, 189–90, 283
Noah, 209
Normans, 170, 190–1, 197
Novatians, Novatianism, 21, 51, 150 n.
Novgorod, 278–81

Oaths, *see* Swearing
Ochrida (Ohrid), 73 n., 78 n., 84 n., 88–90, 102, 148, 153, 156, 163 n., 165 n., 167, 169–71, 172 n., 192, 196 n., 197, 200 n., 284
Old Testament, *see* Testaments
Omortag, Khan of Bulgaria, 61, 64–5, 70 n., 87 n.
Opsikion, Theme of, 174–7
Ormazd, 11–12, 14–15
Orthodox Church, *see* Church (Bulgarian, Byzantine, Russian, Serbian)
Osrhoene, 50
Our Lady, *see* Mary, Virgin

Pachomius, St, 103 n.
Pacific Ocean, 5
Pacurianus, Gregory, founder of Bachkovo monastery, 103 n.
Paganism, pagans, 44 n., 65–72, 74–5, 79 n., 83, 85–8, 95, 110, 144, 186, 247, 259–60, 264, 272
Palamas, Gregory, St, archbishop of Thessalonica, 253–4, 257, 259
Palea, 273, 281–2
Pamphylia, 19, 50, 94
Pank, correspondent of Athanasius of Jerusalem, 271
Pannonia, 88–9
Panoplia Dogmatica, *see* Zigabenus
Panteleimon, St, monastery of, 88–9, 102 n.
Papacy, the, 74–7, 233 n., 234–5, 250; *see also* Church, Roman
Paphlagonia, 17, 22
Papikion, Mount, monastery of, 259
Paraclete, the, 20, 25, 37, 44, 53
Paradise, 128–9, 180, 227, 273, 282
Paraoria (near Adrianople), 257
Paristrium, Theme of, 160
Parsis, 14 n.
Passau, 76
Patarenes, vii, 134 n., 154 n., 157, 162, 226, 243–6, 265, 285–9

Patrocinium (prostasia), 98, 100–1
Patzinaks, *see* Pechenegs
Paul, bishop of Populonia, Papal legate, 76
Paul, St, the Apostle, 19, 33–6, 37 n., 39, 41, 46–7, 49, 53–5, 57 n., 70 n., 131, 182, 183 n., 209 n., 266 n.
Paul, son of Callinice, Manichaean, 32, 43, 54, 57, 114 n.
Paul of Samosata, heretical bishop of Antioch, 55–7, 115–16
Paul the Armenian, Paulician, 114 n.
Paulicians, Paulicianism, ix, 17–18, 20 n., 22, 25–7, Chapter II *passim*, 59–62, 65–6, 68–70, 74, 78–83, 91, 93–5, 102, 105, 109–19, 124–5, 129–30, 131 n., 132–3, 135 n., 136 n., 138–9, 143–4, 146–7, 149 n., 152–3, 158–9, 163 n., 161–2, 163 n., 164 n., 167, 169, 172, 178, 179 n., 181 n., 182–4, 188–96, 198, 201 n., 202, 206–7, 209 n., 211 n., 218, 223–5, 231–3, 236, 238, 251, 262 n., 266, 271, 274 n., 276
Pechenegs (Patzinaks), 104, 106, 170, 172, 189–90, 192–3, 197, 268
Pelagonia (in Macedonia), 163 n., 164 n., 196 n.
Peloponnesus, 63
Περὶ κτίσεως κόσμου καὶ νόημα οὐράνιον ἐπὶ τῆς γῆς, 186 n., 210 n.
Peribleptos, monastery in Constantinople, 174
Persecution:
of Bogomils, 116, 141–3, 153, 196, 203–5, 219, 225, 236–7, 241, 244, 251, 264, 275–6, 284
of Borborites, 52
of Bulgarian Jews, 241, 263
of Massalians, 50
of Paulicians, 28–9, 42, 191, 194–5
of Russian 'Judaizers', 279
Persia, Persians (Iran, Iranians), 3 n., 5, 7, 9–10, 13 n., 15–16, 27, 95 n.
Peter the Abbot, writer, 31
Peter, St, the Apostle, 42, 53, 266 n.
Peter, Bogomil, 145, 240
Peter of Cappadocia, Bogomil 'dyed' of Sredets, 240, 242–5
Peter of Sicily, Byzantine ambassador, 29–33, 35, 37–45, 47–8, 54, 57, 59, 70 n., 81, 113–15, 123, 125–6, 181 n.
Peter, Tsar of Bulgaria, 91, 96–113, 116–20, 125, 136, 138 n., 143, 148, 151–3, 161, 169, 238, 268–9

Petracus, Bogomil, 289
'Philippi', Paulician Church of, 36
Philippians, 36 n.
Philippopolis, 62, 84 n., 144 n., 146, 149 n., 158–61, 188–95, 232–3, 266, 276
Philotheus, Patriarch of Constantinople, 259
Pholus, Paulician, 194–5
Photin, Bulgarian heretic, 240
Photius, St, Patriarch of Constantinople, 25–6, 31, 36 n., 44 n., 51, 72–3, 77, 78 n., 80–1, 102 n.
Phrygia, 19–21, 28, 36, 174, 183 n.
Phundagiagitae (Phundaitae), heretics, 167, 177–83
Plato, 3
Pliska, 76, 85, 87–8, 98 n., 99 n.
Pogodin's Nomocanon, 271
Polabian Slavs, 68 n., 69 n.
Polog, region of (in Macedonia), 165 n., 166 n.
Polovtsi, *see* Cumans
Pomaks (Moslem Bulgarians), 166–7, 265 n.; see also *Torbeshi*
Pontus, 33, 36
Porphyry, 187 n.
Poverty:
  Bogomil view of, 136–8, 177–8, 190, 262
  Manichaean view of, 13
  Massalian view of, 50, 255, 262
  Zoroastrian view of, 13
Prayer, prayers:
  attitude of Bogomils to, 120 n., 134–5, 138, 145, 179, 182–3, 215–17, 239–42, 254, 262, 288
  attitude of Massalians to, 49, 240, 254
Preslav, 85, 87, 89, 97, 102 n., 152
Prespa, 150
Priesthood, priests:
  attitude of Bogomils to, 132–3, 140, 143, 179, 214, 239
  attitude of Bulgarian Jews to, 263
  attitude of Jeremiah to, 273
  attitude of Paulicians to, 41, 53
  attitude of Russian 'Judaizers' to, 279, 281
Prilep, 163 n., 165
Proclus, 187 n.
Procopius, Byzantine historian, 66, 69 n.
Propontis, the, 38
Prostasia, see *Patrocinium*
Provadia, bishopric of, 84 n.

Psellus, Michael, Byzantine philosopher and historian, 94, 123 n., 183–8, 196, 211, 242 n.
Pskov, 278
Pyrenees, 234, 289

Raš (Rash), 283 n.
Raška (Rashka), Rascia, 283–4
Rastislav, Prince of Moravia, 70, 79 n.
Rastko, *see* Sava
Relics:
  attitude of Bogomils to, 130–1, 150–1, 214
  attitude of Bulgarian Jews to, 263 n.
  attitude of Russian 'Judaizers' to, 279
Rendina, gulf of, 163 n.
Renier of Trit, Duke of Philippopolis, 232–3
Renjdane, Slavonic tribe, 163 n.
Resurrection of the body, 12, 26, 134 n., 181–2, 228, 241, 263 n.; *see also* Eschatology
Rhine, river, 8
Rhodope Mountains, 152, 157, 167
Rila Monastery, 103 n.
Rila Mountains, 103
Ritual of the Bogomils, 135, 182–3, 215–17, 219–20, 238–9, 241–2, 244, 288
Robert Guiscard, Duke of Apulia, 190
Roman Catholic Church, *see* Church, Roman
Roman Empire, ancient, 6, 8, 16–17, 18 n.
'Romania', 159–60
Romanus I, Lecapenus, Emperor, 97, 111
Romanus III, Argyrus, Emperor, 174
Rome, 8, 107, 153, 234, 266 n.
Rome, Church of, *see* Church, Roman
Rostov, 277, 280
Russia, Russians, 63, 67, 83 n., 86, 90, 104, 106, 118, 146, 148, 155 n., 234, 268, 272–4, 277–83
Russian Church, *see* Church, Russian
Russian Primary Chronicle, 277–8

Sabbas, St, monastery of, 103 n.
Sabellius, Sabellianism (Modalist Monarchianism), 212
Sabin, Khan of Bulgaria, 64 n.
Sacchoni, Reinerius, Italian Patarene, later Dominican, 157–63, 242–5, 289

INDEX

Sacraments, 24, 40, 44, 47, 51, 53, 129, 134, 140, 142, 145, 181, 214, 239, 254, 279; *see also* Baptism, Eucharist, Marriage
Sahak, Catholicos of Armenia, 52
Saint-Félix de Caraman, dualist Council of, 156, 245–6, 288
St Sophia, church of, at Constantinople, 214 n.
Saints:
  attitude of Bogomils to, 131, 179, 214
  attitude of Bulgarian Jews to, 263 n.
  attitude of Russian 'Judaizers' to, 279
Samosata, 32, 54, 57; *see also* Paul of Samosata
Samuel, Tsar of Bulgaria, 147 n., 148–51, 153, 160, 168–9, 171 n., 233, 269, 283 n.
Samuel of Ani, Armenian historian, 17–18
Santabarenus, Theodore, archbishop of Euchaita, 80–1
Santabarenus, father of Theodore Santabarenus, accused of Manichaeism, 80–1
Šar Planina, *see* Shar Planina
Sardica (Sofia, Sredets), 62, 84 n., 164, 189–90, 240, 242–5, 283 n.
Sassanians (Sassanids), 15, 27
Satan, *see* Devil
Satanael, 129, 162, 185–7, 199, 201 n., 203, 207–10, 213, 240, 273, 281–2; *see also* Devil
Sava, St (Rastko), archbishop of Serbia, 284
Scylitzes, Joannes, Byzantine chronicler, 190
Scythianus, heretic, 114 n.
Sea of Tiberias, The, 128 n., 154 n., 186 n., 209 n., 210 n., 213 n., 282
Secret Book, see *Liber Sancti Johannis*
Sembat, Thonraki, 52
Serbia, Serbian, 8, 90, 162 n., 165 n., 175 n., 229, 249 n., 252, 283–5, 287
Serbian Church, *see* Church, Serbian
Sergius ('Tychicus'), Paulician, 20 n., 29, 35–7, 39, 53, 114 n.
Sermon against the Heretics, see Cosmas, Bulgarian priest
Sexual immorality, 50, 52, 95, 186–7, 199, 201, 242 n., 251–2, 258, 260, 262, 264
Sexual intercourse, *see* Marriage
Shahapivan, Synod of, 51
Shamanism, 67

Shar Planina, mountain range in Macedonia, 165 n., 167, 283
Shutil, Paulician missionary, 81
Sidor the Frank, Bulgarian heretic, 272
Silvanus (Silas), companion of St Paul, 33–4
Sinai, 166, 209, 253–4
Sisinnius II, Patriarch of Constantinople, 271
Sitnica, river, 283
Sivas, 29
Skoplje, 102, 160, 167, 283
Slav, Slavonic, Slavs, 59, 63–72, 77 n., 79 n., 88, 95, 110 n., 160, 175–6, 249–50, 265 *et passim*
Slavonic liturgy, 88, 91–2
Smyrna, 175, 177
Sofia, *see* Sardica
Solovki, monastery of, 272, 282
Sosandra, 220, 221 n.
Spain, 8
Sredets, *see* Sardica
Staria, *see* Colonia
Stephen, Bogomil, 145, 240
Stephen, priest, Bogomil, 260–2
Stephen Dušan (Dushan), Tsar of Serbia, 252, 284–5
Stephen Nemanja (St Symeon), Grand Župan of Serbia, 236, 237 n., 284
Stephen the First-Crowned, King of Serbia, 237 n., 284
Stephen V, Pope, 80
Stoop, E. de, 22, 26
Story of Adam and Eve, 154 n.
'Strigolniki', Russian heretics, 278–9
Struma, river, 157
Studion, monastery of, Studite Rule, 103 n.
Stylianus, bishop of Neocaesarea, 80–1
Subotin, Paulician missionary, 81
Sunday fast, 22, 134
Sursubul, George, Regent of Bulgaria, 108
Svantovit, pagan god of the Slavs, 69 n.
Svyatoslav, Grand Prince of Kiev, 268
Swearing (oaths), Bogomil attitude to, 218
Symeon, archbishop of Thessalonica, 166, 267
Symeon, Patriarch of Trnovo, 261 n.
Symeon ('Titus'), Paulician, 34–6, 114 n.
Symeon, Tsar of Bulgaria, 87–93, 95–7, 99–100, 102 n., 104, 106–9, 149 n., 151, 161, 250, 268–9

INDEX

Synodicon for the Sunday of Orthodoxy, 249
Synodicon of the Tsar Boril, 118–19, 126, 137 n., 145, 146 n., 164, 200, 206 n., 228 n., 235–49, 261
Syria, Syrians, 7–8, 16 n., 17, 27, 32, 50, 60–2, 80, 94, 120 n., 139 n., 146, 203 n.

Tatars, 252
Telerig, Khan of Bulgaria, 63
Telets, Khan of Bulgaria, 61, 63
Tephrice, 29–31, 38, 81
Terebinthus, heretic, 114 n.
Tervel, Khan of Bulgaria, 63, 71
Testaments, New Testament, 16, 32, 39, 44, 47, 49, 58, 68, 127, 130–1, 139–41, 155, 182, 206 n., 209 n., 212, 214 n., 215–18, 220, 278–9, 283, 288; Old Testament, 16, 26, 39, 44, 47, 127, 139 n., 155, 179 n., 209–10, 212, 214 n., 218, 224 n., 228, 239, 273, 278–9, 281, 283
Tetovo (in Macedonia), 165 n.
Theodora, Bulgarian princess, *see* Kosara
Theodora, Empress, wife of Theophilus, 28, 37
Theodore, Bogomil, 145, 240
Theodore, Patriarch of Antioch, 147
Theodore Balsamon, *see* Balsamon
Theodore bar Khonai, Syrian writer, 7 n.
Theodore Santabarenus, *see* Santabarenus
Theodore the Armenian, Paulician, 114 n.
Theodoret, bishop of Cyrus, 46 n.
Theodoret, Bulgarian heretic, 259–60
Theodosiopolis (Erzerum), 60
Theodosius II, Emperor, 52
Theodosius, Bulgarian heretic, 259–60
Theodosius, Patriarch of Trnovo, 257, 261 n.
Theodosius, St, of Trnovo, 255–65
Theophanes, Byzantine historian, 60, 80
Theophilus, Emperor, 28, 84 n.
Theophylact, Patriarch of Constantinople, 111–18, 119 n., 122–3, 125–6, 136, 143, 146, 198, 206, 238
Theophylact of Euboea, archbishop of Ochrida, 73 n., 78 n., 170–1, 172 n., 196 n., 197, 200 n.
Θεοτόκοι, 215–17, 229 n.
Thessalonica, 82, 83 n., 102, 153, 158, 163 n., 166, 241, 255–7, 263, 267, 283 n.
Thessaly, 163 n.

Thomas, rebel against Michael II, 69 n.
Thomas, St, Gospel of, see *Gospel of St Thomas*
Thonraki, Armenian heretics, 52–3
Thrace, 59–63, 69 n., 79–80, 82, 94, 110 n., 131 n., 145–7, 152–3, 158–61, 162 n., 167, 169, 172, 175–7, 184, 187–9, 190 n., 192–3, 196, 231–3, 242 n., 258–9, 274 n., 288
Thracesian Theme, 175
'Three languages heresy', 91–2
Tiberias, The Sea of, see *Sea of Tiberias, The*
Ticha, river, 89
Timok, river, 63
Timothy, St, disciple of St Paul, 19 n., 36 n.
Titus, St, disciple of St Paul, 34
Torbeshi, 166–7, 178 n., 265 n.; see also *Pomaks*
Toulouse, 156, 245
Tragurium, Trau, *see* Trogir
Traulus, Paulician, 191–2
Treskavats, monastery of, 165 n.
Trinity, the, 26, 49, 150–1, 182, 211–12, 215, 242, 279
Trnovo, 161, 164, 223, 230, 233 n., 235–6, 237 n., 240, 247–8, 250, 257–60, 261 n., 264–5
Trnovo, Council of, *see* Council
Trogir (Tragurium, Trau), 159 n.
Turfan discoveries, 7, 10
Turkestan, 7, 16, 193 n.
Turks, Turkish, 63, 84 n., 144 n., 170–1, 189, 193 n., 252, 258, 265–7, 276
Tyana, 220
Tychicus, St, disciple of St Paul, 35, 39 n.
Tzurillas, John, Bogomil, 174–83

Unjust Steward, parable of the, 123–4, 207, 227, 239, 288
Urania, Manichaean woman, 26 n.

Valentinus, Gnostic, 32–3
Van, Lake (in Armenia), 52
Vardar, river, 84 n., 153, 164–5, 193, 255
'Vardar Turks', 84 n.
Veles, 165
Venice, 90 n.
Versinicia, 61
Via Egnatia, 153
Villehardouin, Geoffroy de, the historian, 232–3
Virgin Mary, the, *see* Mary, Virgin
Vision of Isaiah, 154 n., 156 n., 282

INDEX 317

Vladimir, Prince of Bulgaria, 87
Vladimir (John-Vladimir), St, Prince of Dioclea, 93–4, 149 n., 150 n., 283 n.
Vodena, 153, 163 n., 166
Volga Bulgars, *see* Bulgars
Vulgate, the, 288

Wallachia, 160–1
Western Church, *see* Church, Roman
William I, the Conqueror, King of England, 190 n.
William of Apulia, 190 n.
Wine, drinking of, 6 n., 19–20, 26, 44, 127–8, 129 n., 144, 221
Women:
  in Bogomil sect, 135, 199, 201
  in Marcionite, Massalian and Montanist sects, 50

Xantas, Paulician, 190 n.

Yaroslavl, 280

Zacharias, Paulician, 114 n.
Zarvan, Zarvanism, 14–15, 95 n.
Zcerneboch, *see* Chernebog
Zeta, 283
Zigabenus, Euthymius, Byzantine theologian, 31, 119–20, 123 n., 174 n., 177 n., 179 n., 185–6, 197, 199–200, 201 n., 202, 205–19, 222, 224, 225 n., 226, 229 n., 239–41, 249, 282
Zonaras, Joannes, Byzantine historian, 75, 147, 200
Zoroaster, Zoroastrianism, 3 n., 10–15, 193 n.
Zosima, metropolitan of Moscow, 272

For EU product safety concerns, contact us at Calle de José Abascal, 56–1°,
28003 Madrid, Spain or eugpsr@cambridge.org.

www.ingramcontent.com/pod-product-compliance
Lightning Source LLC
LaVergne TN
LVHW091933070526
838200LV00068B/545